Reading Rawls

Reading Rawls

CRITICAL STUDIES ON RAWLS' *A THEORY OF JUSTICE*

EDITED WITH AN INTRODUCTION BY

Norman Daniels

BASIC BOOKS, INC., PUBLISHERS
NEW YORK

Library of Congress Catalog Card Number: LC 74–25908

ISBN 465–06854–5 (Cased)
465–06855–3 (Paper)

10 9 8 7 6 5 4 3

Contents

Acknowledgments

I owe special gratitude to Professors Hugo Adam Bedau and John Rawls for many helpful discussions of the general idea and scope, as well as the specific contents and organization, of this collection. Without their constant advice, this collection might never have been put together. I would also like to thank the Tufts University graduate students who participated in my seminars on 'Justice' and 'Equality' in the years 1972 and 1973. Their comments helped me see more clearly what kinds of problems and arguments might most usefully be discussed in contributions to the collection. Tufts University provided funds for a lecture series on 'Justice' (Fall 1972) in which earlier versions of two papers included here (David Lyons' and Richard Miller's) were read. Finally, I would like to thank Eugene Warfel for many hours spent in the preparation of the manuscript. I wish to acknowledge permission to use material contained in this volume as follows:

1 Thomas Nagel's 'Rawls on Justice', *Philosophical Review*, Vol. LXXXII, No. 2 (April 1973), pp. 220–34, is reprinted by permission of the *Philosophical Review* and Professor Nagel.

2 Ronald Dworkin's 'The Original Position', *University of Chicago Law Review*, Vol. 40, No. 3 (Spring 1973), pp. 500–33, is reprinted by permission of the *Chicago Law Review* and Professor Dworkin.

3 Milton Fisk's 'History and Reason in Rawls' Moral Theory' was originally prepared for this collection.

4 R. M. Hare's 'Rawls' Theory of Justice' is slightly revised from his critical review by the same title which appeared in *Philosophical Quarterly*, Vol. 23 (April 1973), pp. 144–55, and Vol. 23 (July 1973), pp. 241–51, © R. M. Hare, 1973, and is reprinted by permission of Professor Hare and the *Philosophical Quarterly*.

5 Joel Feinberg's 'Rawls and Intuitionism' is specially prepared

for this collection from material originally contained in (a) 'Justice, Fairness, and Rationality', *Yale Law Journal*, Vol. 81 (1972), p. 1004, reprinted by permission of the Yale Law Journal Company and Fred B. Rothman & Company; and (b) 'Duty and Obligation in a Non-Ideal World', *Journal of Philosophy*, Vol. LXX, No. 9 (10 May 1973), pp. 263–75, reprinted by permission of the *Journal of Philosophy*.

6 Gerald Dworkin's 'Non-Neutral Principles', originally prepared for this collection, has since appeared in the *Journal of Philosophy*, Vol. LXXI, No. 14 (15 August 1974), and is reprinted by permission of the *Journal of Philosophy* and Professor Dworkin.

7 David Lyons' 'Nature of Soundness of the Contract and Coherence Arguments' was specially prepared for this collection. It is based on material from (a) 'Rawls versus Utilitarianism', *Journal of Philosophy*, Vol. LXIX, No. 18 (5 October 1972), pp. 535–45, reprinted by permission of the *Journal of Philosophy*; and (b) 'The Nature of the Contract Argument', *Cornell Law Review*, Vol. 59, No. 6 (1974), © 1974 by Cornell University, reprinted by permission of *Cornell Law Review*.

8 T. M. Scanlon's 'Rawls' Theory of Justice' is specially revised for this collection from Parts II and IV of his 'Rawls' Theory of Justice', *University of Pennsylvania Law Review*, Vol. 121, No. 5 (May 1973), pp. 1020–69, © 1973 *University of Pennsylvania Law Review*, with permission to reprint from *The University of Pennsylvania Law Review* and Fred B. Rothman & Co.

9 Richard W. Miller's 'Rawls and Marxism' was originally prepared for this collection. It has since appeared as 'Rawls and Marxism', *Philosophy and Public Affairs*, Vol. 3, No. 2 (Winter 1974), pp. 167–91, © by Princeton University Press; material used in this slightly revised version is reprinted by permission of Princeton University Press and Professor Miller.

10 H. L. A. Hart's 'Rawls on Liberty and its Priority' is reprinted from *University of Chicago Law Review*, Vol. 40, No. 3 (Spring 1973), p. 534–55, by permission of *Chicago Law Review* and Professor Hart.

11 Norman Daniels' 'Equal Liberty and Unequal Worth of Liberty' was specially prepared for this collection.

12 A. K. Sen's 'Rawls versus Bentham: An Axiomatic Examination of the Pure Distribution Problem' is slightly revised from his article by the same title in *Theory and Decision*, Vol. IV, No. 3–4, pp. 301–10, and is reprinted by permission of *Theory and Decision* and Professor Sen.

13 Benjamin Barber's 'Justifying Justice: Problems of Psychology, Politics, and Measurement' was specially prepared for this collection and is forthcoming in the *American Political Science Review*, Spring 1975, as well.
14 Frank Michelman's 'Constitutional Welfare Rights and *A Theory of Justice*' is specially prepared for this collection from material in 'In pursuit of Constitutional Welfare Rights: One View of Rawls' Theory of Justice', *University of Pennsylvania Law Review*, Vol. 121, No. 5, pp. 962–1019, © *University of Pennsylvania Law Review*; permission to reprint previously published material is from the *Pennsylvania Law Review* and Fred B. Rothman & Co.

Introduction

I. *The Response to 'A Theory of Justice'*

No one would have dared predict the broad critical acclaim, even fame, John Rawls' *A Theory of Justice* was to receive in the non-academic press upon publication in 1971. Certainly no one in the circle of moral and political philosophers familiar with the development of Rawls' theory, from the initial journal presentation of 'Justice as Fairness' in 1958 through several stages of privately circulated mimeo drafts, would have ventured such a prediction. However highly regarded by academics, scholarly works in moral and political philosophy, especially those emerging from the Anglo-American 'analytic' tradition, *never* receive rave, lead reviews in the *New York Review of Books*, the *New York Times Book Review*, or *The Times Literary Supplement*, as *A Theory of Justice* did. Nor are they enthusiastically greeted by reviews in *The Economist*, the *Spectator*, *Nation*, *New Republic*, the *Listener*, the *New Statesman*, the *Washington Post*, the *Observer*, *The Times Higher Education Supplement*, and many others.

What explains the unusually wide interest shown in Rawls' work? One obvious factor is that many readers and editors found in Rawls' work a welcome return to an older tradition of substantive, rather than semantic moral and political philosophy. Rawls' approach stands in sharp contrast to the work of the logical positivists and the analytic school in general. To the non-professional, logical positivism and the analytic approach seemed to abandon crucial issues of right and wrong in favor of technical questions about the 'emotive' function of moral language and the meanings of moral and political terms. Rawls, at least, asks the substantive questions. As *The Times Literary Supplement* put it,

> It is a convincing refutation, if indeed one is needed, of any lingering suspicions that the tradition of English-speaking

> political philosophy might be dead. Indeed his book might plausibly be claimed to be the most notable contribution to that tradition to have been published since Sidgwick and Mill.[1]

There is irony in the perception that Rawls' work marks a major break with the recent trend of technical philosophy. The book is far more philosophically and technically sophisticated than it might seem to be at first reading. To be sure, *A Theory of Justice* uses a social contract model, familiar to all who have ever read any political philosophy, to discuss issues of justice which are of interest to many. It has, in a sense, a plot that is readily grasped in its broad outlines. On this level it is far more available to a broad audience than most recent Anglo-American moral and political philosophy and so seems to be a refreshing return to a type of philosophy the non-professional can understand.

But *A Theory of Justice* operates on another level as well. In his effort to be 'substantive', Rawls does not at all abandon the sophisticated apparatus and techniques of the professional philosopher, though the apparatus is not that of the logical positivist. It is precisely because those closest to Rawls' work viewed it as a contribution to the more technical tradition in moral philosophy that they could not anticipate its popular reception. Its broader relevance to the non-professional may have been obscured by the extensive professional criticism and continuous refinement it received in academic journals and graduate seminars. In a sense it was hard to see the forest for the trees. Nevertheless, in spite of the high degree of professionalism, and in spite of the fact that the work is a product of the 'philosophical academy', *A Theory of Justice* discusses general issues many people relate to and find of deep interest. Ronald Dworkin suggests that these two levels, the 'significant' and the professional, are not by any means an either-or proposition:

> It is also a refutation of the popular but silly view that Anglo-American philosophy has recently sacrificed significance for clarity. Rawls demonstrates that clarity is a necessary condition of genuine intellectual significance, not a commodity that may be traded off to purchase it.[2]

[1] 'The Good of Justice as Fairness', *The Times Literary Supplement*, 5 May 1972, p. 1505.

[2] Ronald Dworkin, 'Social Contract', *Sunday Times*, London, 9 July 1972.

There is another reason for the broad appeal this scholarly work in philosophy has had. In drawing on material from the social sciences, Rawls opens the pages of his book to a far wider audience. As Marshall Cohen remarks, Rawls 'revives the English tradition of Hume and Adam Smith, of Bentham and of John Stuart Mill, which insists on relating its political speculations to fundamental research in moral psychology and political economy'.[3] Many more people are familiar with concepts and theories in the social sciences than are familiar with the machinery of the professional philosopher.

There is more than increased familiarity at work here, however. A work of the scope of *A Theory of Justice* also holds the promise of cross-fertilizing and enriching the social sciences, not just depending on them. This possibility may be what Ronald Dworkin had in mind when he noted, 'It is one of those books, rare but essential to philosophy, that define the state of the art by describing the theories in the field in a fresh way, in the light of developments in economics and the social sciences, and thus providing a new vocabulary that will no doubt be used even in criticism of its own arguments.'[4] This promise no doubt must have interested many editors of the literary, non-academic press. In any event, it certainly interested professional social scientists. *A Theory of Justice* has been the subject of symposia in journals and at conferences of political scientists, economists and legal theorists. As a result, Rawls' work was precipitated into still another arena usually denied to contemporary works in moral and political philosophy.

But Rawls' return to the tradition of 'substantive' moral and political philosophy also adds another dimension to his work, one which I believe has had the most to do with its unusual reception. Most major works in the substantive tradition Rawls writes in have had great ideological importance. They have set the dominant—or rising—moral and political ideas of their time in a theoretical framework that lent them great coherence and power, often for centuries to come. By appealing to contemporary models of scientific method and findings in the social sciences, they have tried to show that the dominant—or rising—views of a period should be acceptable to all reasonable men.

Rawls' goal in *A Theory of Justice* has similar ideological

[3] Marshall Cohen, 'The Social Contract Explained and Defended', *New York Times Book Review*, 16 July 1972, p. 1.

[4] Ronald Dworkin, 'Social Contract', cited above.

import. Rawls intends to do three things. He wants to reveal the principles of justice which underlie the dominant moral and political views of our period. He wants to show that these principles can be viewed as the result of a selection procedure that all people can agree is fair (thus, 'justice as fairness'). And he wants to show that these principles describe a workable social arrangement, given everything we know from the social sciences. But, the dominant moral and political ideology of our time, reflected in these principles, is, of course, a form of liberalism. Perhaps it is a more egalitarian liberalism than dominated the eighteenth and nineteenth centuries, but it is liberalism nonetheless. Rawls' goal, then, is to produce a persuasive, coherent framework for this liberalism.

This ideological import of Rawls' work may well have been obscured to the view of those closest to it in its earlier stages. It is another feature of the forest not seen for the trees. But once the work as a whole was available, reviewers were quick to comment on it. Hugo Adam Bedau remarked in *Nation*, 'As a work of original scholarship in the services of the dominant moral and political ideology of our civilization, Rawls' treatise is simply without a rival.'[5] Marshall Cohen also notes this ideological feature of Rawls' work in the *New York Times Book Review*:

> For too long now, the main tradition of moral philosophy has been utilitarian in its broad assumptions: people ought to work for 'the greatest happiness of the greatest number' of their fellow-men; minorities should submit to the interests of the majority. But the utilitarian attitudes are incompatible with *our moral judgment and with the principles on which our Constitution rests.* It is, therefore, a crucial task of moral and political philosophy to make clear the inadequacy of utilitarian concepts, and, more important, to provide a persuasive alternative to them.[6]

Rawls' task is to provide just such an alternative theoretical framework for these dominant views.

Rawls' goal of providing such an alternative is of particular interest because *A Theory of Justice* was published on the heels of a period of intense political struggle and questioning, a period of

[5] Hugo Adam Bedau, 'Rawls: A Theory of Justice', *Nation*, 11 September 1972, p. 180.

[6] Marshall Cohen, 'Social Contract Explained', p. 1. (My emphasis.)

serious challenge to liberalism. In the United States in particular, the Civil Rights and Black Liberation movements, followed by the Anti-Vietnam War movement, brought millions of people into conflict with existing political institutions and policies. These movements raised, in sharp form, fundamental questions about the justice of basic political and social institutions, questions about the distribution of liberties and other social goods, and questions about the just use of political power. Liberal moral and political judgments were pitted against liberal political institutions, and these institutions were in turn defended by liberal political arguments. That Rawls worked on his book throughout this period, and may have been responded to it (as in his thoughtful discussion of civil disobedience) did not go unnoticed by reviewers:

> All the great political philosophies of the past—Plato's, Hobbes's Rousseau's—have responded to the realities of contemporary politics, and it is therefore not surprising that Rawls' penetrating account of the principles to which our public life is committed should appear at a time when these principles are persistently being obscured and betrayed.[7]

Of course it would be quite wrong to conclude that, because Rawls' work has this timely, ideological significance, he is merely a polemicist for the *status quo*. His principles of justice and his 'ideal theory' in general go some way toward challenging existing political and social institutions. Moreover, his aproach to, and execution of, his project is very far from the manner of a polemical ideologue. Indeed, Rawls even suggests that this project produces results which are, at least in principle, compatible with such different socio-economic systems as capitalism and socialism.

Nevertheless, in providing a theoretical foundation for the dominant moral and political ideas of our time, if not for many of the dominant practices or institutions, and in attempting this on the heels of a period many thought proved the ideological bankruptcy of liberalism,[8] Rawls invites a special form of con-

[7] Ibid., p. 1.

[8] In his challenging critique of Rawls, Brian Barry remarks, 'Both *A Theory of Justice* and (Sidgwick's) *The Methods of Ethics*, which appeared just two years short of a century before it, are comprehensive and systematic statements of a thorough-going liberal position; and both, it might be added, appear at a time when liberalism is becoming unfashionable, dismissed in smart circles as shallow compared with the deep (not to say unfathomable) truths of Hegel or a Hegelianized Marx.' *The Liberal Theory of Justice:*

troversy. Steven Lukes signals this controversy in the *Observer Review* when he suggests that Rawls' defense of liberalism is not itself without ideological bias:

> In the end, the 'Archimedean point for judging the basic structure of society' that Rawls seeks eludes him. Every political theory, and every theory of justice, expresses a particular political and moral perspective. Rawls' achievement, which is considerable, is indeed to have produced *a* theory of justice—a theory of liberal democratic justice.[9]

The controversy will take different forms. Some will argue that the particular version of liberalism that Rawls' book is concerned with is wrong, believing it too egalitarian or not egalitarian enough. Some will claim it shows too little concern for other classical notions, like merit or entitlement, or as Stuart Hampshire suggests, for the notions of guilt and innocence and the retributive background to the concept of justice.[10] They will be challenged to show where Rawls' work is at fault. Others, who think that all forms of liberalism are inadequate, will now have the task of attacking a powerful theoretical defense of it. Bernard Crick remarks on this challenge in the *New Statesman*,

> Rawls is profoundly wrong but almost perfectly relevant. Truly he forces us to see that theories of socialism without a critical moral philosophy are as undesirable as they are impossible. Who can answer him fully and go beyond?[11]

Rawls' work, which avoids discussing possible Marxist objections, nevertheless provides Marxists with a serious and sophisticated challenge.

II. *The Purpose of this Collection*

Because Rawls' work has such a broad audience and such ideological importance, its systematic critical evaluation is par-

A Critical Examination of the Principal Doctrines in A Theory of Justice *by John Rawls* (Oxford: Clarendon Press, 1973), p. 4.

[9] Steven Lukes, 'An Archimedean Point', *Observer Review*, 4 June 1972.

[10] Cf. Stuart Hampshire, 'A New Philosophy of the Just Society', *New York Review of Books*, 24 February 1972, pp. 38–9.

[11] Bernard Crick, 'On Justice', *New Statesman*, 5 May 1972, p. 602.

ticularly important. Fortunately, the first wave of academic criticism of *A Theory of Justice* has not been restricted to general review articles. Instead, it includes extended critical studies which analyze in some depth the special contributions and important problem areas in Rawls' work. Taken together, these first studies still do not permit any full scholarly assessment of Rawls' achievement, something which may take many years.[12] Nevertheless, they do provide a useful indication or guide to the problems and lines of criticism which must be pursued further if such an assessment is to be possible. The goal of this collection is to bring together just such an early guide, both to further critical work and to the education of advanced students.

Indeed, I first became aware of the need for a topically oriented collection of studies on *A Theory of Justice* after using the book in a series of graduate seminars and advanced courses beginning in the summer of 1972. Confronted with over 600 pages of close argument, including extensive digressions into related material, students fell into either of two traps. Some students became so enamored of the elegant, central treatment of the social contract that they were unable to look objectively at the book. Engulfed by the seemingly simple plot, they ignored possible lines of criticism and were often uninterested in all the subsidiary material that adds to Rawls' complex argument. Other students had the opposite reaction. They were unable to grasp the general outlines of Rawls' theory because they became bogged down in the many detailed points of Rawls' exposition. In the desire to see how all of it fit together, they were unable to distinguish the peripheral and supplementary from the central. They, too, were in no position to assess possible lines of criticism. Nor, to tell the whole story, was the problem one that afflicted only my students. Teaching Rawls' complex theory without a supplementary, topically organized set of critical readings was not always very successful.

In assembling this collection, I have tried to keep these difficulties in mind. The contributions are organized around four main topic areas: Rawls' contractarian approach, epistemological questions and other questions of method, the analysis and derivation of the principles of justice, and the bearing of Rawls' work on several areas of the social sciences. I have avoided general review articles and I have solicited articles written from a

[12] One cannot help admire Brian Barry's *tour de force* effort at such an assessment. See his *Liberal Theory of Justice*.

variety of methodological and ideological perspectives. Since I felt it important to bring several studies to bear on each of these central topic areas, I have had to restrict the scope of the collection. In particular, I have not included papers on Rawls' moral psychology, his theory of goodness, or his material on rationality and duty. Most painfully, I have had to resist the temptation to keep adding to the collection. I was aware of many excellent articles which had to be omitted even though they would have improved the collection.

By no means, then, does this volume pretend to be the 'final word' on Rawls. It is intended as a working guide to the ongoing critical assessment of fundamental issues in Rawls' work. Just as important, it is intended as an aid to the education of advanced students. It is also my hope that its topical organization will even help the general reader attempting a first reading of Rawls' book.

III. *Rawls' Contractarianism*

The papers in Part One of this volume focus on the most distinctive feature of *A Theory of Justice*, Rawls' attempt to revive a version of the social contract. In Rawls' version of the contract, parties select principles of justice from a hypothetical 'original position', much as parties contracted to form civil society in Hobbes, Locke, and Rousseau. The principles are to apply to basic social and political institutions and they are to meet certain formal conditions: they are to be general, to be publicly known, and to constitute a final choice. Moreover, the principles must be compatible with our considered judgments about what is just in particular situations. This 'coherence' constraint, which raises important epistemological problems, is discussed in papers in Part Two of this volume.

The parties making this selection of principles are not savages in a state of nature, but people just like us. They are, however, subject to certain constraints on their knowledge and motivation. Though self-interested, presumably like us, these rational agents are mutually disinterested and are concerned only with their (family's) share of the primary social goods (wealth, income, powers, authority, self-respect and liberty). They know all there is to know about the general laws of social theory, but they operate behind a veil of ignorance which prevents them from knowing their natural characteristics and endowments, their position in society or the historical period they are born into. Given these various conditions on both the principles and the parties choosing them, Rawls believes two principles of justice will be selected. The papers in Part Three of this collection discuss the derivation of Rawls' First Principle, which guarantees equal liberty, and his Second Principle, which permits inequalities in other goods only if they help the least advantaged.

Rawls uses his original position in two capacities. In its analytic role it helps us to understand the degree to which our conception of justice depends on various formal requirements like the conditions of generality, finality, and publicity that apply to the principles. Also, in its analytic capacity, it provides a concrete model for reducing a relatively complex problem, the social choice of principles of justice, to a more manageable problem, the rational individual choice of principles. If this reduction succeeds, it means the original position is of considerable importance as an analytic device.

But Rawls places great emphasis on another role of the original position, its role as a justificatory device. It is this justificatory role of Rawls' contractarianism which is discussed by the papers in Part One of this volume. Rawls wants the choice of a set of principles in the original position to constitute a justification of them. But, as Thomas Nagel and Ronald Dworkin ask in the opening sections of their papers, why should people not in the original position believe it important that those subject to its stringent constraints happen to choose certain principles? Why should *their* choice 'justify' *our* adopting them?

Rawls' answer is that the conditions and constraints embodied in the original position constitute a model of procedural fairness and, as such, they should be acceptable to everyone on due consideration. In other words, Rawls believes he can show there is a justificatory role played by the original position if he can show the moral persons who are parties in it are treated fairly. Two main critical responses to this position are taken up by papers in this volume. Thomas Nagel and Milton Fisk try to demonstrate that some form of bias exists in the original position and that this bias undermines its ability to serve as a fair device for selecting principles. Ronald Dworkin, on the other hand, tries to show that there is a deeper, presupposed theory behind the original position which gives it the justificatory power it has.

Nagel claims Rawls' original position contains a kind of bias which is an unavoidable drawback of all contract theories. Since parties in the original position must arrive at a unanimous choice, a feature of all contract theories, Rawls is forced to assume that the 'thin' theory of primary social goods is a sufficient basis for parties to act on choosing a conception of justice. The theory is 'thin' because parties do not know their full conception of the good but know only that income, wealth, power and self-respect, for example, are desirable, whatever else they desire. If differing, full conceptions of the good were allowed, no unanimity in principles would result. But it is one thing to rule out knowledge of morally irrelevant features like sex, race and class position. It is another, suggests Nagel, to eliminate morally relevant knowledge, like one's conception of the good. The 'thin' theory is more compatible with some conceptions of the good than with others and, so, the original position treats parties in it unequally and unfairly. Nagel also argues that more particular forms of bias underlie the arguments for the particular principles of justice. For example, he claims behind the Second Principle there is a presupposition that sacrifices by the higher levels of society are preferable to sacrifices by the lower levels, a point which Richard Miller develops in his contribution to Part Three, but which T. M. Scanlon disagrees with in his paper. Nagel's point here is not that biases lead to the wrong substantive principles but rather that they undermine the claim that the contract is a fair basis for selecting and thus justifying principles.

Fisk also tries to show that Rawls' biases in the original position are a general feature of contract arguments, arguing his claim from a Marxist point of view. Fisk claims that Rawls, like earlier theorists, thinks he can earn justificatory force for the contract position by stripping men down to their 'natural' characteristics of 'freedom' and 'equality'. These characteristics, however, reflect a particular ideological bias. For example, Rawls supposes men are naturally 'free', by which he means they choose plans of life without antecedent restrictions, like those imposed by a concern for the interests of others. Fisk argues, however, that abstracting man from his real condition, which always involves group interests, especially class interests, is an unfortunate distortion. This distortion has three bad effects. It leads to selecting ideologically biased principles; the liberal principle of freedom of thought is Fisk's example of a principle which really favors the dominant class's interests. It results in a misleading description of the state, which is viewed as a check to individual selfishness rather than as an instrument of class oppression. And finally, it makes genuine 'community' impossible, a point Benjamin Barber also makes in his contribution to Part Four. Fisk suggests that a second major distortion of human nature is embodied in the claim that people are 'equal'. In reality, Fisk claims, people are 'equally' members of classes, not members of society as a whole. Adopting the view of society as a whole leads to principles which seem to treat all people equally, but which really have different effects on different classes.

Fisk's overall argument has the form, then, of granting the analytic function of the original position but denying its justificatory function. Rawls reduces liberal moral and political theory to a problem of individual rational choice. But Fisk rejects the model of the individual who makes the choice as itself ideologically biased.

Ronald Dworkin is less interested in challenging the justificatory role of Rawls' original position than he is in showing just what presupposed deep theory gives it its justificatory power. Distinguishing three types of theories, goal-based, duty-based and right-based, Dworkin first argues that only right-based deep theories are compatible with the contract model. He then suggests that the particular right which lies at the heart of Rawls' 'deep' theory is the right of each individual to equal concern and respect. This right is not a product of the contract but a presupposition of Rawls' use of the contract. The role of the contract is to decide which of several ways of granting such equal respect and concern is to be preferred. Moreover, since Rawls presupposes this fundamentally egalitarian right, and not the particular liberties Locke and Mill thought important, then those who accuse Rawls of simply assuming liberalism and then idealizing it are wrong. They must show just what is wrong with the egalitarian assumption.

Other contributors to the collection also make important remarks about Rawls' contractarianism, though for various reasons I have grouped their papers in other sections of this volume. R. M. Hare comments on the analytic role of the original position. He argues that Rawls' contract theory is equivalent, in the formulation of its main conditions on choice, to other general approaches to justifying moral theory, in particular to ideal observer theories and to the formal conditions of universalizability and prescriptivity used in his own theory. Where the formulation of conditions differ, as in Rawls' insistence on a 'thick' veil of ignorance, he claims there is no reason to accept Rawls' special conditions.

Two other contributors challenge the contractarian attempt to reduce the social choice of principles of justice to a problem of rational individual choice. David Lyons, in Section III of his paper, and Benjamin Barber, in his contribution to Part Four, both ask why we should view the results of choice in the original position as the selection of principles of justice, rather than as the selection of principles governing self-interested departures from just egalitarian arrangements. Barber further questions whether there is any solution to the choice problem as Rawls constructs it. Do parties in the original position know enough about their interests to make any rational choice? Lyons asks a further question about the central feature of the original position, its commitment to procedural fairness and impartiality. Why, he asks, does the presence of this feature turn choice into justification?

T. M. Scanlon, in Section IA of his paper, also tries to answer the question why we should be motivated to accept choice behind a veil of ignorance as a procedure for selecting principles of justice. He thinks the answer lies in Rawls' view of what constitutes the good for an individual. Rather than identifying the good with an individual's particular commitments and goals, Rawls sees the good for an individual in his protected status as a rational chooser, one capable of modifying his goals throughout his life. The intuitive appeal of this model of the individual is what makes Rawls' contractarianism attractive, in Scanlon's view. Scanlon's analysis, it might be noted, has a certain similarity to the egalitarian principle of equal respect which Ronald Dworkin isolates as Rawls' 'deep theory', though Scanlon does not suggest a presupposed set of rights, only a presupposed ideal of the individual. Scanlon's description of Rawls' model of the individual should be compared to the rather different analysis Fisk makes.

IV. *Other general features of Rawls' Approach*

The revival of the social contract is not the only general feature of Rawls' approach that has prompted critical discussion. Papers in Part Two of this volume focus on four other general aspects of Rawls' approach: the 'coherence' constraint on the selection of moral principles, the attempt to avoid intuitionism in the ordering of moral principles, the concern to rule out moral principles whose applicability is not determined by procedures all can agree to, and the general goal of showing that a variant of utilitarianism does not emerge from the contract position.

One way to see the importance of the coherence constraint is to ask what assurances we have that the principles selected in the contract situation are really principles of justice, rather than some other sort of principle. Rawls' answer is that these general principles can be seen as generating moral judgments which more or less match—that is, which cohere with—our considered judgments about what is just. In a sense, a test of whether we have the correct moral theory, construed as a set of moral principles, is whether our theory matches the moral data, construed as our moral judgments. Since we are concerned with the best overall fit of theory and data, we may have to make certain adjustments both at the theory end and at the data end, just as is done in science. Rawls calls this best fit a 'reflective equilibrium'.

R. M. Hare and David Lyons both treat Rawls' coherence constraint as part of his justificatory apparatus for the principles. Can the coherence constraint add justificatory force? In the opening section of his paper, Hare suggests that the procedure of finding principles to match our intuitions is ultimately a highly subjectivist approach. It can provide no more justificatory support for the principles of justice than is found in intuitionism itself, where intuitionism appeals directly to the intuitive validity of the principles. (Thomas Nagel, in his contribution to Part One, disagrees with the assimilation of the coherence argument to intuitionism.) Lyons also challenges the justificatory power of the coherence argument, and suggests that any justificatory power of Rawls' theory must be found in the contract argument. He carries the challenge one step further and suggests that, on one interpretation of the contract argument, it simply assumes the importance of fairness as if fairness were a 'fixed point' in our intuitions. On this interpretation Rawls' contract argument is subsumed under the coherence argument and has no more justificatory power than it.

A slightly different challenge to the analogy between scientific and moral theories is provided by Section II of Ronald Dworkin's paper, found in Part One of this volume. Dworkin suggests that we can give a better account of the process of arriving at a 'reflective equilibrium' if we drop the analogy to fitting scientific laws with scientific data and substitute for it a 'constructive' model derived from legal theory. His remarks should be compared to

Nagel's comment on the disanalogy between moral theories and linguistic theories.

A second general feature of Rawls' approach is his effort to avoid an intuitionistic pluralism of basic principles. He hopes to avoid the problem, unsolved by classical intuitionists, of ordering a set of principles so they do not conflict, but rather, have determinate priorities among them. Joel Feinberg argues that Rawls falls down in two ways in his attack on intuitionism. First, Rawls fails to establish that his—or any—principles of justice must always take precedence over other considerations of rightness. Even if justice is weighted very heavily as a factor in determining the rightness of actions, Rawls fails to show that other considerations—even utilitarian ones—don't sometimes outweigh the demands of justice. But in failing to show the uniform priority of justice, Rawls fails to show that intuitionism is wrong. Moreover, Feinberg argues that Rawls himself admits intuitionist considerations into his own substantive theory, notably in his discussion of civil disobedience. Feinberg's accusation that Rawls fails to avoid intuitionist flexibility in the ordering of principles must be distinguished from Hare's claim that the epistemological grounding for Rawls' two principles is intuitionist.

A third general feature of Rawls' contractarian approach is that it requires the selection of principles to which all can agree in the original position. An important question to ask is whether or not this contractarian approach forces us to eliminate certain types of general moral principles whose elimination is not justified by other formal considerations. If the original position is weighted in this way against otherwise formally acceptable moral principles, what accounts for the weighting?

Gerald Dworkin asks both of these questions in his contribution to Part Two. He distinguishes two general types of moral principles, 'neutral' and 'non-neutral' principles. A 'non-neutral' principle is one whose application to particular cases is a matter of controversy for the parties whose conduct is supposed to be regulated by the principle. Thus 'Punish those who are guilty' is a non-neutral principle since it is a matter of controversy who is guilty. In contrast, 'Punish those found guilty in a trial by their peers' is a neutral principle, since, by specifying a procedure (even an imperfect one) for determining guilt, the principle will not be controversial in its application. Dworkin argues, first, that non-neutral principles cannot be eliminated by formal arguments alone. Then he argues that Rawls requires certain controversial epistemological assumptions in order to produce agreement on the selection of neutral principles, like the principle of toleration. Ruling out the choice of non-neutral principles is only as justifiable as those epistemological assumptions.

The fourth general feature involves one of Rawls' most important goals in *A Theory of Justice*. Rawls directs much of his argument to showing that no variant of utilitarianism, which he regards as the dominant theoretical framework now supporting liberal moral and political views, would be selected in the original position. Instead, he tries to demonstrate that his two principles of justice will be a more rational choice than the principle of utility, given the fair procedures and construction of the contract situation. The papers by R. M. Hare and, particularly, by David Lyons challenge Rawls' arguments against utilitarianism. They question certain general features of the original position which are designed to weight it in favor of Rawls' two principles and they question the most general features of Rawls' argument against utility. Thus, they constitute a transition from the consideration of general questions about Rawls' method, the concern of Part Two of our collection, to the consideration of Rawls' derivation of his own principles, the concern of Part Three.

Taken together, Hare and Lyons raise three main lines of criticism. Lyons argues, in sections 1 and 2 of his contribution, that Rawls' two principles and utilitarianism are not equivalent, but that there is an unfortunate tendency to exaggerate the differences. In fact, he suggests the two sets of principles may not be so different with regard to what actual or likely real-world arrangements they justify. Consequently, there may not be so clear-cut a difference in the degree to which utilitarianism and Rawls' principles of justice match our considered judgments when we apply the 'coherence' constraint.

A second line of criticism is taken up by both Hare and Lyons. The 'Difference Principle', which is the main clause in Rawls' Second Principle, allows inequalities only if they act to benefit the least advantaged. The principle of utility, on the other hand, places no such constraints on inequality. Rawls argues that the difference principle would be adopted in the original position because it follows from the uniquely rational strategy of minimizing risks—the 'maximin' strategy. To know that a maximin strategy is the rational one, however, Rawls has to show that certain special conditions obtain, particularly that choices based on knowing the probabilities of arrangements outside the veil of ignorance are impossible. Hare attacks Rawls' use of a 'thick' veil of ignorance, one that blocks knowledge of such probabilities. He contends that a thinner veil, permitting probability calculations, would still provide for the impartiality of principles being selected. A thinner veil is thus all that is needed to meet the formal constraints on moral principles. A thinner veil, however, permits selection of the utilitarian principle.

Lyons also challenges the several components of Rawls' crucial argument in favor of a maximin strategy. But Lyons carries the argument one step further. Drawing on his earlier contention, that justice as fairness and utilitarianism are not so different as is often thought, Lyons then argues that choosing a risk-minimizing strategy does not determine selection of the difference principle over the principle of utility.

Other papers in the collection also remark on points relevant to the rejection of utilitarianism or to the contrast between utilitarianism and justice as fairness. T. M. Scanlon discusses Rawls' crucial difference principle and the justification for it in Section IIB of his paper. Benjamin Barber raises several questions about the special psychological assumptions implicit in claiming that a low-risk choice strategy would be selected in the original position. He argues it is this conservative psychology and not the veil of ignorance that leads to a maximin strategy. A. K. Sen attempts to show in formal terms where Rawls' solution to the pure distribution problem differs from Bentham's. He contends that there are solutions preferable both to Rawls' difference principle and to the principle of utility, though they are not considered by Rawls when he rejects utilitarianism. Finally, Frank Michelman argues that utilitarianism and Rawls' Second Principle have quite different implications for legal theorists interested in the theoretical grounding of welfare rights.

V. *The Principles of Justice as Fairness*

We turn, in Part Three, from the more general features of Rawls' methods and goals to discussion of the principles of justice he argues would be chosen in the original position. Actually, Rawls distinguishes between a General and a Special Conception of justice as fairness. The General Conception consists of one principle which distributes all social goods, including liberty, so that inequalities benefit the worst-off. The General Conception applies whenever a certain basic level of material well-being, necessary to the maintenance of effective equal wealth, is unavailable. Whenever appropriate conditions are reached, however, the Special Conception, consisting of two principles, applies. The First Principle, which has priority over the Second, guarantees a maximal system of equal basic liberties. The Second Principle distributes all social goods, other than liberty, allowing inequalities in them provided they benefit the least advantaged and provided equality of opportunity is present. It is this Special Conception which has received most of the critical attention to date, a fact reflected in the contributions to Part Three.

We may begin with the Second Principle. It is the principle Rawls argues for first and we have already touched on arguments relevant to it in our discussion of Rawls' rejection of utilitarianism. Hare and Lyons, as we have seen, raise one central line of criticism of the difference principle, which is the main clause of Rawls' Second Principle. They argue that there is no reason to accept the features of the original position that make risk-minimization a preferred strategy. Barber has a different argument with a similar conclusion. T. M. Scanlon, however, following a careful attempt to analyze how the difference principle is to be understood and applied (Section IIA of his paper), suggests that there is another basis to Rawls' argument for the difference principle.

The difference principle requires that those with various natural talents and endowments must accept restrictions on how much they will benefit from their natural advantages. Their talents must be harnessed to benefit the least advantaged. Why would anyone agree to such a restriction in the first place or to abide by it once they discovered their true talents after the veil of ignorance was raised? Scanlon suggests, in Section IIB of his paper, that agreement on the difference principle presupposes a Rawlsian idea of social cooperation. Social and political institutions, especially economic institutions, are viewed as reciprocal arrangements which operate to the mutual advantage of those who cooperate, as equal partners, in them. Scanlon contrasts this ideal to the views of social cooperation which are embodied in the systems of 'natural liberty' and utilitarianism which Rawls rejects. Scanlon's interpretation stands opposed to the position taken by Nagel, noted earlier. Nagel suggested that sacrifices by the best-off for the worst-off reflected a bias in Rawls' original position. Scanlon sees an ideal of cooperation between equals, not a bias against the best-off, as the source of agreement on the difference principle. Of course, what Rawls would have to show is that the whole product of a society is to be viewed as its common product, rather than just some portion of that product, the remainder to be viewed as the product of particular individuals and groups.

Where Scanlon finds an appealing ideal of social cooperation behind Rawls' support for the difference principle, Richard Miller finds a controversial body of social theory. Miller, like Nagel and Scanlon, believes the problematic feature of the difference principle is its apparent lack of appeal to those who will end up in the best-off levels of a society. Suppose, Miller argues, that some societies have these three features: (1) no institutional arrangement acceptable to the best-off class is acceptable to the worst-off; (2) the best-off class is a ruling class; and (3) the best-off class has a more acute need for wealth and power than is typical of the worst-off class. If these features of class-divided society, as depicted in Marxist social theory, are ever found, then Rawls will not be able to show that representatives of the best-off classes would be able to tolerate the contraints of the difference principle. Nor would he be able to show that the difference principle constituted minimization of risk: if parties in the original position knew the real needs of the best-off class, in this Marxist view, they would not think the difference principle really minimized risk. So, Rawls must be presupposing that societies have only moderate class conflict, no ruling class, and no class differentiated basic desires, at the very least a controversial, if not false, set of assumptions.

Rawls' contribution to the theory of distributive justice, his Second Principle, applies only to social goods other than liberty. There remains the traditional problem of how liberty and justice are related. It is the First Principle, including the priority Rawls grants to it, which elevates liberty to a special importance and which constitutes Rawls' distinctive contribution to this problem area. The First Principle calls for a maximally extensive system of equal basic liberties. The 'priority of liberty' means that liberty is not to be traded off for other social goods once a certain level of material well-being has been achieved. Instead, liberty may be restricted only for the sake of liberty. Discussion of Rawls' First Principle falls into four problem areas.

The first problem area is the analysis of Rawls' concept of liberty and his constraints on attempts to restrict liberty. What notion of liberty does the First Principle govern? Hart suggests that Rawls shifts away from a broad notion of liberty used in his early writing and adopts a narrower list of 'basic' liberties in *A Theory of Justice*, in particular the liberties of conscience, person and political participation. By avoiding the broad notion, Hart claims, Rawls successfully avoids the traditional problem of trying to reconcile equality of liberty with various inequalities in property rights. On the other hand, a new problem arises: Hart believes Rawls' narrow list of basic liberties may not be comprehensive. In my own contribution to the volume, I argue that Rawls arbitrarily excludes economic factors from among the constraints definitive of basic liberties. I argue that economic factors function much like the legal restrictions and public pressures which Rawls accepts as defining constraints. The First Principle calls for rather different things depending on which definition of liberty is used.

Assuming we understand what the basic liberties are, we still must understand how they are to be balanced against each other in cases of conflict and how they are to be restricted, if need be. Scanlon believes there may be a conflict between two criteria Rawls uses in discussing restrictions on the system of liberties. On the one hand, liberty is to be restricted only for the sake of protecting the overall system of liberties. On the other, a 'Principle of Common Interests' is employed which says liberty may be restricted provided all can agree that a lack of restriction will be harmful to all. Are these criteria going to agree with each other? Scanlon thinks not. Hart also believes the Principle of Common Interest will break down in important cases. What is more, Hart thinks Rawls underestimates the difficulty of balancing conflicting liberties. He fails to provide criteria for their relative importance. He also underestimates the complexity of balancing those advantages and disadvantages that accompany the conflicting interests people have in liberty. Milton Fisk's discussion of the class bias behind the liberal principle of toleration might be viewed as an elaboration of this problem from a Marxist point of view.

A second problem area concerns the rationale for granting priority to liberty over other social goods. Rawls argues that once a certain level of well-being has been reached in a society, even the worst-off members of society will prefer increments in their liberty to increments in other social goods. Those in the original position recognize this preference by refusing to allow liberty to be traded off for other goods, provided the basic level of well-being obtains. Hart challenges this general argument for the priority of liberty. He suggests that no such permanent ordering of preferences is possible behind the veil of ignorance because parties do not know enough about their preferences. Benjamin Barber, in his contribution to Part Four, provides some relevant examples which illustrate this line of criticism of the general priority argument.

Both Scanlon and I raise a further question, does the priority argument work equally well for the different liberties covered by the First Principle? In his argument for the priority of equal liberty of conscience, Rawls appeals to the importance that some people may place on moral and religious obligations once outside the veil of ignorance. They may feel that meeting these obligations is far more important than increments in other social goods. Parties behind the veil will want to minimize the risk of not being able to meet such obligations and would choose a principle of equal toleration, giving it priority over other goods. How generalizable is this argument to other liberties? I suggest that there are difficulties with the argument itself, and, further, that it is not generalizable. Scanlon contends that a 'Principle of Limited Authority' is really needed to give an account of freedom of expression and that such a principle needs a different rationale.

A third problem area concerns the general rationale Rawls offers for treating liberty with such special importance. Scanlon argues that a particular model of the individual—as a free chooser of his goals of life-plans—is the ideal that motivates Rawls' special treatment of liberty as well as his rejection of 'perfectionism'. A 'perfectionist' moral theory might justify various incursions on individual liberty in the name of the perfectionist ideal. The ideal of the individual as an unfettered choice-maker provides a common thread linking the attack on perfectionism with the defense of the priority of equal liberty. A somewhat similar 'deep theory' is suggested, as I have remarked, by Ronald Dworkin, although Dworkin couches his version in the form of a presupposed 'right' of the individual to equal respect.

Finally, my own contribution to the volume suggests one more problem area. I argue that there may be a serious incompatibility between the First Principle demand for equal liberty and the Second Principle justification of inequalities, even significant inequalities, in wealth and liberty. An important fact of empirical social theory is in question: do inequalities in wealth and power always produce inequalities in basic liberty? I argue that if they do, Rawls may not be able to reconcile the two principles. He tries to avoid this problem by distinguishing between liberty and worth of liberty, but I believe the distinction, aside from being arbitrary, does not help. Parties in the original position have equally good reason for choosing equality with regard to both notions. My argument suggests that Rawls may have underestimated the egalitarian thrust of his First Principle.

VI. *'A Theory of Justice' and the Social Sciences*

Rawls' return to the tradition of substantive political philosophy is marked by his attempt to relate his theory to major areas of the social sciences. The relation operates in two ways. On the one hand, Rawls plausibly contends that a moral and political theory of justice must draw on what we believe to be true about people and social systems. It is not unrelated to the facts of life. Thus, our knowledge of the social sciences is one important empirical constraint on theories of justice. On the other hand, by showing how specific bodies of knowledge from different areas of the social sciences are related to a comprehensive, systematic theory of justice, Rawls may reveal implications for the social sciences themselves. After all, the problems of distributing social goods, including powers and liberties, has its application in economics, political science, and legal theory, three social sciences represented by contributions to Part Four of this volume.

Before discussing the contributions to Part Four, it is worth noting a common thread linking several papers in this volume. These papers argue that Rawls is correct in his desire to relate the theory of justice to knowledge of the social sciences but they claim that he appeals to a false, or at least controversial, body of social theory. *A Theory of Justice* says very little about Marxism, either in its social, political, or economic implications. Instead it treats utilitarianism as the main opponent. The papers by Fisk, Miller, Barber and myself fill this gap. They raise various objections to Rawls' background assumptions about social theory which are either explicitly Marxist or compatible with Marxist social theory. Fisk argues that Rawls' model of the free and equal rational agent ignores what we know about the group—or class—

nature of people and so introduces a liberal bias into the contract situation itself. Miller argues that Rawls presupposes the falsity of important features of Marx's theory of class struggle in his arguments for the Second Principle. Barber suggests the incompatibility of Marxist theory with the First Principle. He suggests that on a Marxist-Leninist theory of the dictatorship of the proletariat, sacrifices of equal liberty would be justifiable, even if basic levels of well-being sufficient for the operation of the Special Conception of Justice as fairness are present. Accordingly, the principles of justice fail to meet everyone's considered notions about justice. I suggest in my own contribution that Rawls shares with other liberal theorists a serious underestimate of the effect of economic inequalities on political liberties. As a result, his two principles may in fact be incompatible. Even taken together, these contributions hardly constitute a systematic Marxist response to Rawls' but they do suggest that there is need for further work along these lines. The need for a systematic Marxist response is especially great, since Rawls' work, when viewed as a defense of liberal, moral and political theory, has such ideological importance.

Rawls' concern to establish an alternative theory of justice to utilitarianism is, as we have seen, a major feature of *A Theory of Justice*. In view of this goal, A. K. Sen's paper is of special interest. It suggests that both the difference principle and utilitarianism leave something to be desired when viewed as solutions to the 'pure distribution problem', the problem typified by dividing a cake among an arbitrary number of persons. Sen sets out three intuitively appealing 'rules of choice', each of which seems to capture some of our inhibitions about just distributions. He then argues that a utilitarian solution to the cake-dividing problem, maximizing the sum of individual welfares, violates a 'weak equity axiom'. The weak equity rule of choice says that a person with a relatively small share of the cake should not be given an even smaller one on further redivisions of the cake. Rawls' difference principle, on the other hand, also violates intuitively appealing rules. For example, it violates the 'symmetry preference axiom', which favors reductions of inequality provided persons have identical needs. The difference between the two theories comes from the fact that utilitarianism is solely concerned with individual gains and losses, whereas the Rawlsian approach compares levels of welfare. Sen suggests there are other solutions to the distribution problem which are not 'one-sided' in either of these two ways and which therefore do not violate any of the intuitively appealing rules of choice. The implication is that neither the difference principle nor utilitarianism provides a *sufficient* basis for guiding moral judgments about distributions in public policy.

We have already seen that Benjamin Barber shares with other critics questions about conservative, risk-averting psychological assumptions Rawls may have made in the design of the original position. But Barber's main criticism of *A Theory of Justice* is that Rawls' work is divorced in certain crucial ways from the historical and political realities any theory of justice must face. The problem shows up, Barber contends, in three ways. First, Rawls underestimates the difficulty of designing the index of primary social goods. The index is needed if the relative levels of individuals in society are to be made comparable. Barber contends that basing the index primarily on income is likely to prove inadequate, especially since Rawls has not suggested how relative values among the different social goods are to be determined. Second, even if we assume there is an appropriate index, Rawls drastically oversimplifies, in an ahistorical and apolitical way, his discussion of how incomes are likely to vary with each other given different modifications of social structures. Barber contends that far more complicated relations will hold (depicted in his graphs of 'contribution' or OP curves) and that Rawls' principles fail to tell us how to behave justly in view of these complexities. Finally, Barber suggests that important political remarks made by Rawls are also out of touch with political realities.

Barber's claims about the apoliticity and ahistoricity of Rawls' theory raises an important general question not systematically explored by any paper in the collection though touched on by several: how is 'ideal' theory related to judgments about justice in the real, non-ideal world? There are at least two main issues here. One is epistemological: can an 'ideal' theory be based on arguments abstracted in certain ways from our knowledge of existing political systems and still be 'relevant', even as an ideal? The second is normative: given that the real world deviates from the conditions Rawls believes obtain in a just, well-ordered society, how much does concern for the 'ideal' determine judgment about strategies for reaching justice from conditions lacking both justice and the social forces Rawls postulates in ideal theory? These two problem areas deserve systematic study.

Frank Michelman's contribution to Part Four investigates what basis may be found in justice as fairness for constructing an adequate legal framework for welfare rights. Justice as fairness, he suggests, in contrast to utilitarianism, seems to provide for distributive claims which the disadvantaged can make on their own behalf rather than as 'happenstance advocates for the general public interest'. The difficulty for justice as fairness comes in relating the right to a social minimum (measured primarily in terms of income) to specific rights to the insurance of basic needs, like health, housing and education and old age support. For example, how does justice as fairness respond to the needs of a worst-off individual who also happens to be a diabetic? What is the basis, in Rawls' distributive theory, for making insurance claims on behalf of the diabetic? Michelman tries to explain why Rawls did not select a more need-based version of the difference principle. Finally, he assesses how far the difference principle or justice as fairness as a whole can go toward justifying an insurance-rights package if the notion of self-respect is emphasized.

It should be obvious, from the highly suggestive contributions to Part Four of this collection, that we are very far from a complete assessment of the relation between *A Theory of Justice* and the various social sciences. If other parts of the volume seem more complete or final, it is only because I have provided more structure to them by suggesting what the key lines of criticism have been so far. I should like to warn the reader, by way of conclusion, that further work may well require revision of my assessments of the key problems. This volume is intended as a working guide to the scholar, student, and general reader, not as a final word on reading Rawls.

THE ORIGINAL POSITION

1 Rawls on Justice

THOMAS NAGEL*

A *Theory of Justice*[1] is a rich, complicated, and fundamental work. It offers an elaborate set of arguments and provides many issues for discussion. This review will focus on its contribution to the more abstract portions of ethical theory.

The book contains three elements. One is a vision of men and society as they should be. Another is a conception of moral theory. The third is a construction that attempts to derive principles expressive of the vision, in accordance with methods that reflect the conception of moral theory. In that construction Rawls has pursued the contractarian tradition in moral and political philosophy. His version of the social contract, a hypothetical choice situation called the original position, was first presented in 1958 and is here developed in great and explicit detail. The aim is to provide a way of treating the basic problems of social choice, for which no generally recognized methods of precise solution exist, through the proxy of a specially constructed parallel problem of individual choice, which can be solved by the more reliable intuitions and decision procedures of rational prudence.

If this enterprise is to succeed, and the solution to the clearer prudential problem is to be accepted as a solution to the more obscure moral one, then the alleged correspondence between the two problems must bear a great deal of weight. Critics of the theory have tended to take issue with Rawls over what principles would be chosen in the original position, but it is also necessary

* Professor of Philosophy, Princeton University.

[1] John Rawls, *A Theory of Justice* (Cambridge, Mass., Harvard University Press, 1971), pp. xv, 607. All page references to *Theory* will appear in text in parentheses.

to examine those features of the position that are thought to support the most controversial choices and to ask why the results of a decision taken under these highly specific and rather peculiar conditions should confirm the justice of the principles chosen. This doctrine of correspondence is both fundamental and obscure, and its defense is not easy to extract from the book. A proper treatment of the subject will have to cover considerable ground, and it is probably best to begin with Rawls' moral epistemology.

Rawls believes that it will be more profitable to investigate the foundations of ethics when there are more substantive ethical results to seek the foundations of. Nevertheless, in Section 9 he expounds a general position that helps to explain his method of proceeding. Ethics, he says, cannot be derived from self-evident axioms nor from definitions, but must be developed, like any other scientific subject, through the constant interaction between theoretical construction and particular observation. In this case the particular observations are not experiments but substantive moral judgments. It is a bit like linguistics: ethics explores our moral sense as grammar explores our linguistic competence.[2]

Intuitionism attempts to capture the moral sense by summarizing our particular moral intuitions in principles of maximum generality, relying on further intuitions to settle conflicts among those principles. This is not what Rawls means. He intends rather that the underlying principles should possess intuitive moral plausibility of their own, and that the total theory should not merely summarize but illuminate and make plausible the particular judgments that it explains. Moreover, its intrinsic plausibility may persuade us to modify or extend our intuitions, thereby achieving greater theoretical coherence. Our knowledge of contingent facts about human nature and society will play a substantial part in the process.

When this interplay between general and particular has produced a relatively stable outcome, and no immediate improvements on either level suggest themselves, then our judgments are said to be in a state of *reflective equilibrium*. Its name implies

[2] This seems to me a false analogy, because the intuitions of native speakers are decisive as regards grammar. Whatever native speakers agree on is English, but whatever ordinary men agree in condemning is not necessarily wrong. Therefore the intrinsic plausibility of an ethical theory can impel a change in our moral intuitions. Nothing corresponds to this in linguistics (*pace* Rawls' suggestion on p. 49), where the final test of a theory is its ability to explain the data.

that the state is always subject to change, and that our current best approximation to the truth will eventually be superseded. The indefinite article in Rawls' title is significant: he believes that all present moral theories 'are primitive and have grave defects' (p. 52). His own results are provisional: 'I doubt', he says (p. 581), 'that the principles of justice (as I have defined them) will be the preferred conception on anything resembling a complete list.'

If the principles and judgments of a theory are controversial and do not command immediate intuitive assent, then the support they receive from the underlying moral conception assumes special importance. To a certain extent that conception may reveal itself directly in the basic principles of the theory, but it is more clearly visible when the theory contains a model or construction that accounts for the principles and for their relation to one another. Alternative theories of justice are intuitively represented by different models (utilitarianism, for example, by the impartial sympathetic observer). Rawls' model is the original position, and the principles it is used to support are controversial. To enhance their appeal, the construction must express an intuitive idea that has independent plausibility. Before turning to the model itself, it will be useful to review briefly the substantive conclusions of the theory, identifying their controversial elements and thus the respects in which they are most in need of independent support.

Rawls' substantive doctrine is a rather pure form of egalitarian liberalism, whose controversial elements are its egalitarianism, its anti-perfectionism and anti-meritocracy, the primacy it gives to liberty, and the fact that it is more egalitarian about liberty than about other goods. The justice of social institutions is measured not by their tendency to maximize the sum or average of certain advantages, but by their tendency to counteract the natural inequalities deriving from birth, talent, and circumstance, pooling those resources in the service of the common good. The common good is measured in terms of a very restricted, basic set of benefits to individuals: personal and political liberty, economic and social advantages, and self-respect.

The justice of institutions depends on their conformity to two principles. The first requires the greatest equal liberty compatible with a like liberty for all. The second (the difference principle) permits only those inequalities in the distribution of primary economic and social advantages that benefit everyone, in particular the worst-off. Liberty is prior in the sense that it cannot be sacrificed for economic and social advantages, unless they are so scarce or unequal as to prevent the meaningful exercise of equal liberty until material conditions have improved.

The view is firmly opposed to mere equality of opportunity, which allows too much influence to the morally irrelevant contingencies of birth and talent; it is also opposed to counting a society's advanced cultural or intellectual achievements among the gains which can make sacrifice of the more primary goods just. What matters is that everyone be provided with the basic conditions for the realization of his own aims, regardless of the absolute level of achievement that may represent.

When the social and political implications of this view are worked out in detail, as is done in Part Two of the book, it is extremely appealing, but far from self-evident. In considering its theoretical basis, one should therefore ask whether the contractarian approach, realized in terms of the original position, depends on assumptions any less controversial than the substantive conclusions it is adduced to support.

The notion that a contract is the appropriate model for a theory of social justice depends on the view that it is fair to require people to submit to procedures and institutions only if, given the opportunity, they could in some sense have agreed in advance on the principles to which they must submit. That is why Rawls calls the theory 'justice as fairness'. (Indeed, he believes that a similar contractual basis can be found for the principles of individual morality, yielding a theory of rightness as fairness.) The fundamental attitude toward persons on which justice as fairness depends is a respect for their autonomy or freedom.[8] Since social institutions are simply there and people are born into them, submission cannot be literally voluntary, but

[8] Expanding on this point, Rawls submits that his view is susceptible to a Kantian interpretation, but the details of the analogy are not always convincing. See, e.g., the claim on p. 253 that the principles of justice are categorical imperatives, because the argument for them does not assume that the parties to the agreement have particular ends, but only that they desire those primary goods that it is rational to want whatever else one wants. First of all, the desire for those primary goods is not itself the motive for obeying the principles of justice in real life, but only for choosing them in the original position. Secondly, imperatives deriving from such a desire would be hypothetical and assertoric in Kant's system, not categorical. But since our adherence to the two principles is supposed to be motivated by a sense of justice growing out of gratitude for the benefits received from just institutions, the imperatives of justice as fairness would in fact appear to be hypothetical and problematic (*Foundation of the Metaphysics of Morals*, pp. 415–16 of the Prussian Academy Edition).

(p. 13) 'A society satisfying the principles of justice as fairness comes as close as a society can to being a voluntary scheme, for it meets the principles which free and equal persons would assent to under circumstances that are fair.'

Before considering whether the original position embodies these conditions, we must ask why respect for the freedom of others, and the desire to make society as near to voluntary as possible, should be taken as the mainspring of the sense of justice. That gives liberty a position of great importance from the very beginning, an importance that it retains in the resulting substantive theory. But we must ask how the respect for autonomy by itself can be expected to yield further results as well.

When one justifies a policy on the ground that the affected parties would have (or even have) agreed to it, much depends on the reasons for their agreement. If it is motivated by ignorance or fear or helplessness or a defective sense of what is reasonable, then actual or possible prior agreement does not sanction anything. In other cases, prior agreement for the right reasons can be obtained or presumed, but it is not the agreement that justifies what has been agreed to, but rather whatever justifies the agreement itself. If, for example, certain principles would be agreed to because they are just, that cannot be what makes them just. In many cases the appeal to hypothetical prior agreement is actually of this character. It is not a final justification, not a mark of respect for autonomy, but merely a way of recalling someone to the kind of *moral* judgment he would make in the absence of distorting influences derived from his special situation.

Actual or presumable consent can be the *source* of a justification only if it is already accepted that the affected parties are to be treated as certain reasons would incline each of them to want to be treated. The circumstances of consent are designed to bring those reasons into operation, suppressing irrelevant considerations, and the fact that the choice would have been made becomes a further reason for adhering to the result.

When the interests of the parties do not naturally coincide, a version of consent may still be preserved if they are able to agree in advance on a procedure for settling conflicts. They may agree unanimously that the procedure treats them equally in relevant respects, though they would not be able to agree in advance to any of the particular distributions of advantages that it might yield. (An example would be a lottery to determine the recipient of some indivisible benefit.)

For the result of such a choice to be morally acceptable, two things must be true: (*a*) the choice must be unanimous; (*b*) the circumstances that make unanimity possible must not undermine the equality of the parties in other respects. Presumably they must be deprived of some knowledge (for example, of who will win the lottery) in order to reach agreement, but it is essential that they not be unequally deprived (as would be the case, for example, if they agreed to submit a dispute to an arbitrator who, unknown to any of them, was extremely biased).

The more disparate the conflicting interests to be balanced, however, the more information the parties must be deprived of to insure unanimity, and doubts begin to arise whether any procedure can be relied on to treat everyone equally in respect of the relevant interests. There is then a real question whether hypothetical choice under conditions of ignorance, as a representation of consent, can by itself provide a moral justification for outcomes that could not be unanimously agreed to if they were known in advance.

Can such a procedure be used to justify principles for evaluating the basic structure of social institutions? Clearly the preferences of individuals are so divergent that they would not voluntarily agree on a common set of principles if all were given an equal voice. According to the theory of the original position, the appeal to prior agreement can be utilized, nevertheless, by requiring the hypothetical choice to be made on the basis of reasons that all men have in common, omitting those which would lead them to select different principles and institutions. By restricting the basis of the hypothetical agreements in this way, however, one may lose some of its justifying power. We must therefore look carefully at the conditions imposed on a choice in the original position. Since Rawls does not, in any case, offer an abstract argument for the contractarian approach, its defense must be found in its application.

The original position is supposed to be the most philosophically favored interpretation of a hypothetical initial *status quo* in which fundamental agreements would be fair. The agreements can then be appealed to in disputes over the justice of institutions. The parties have an equal voice and they choose freely: in fact, they can all arrive independently at the same conclusions. Each of us, moreover, can enter the original position at any time simply by observing its rather special restrictions on arguments, and choosing principles from that point of view.

All this is possible because the grounds of choice are severely restricted as follows. The parties are mutually disinterested—that is, neither altruistic nor envious. About their own desires they know only what is true of everyone: that they have some life plan or conception of the good and a personal commitment to certain other individuals. Whatever the details, they know these interests can be advanced by the employment of very basic primary goods under conditions of liberty. They also possess general knowledge about eonomics, politics, and sociology and they know that the circumstances of justice, conflicting interests and moderate scarcity, obtain. Finally, they believe that they have a sense of justice which will help them to adhere to the principles selected, but they know enough about moral psychology to realize that their choices must take into account the strains of commitment which will be felt when the principles are actually adopted, and the importance of choosing principles that will, when put into application, evoke their own support and thereby acquire psychological stability. Everything else—their talents, their social position, even the general nature or stage of development of their particular society—is covered over with a thick veil of ignorance on the ground that it is morally irrelevant. The choice should not be influenced by social and natural contingencies that would lead some parties to press for special advantages, or give some of them special bargaining power.

Rawls contends (p. 21) that these restrictions 'collect together into one conception a number of conditions on principles that we are ready upon due consideration to recognize as reasonable . . . One argues', he says (p. 18), 'from widely accepted but weak premises to more specific conclusions. Each of the presumptions should by itself be natural and plausible; some of them may seem innocuous or even trivial. The aim of the contract approach is to establish that taken together they impose significant bounds on acceptable principles of justice.'

I do not believe that the assumptions of the original position are either weak or innocuous or uncontroversial. In fact, the situation thus constructed may not be fair. Rawls says that the aim of the veil of ignorance is 'to rule out those principles that it would be rational to propose for acceptance, however little the chance of success, only if one knew certain things that are irrelevant from the standpoint of justice' (p. 18). Let us grant that the parties should be equal and should not be in possession of information which would lead them to seek advantages on morally irrelevant grounds like race, sex, parentage, or natural endowments. But they are deprived also of knowledge of their particular conception of the good. It seems odd to regard that as morally irrelevant from the standpoint of justice. If someone favors certain principles because of his conception of the good, he will not be seeking special advantages for himself so long as he does not know who in the society he is. Rather he will be opting for principles that advance the good for everyone, as defined by that conception. (I assume a conception of the good is just that, and not simply a system of tastes or preferences.) Yet Rawls appears to believe that it would be as unfair to permit people to press for the realization of their conception of the good as to permit them to press for the advantage of their social class.

It is true that men's different conceptions of the good divide them and produce conflict, so allowing this knowledge to the parties in the original position would prevent unanimity. Rawls concludes that the information must be suppressed and a common idea substituted which will permit agreement without selecting any particular conception of the good. This is achieved by means of the class of primary goods that it is supposedly rational to want whatever else one wants. Another possible conclusion, however, is that the model of the original position will not work because in order to secure spontaneous unanimity and avoid the necessity of bargaining one must suppress information that is morally relevant, and moreover suppress it in a way that does not treat the parties equally.

What Rawls wishes to do, by using the notion of primary goods, is to provide an Archimedean point, as he calls it, from which choice is possible without unfairness to any of the fuller conceptions of the good that lead people to differ. A *theory* of the good is presupposed, but it is ostensibly neutral between divergent particular conceptions, and supplies a least common denominator on which a choice in the original position can be based without unfairness to any of the parties. Only later, when the principles of justice have been reached on this basis, will it be possible to rule out certain particular interests or aims as illegitimate because they are unjust. It is a fundamental feature of Rawls' conception of the fairness of the original position that it should not permit the choice of principles of justice to depend on a particular conception of the good over which the parties may differ.

The construction does not, I think, accomplish this, and there are reasons to believe that it cannot be successfully carried out. Any hypothetical choice situation which requires agreement among the parties will have to impose strong restrictions on the grounds of choice, and these restrictions can be justified only in terms of a conception of the good. It is one of those cases in which there is no neutrality to be had, because neutrality needs as much justification as any other position.

Rawls' minimal conception of the good does not amount to a weak assumption: it depends on a strong assumption of the sufficiency of that reduced conception for the purposes of justice. The refusal to rank particular conceptions of the good implies a very marked tolerance for individual inclinations. Rawls is opposed not only to teleological conceptions according to which justice requires adherence to the principles that will maximize the good. He is also opposed to the natural position that even in a nonteleological theory what is just must depend on what is good, at least to the extent that a correct conception of the good must be used in determining what counts as an advantage and what as a disadvantage, and how much, for purposes of distribution and compensation. I interpret him as saying that the principles of justice are objective and interpersonally recognizable in a way that conceptions of the good are not. The refusal to rank individual conceptions and the reliance on primary goods are intended to insure this objectivity.

Objectivity may not be so easily achieved.[4] The suppression of knowledge required to achieve unanimity is not equally fair to all the parties, because the primary goods are not equally valuable in pursuit of all conceptions of the good. They will serve to advance many different individual life plans (some more efficiently than others), but they are less useful in implementing views that hold a good life to be readily achievable only in certain well-defined types of social structure, or only in a society that works concertedly for the realization of certain higher human capacities and the suppression of baser ones, or only given certain types of economic relations among men. The model contains a strong individualistic bias, which is further strengthened by the motivational assumptions of mutual disinterest and absence of envy. These assumptions have the effect of discounting the claims of conceptions of the good that depend heavily on the relation between one's own position and that of others (though Rawls is prepared to allow such considerations to enter

[4] For the ideas in this paragraph I am indebted to Mary Gibson.

in so far as they affect self-esteem). The original position seems to presuppose not just a neutral theory of the good, but a liberal, individualistic conception according to which the best that can be wished for someone is the unimpeded pursuit of his own path, provided it does not interfere with the rights of others. The view is persuasively developed in the later portions of the book, but without a sense of its controversial character.

Among different life plans of this general type the construction is neutral. But given that many conceptions of the good do not fit into the individualistic pattern, how can this be described as a fair choice situation for principles of justice? Why should parties in the original position be prepared to commit themselves to principles that may frustrate or contravene their deepest convictions, just because they are deprived of the knowledge of those convictions?

There does not seem to be any way of redesigning the original position to do away with a restrictive assumption of this kind. One might think it would be an improvement to allow the parties full information about everyone's preferences and conception of the good, merely depriving them of the knowledge of who they were. But this, as Rawls points out (pp. 173–4), would yield no result at all. For either the parties would retain their conceptions of the good and, choosing from different points of view, would not reach unanimity, or else they would possess no aims of their own and would be asked to choose in terms of the aims of all the people they might be—an unintelligible request which provides no basis for a unified choice, in the absence of a dominant conception. The reduction to a common ground of choice is therefore essential for the model to operate at all, and the selection of that ground inevitably represents a strong assumption.

Let us now turn to the argument leading to the choice of the two principles in the original position as constructed. The core of this argument appears in Sections 26–9, intertwined with an argument against the choice of the principle of average utility. Rawls has gone to some lengths to defend his controversial claim that in the original position it is rational to adopt the maximin rule which leads one to choose principles that favor the bottom of the social hierarchy, instead of accepting a greater risk at the bottom in return for the possibility of greater benefits at the top (as might be prudentially rational if one had an equal chance of being anyone in the society).

Rawls states (p. 154) that three conditions which make maximin plausible hold in the original position to a high degree. (1) 'There must be some reason for sharply discounting estimates of ... probabilities'. (2) 'The person choosing has a conception of the good such that he cares very little, if anything, for what he might gain above the minimum stipend that he can, in fact, be sure of by following the maximin rule'. (3) 'The rejected alternatives have outcomes that one can hardly accept'. Let us consider these in turn.

The first condition is very important, and the claim that it holds in the original position is not based simply on a general rejection of the principle of insufficient reason (that is, the principle that where probabilities are unknown they should be regarded as equal). For one could characterize the original position in such a way that the parties would be prudentially rational to choose as if they had an equal chance of being anyone in the society, and the problem is to see why this would be an inappropriate representation of the grounds for a choice of principles.

One factor mentioned by Rawls is that the subject matter of the choice is extremely serious, since it involves institutions that will determine the total life prospects for the parties and those close to them. It is not just a choice of alternatives for a single occasion. Now this would be a reason for a conservative choice even if one knew the relative probabilities of different outcomes. It would be irresponsible to accept even a small risk of dreadful life prospects for oneself and one's descendants in exchange for a good chance of wealth or power. But what is needed is an account of why probabilities should be totally discounted, and not just with regard to the most unacceptable outcomes. The difference principle, for example, is supposed to apply at all levels of social development, so it is not justified merely by the desire to avoid grave risks. The fact that total life prospects are involved does not seem an adequate explanation. There must be some reason against allowing probabilities (proportional, for instance, to the number of persons in each social position) to enter into the choice of distributions above an acceptable minimum. Let me stress that I am posing a question not about decision theory but about the design of the original position and the comprehensiveness of the veil of ignorance. Why should it be thought that a just solution will be reached only if these considerations are suppressed?

Their suppression is justified, I think, only on the assumption that the proportions of people in various social positions are regarded as morally irrelevant, and this must be because it is not thought acceptable to sum advantages and disadvantages over persons, so that a loss for some is compensated by a gain for others. This aspect of the design of the original position appears, therefore, to be motivated by the wish to avoid extending to society as a whole the principle of rational choice for one man. Now this is supposed to be one of the *conclusions* of the contract approach, not one of its presuppositions. Yet the constraints on choice in Rawls' version of the original position are designed to rule out the possibility of such an extension,[5] by requiring that probabilities be discounted. I can see no way to avoid presupposing some definite view on this matter in the design of a contract situation. If that is true, then a contract approach cannot give any particular view very much support.

Consider next the second condition. Keeping in mind that the parties in the original position do not know the stage of development of their society, and therefore do not know what minimum will be guaranteed by a maximin strategy, it is difficult to understand how an individual can know that he 'cares very little, if anything, for what he might gain above the minimum'. The explanation Rawls offers (p. 156) seems weak. Even if parties in the original position accept the priority of liberty, and even if the veil of ignorance leaves them with a skeletal conception of the good, it seems impossible that they should care very little for increases in primary economic and social goods above what the difference principle guarantees at any given stage of social development.

Finally, the third condition, that one should rule out certain possibilities as unacceptable, is certainly a ground for requiring a social minimum and the priority of basic personal liberties, but it is not a ground for adopting the maximin rule in that general form needed to justify the choice of the difference principle. That must rely on stronger egalitarian premises.[6]

[5] I.e. they do not just refuse to assume that the extension is acceptable: they assume that it is unacceptable.

[6] A factor not considered in Rawls' argument, which suggests that the difference principle may be too weak, is the following. If differential social and economic benefits are allowed to provide incentives, then the people at the top will tend to be those with certain talents and abilities, and the people at the bottom, even though they are better off than they would be otherwise, will tend to lack those qualities. Such a consistent schedule of rewards inevitably affects people's sense of their intrinsic worth, and any society operating on the difference principle will have a meritocratic flavor. This is very different from the case where an unequal distribution that benefits the

Some of these premises reveal themselves in other parts of the argument. For example, the strongly egalitarian idea that sacrifice at the bottom is always worse than sacrifice at the top plays a central role in the appeal to strains of commitment and psychological stability. It is urged against the utilitarian alternative to the difference principle, for example, that the sacrifices utilitarianism might require would be psychologically unacceptable.

> The principles of justice apply to the basic structure of the social system and to the determination of life prospects. What the principle of utility asks is precisely a sacrifice of these prospects. We are to accept the greater advantages of others as a sufficient reason for lower expectations over the whole course of our life. This is surely an extreme demand. In fact, when society is conceived as a system of cooperation designed to advance the good of its members, it seems quite incredible that some citizens should be expected, on the basis of political principles, to accept lower prospects of life for the sake of others [p. 178].

Notice that if we substitute the words 'difference principle' for 'principle of utility', we get an argument that might be offered against the difference principle by someone concentrating on the sacrifices it requires of those at the top of the social order. They must live under institutions that limit their life prospects unless an advantage to them also benefits those beneath them. The only difference between the two arguments is in the relative position of the parties and of their sacrifices.[7] It is of course a vital difference, but that depends on a moral judgment—namely, that sacrifices which lessen social inequality are acceptable while sacrifices which increase inequality are not.

This apeal to psychological stability and the strains of commitment therefore adds to the grounds of choice in the original

worst off is not visibly correlated with any independent qualities. Rawls does suggest (p. 546) that 'excusable envy' may be given its due in the operation of the difference principle by including self-esteem among the primary goods. But he does not stress the *bases* of income inequality. The phenomenon I have described is not *envy*. Rawls is too willing to rely on equal liberty as the support of self-esteem; this leads him to underrate the effect of differential rewards on people's conception of themselves. A reward that is consistently attached to a certain quality stops being perceived as mere good luck.

[7] Exactly the same sacrifice could, after all, be either at the bottom or at the top, depending on the stage of advancement of the society.

position a moral view that belongs to the substance theory. The argument may receive some support from Rawls' idea about the natural development of moral sentiments, but they in turn are not independent of his ethical theory. If a hypothetical choice in the original position must be based on what one can expect to find morally acceptable in real life, then that choice is not the true ground of acceptability.[8]

Another strong conclusion of the theory is the priority of equal liberty, expressed in the lexical ordering of the two principles. The argument for *equal* liberty as a natural goal is straightforward. No analogue of the difference principle can apply permanently to liberty because it cannot be indefinitely increased. There will come a point at which increases in the liberty of the worst off can be achieved not by further increasing the liberty of the best off, but only by closing the gap. If one tries to maximize for everyone what really has a maximin, the result is equality.

The priority of liberty over other goods, however, is chosen in the original position on the basis of a judgment that the fundamental interest in determining one's plan of life assumes priority once the most basic material needs have been met, and that further increases in other goods depend for their value primarily on the ability to employ them under conditions of maximum liberty. 'Thus the desire for liberty is the chief regulative interest that the parties must suppose they all will have in common in due course. The veil of ignorance forces them to abstract from the particulars of their plans of life thereby leading to this conclusion. The serial ordering of the two principles then follows' (p. 543). The parties also reflect that equal liberty guarantees

[8] A similar objection could be made to Rawls' claim that the difference principle provides a condition of reciprocal advantage that allows everyone to cooperate willingly in the social order. Obviously, those at the bottom could not prefer any other arrangement, but what about those at the top? Rawls says the following:

> To begin with, it is clear that the well-being of each depends on a scheme of social cooperation without which no one could have a satisfactory life. Secondly, we can ask for the willing cooperation of everyone only if the terms of the scheme are reasonable. The difference principle, then, seems to be a fair basis on which those better endowed, or more fortunate in their social circumstances, could expect others to collaborate with them when some workable arrangement is a necessary condition of the good of all (p. 103).

But if some scheme of social cooperation is necessary for *anyone* to have a satisfactory life, everyone will benefit from a wide range of schemes. To assume that the worst off need further benefits to cooperate willingly while the best off do not is simply to repeat the egalitarian principle.

them all a basic self-esteem against the background of which some differences in social position and wealth will be acceptable. Here again an explicitly liberal conception of individual good is used to defend a choice in the original position.

I have attempted to argue that the presumptions of the contract method Rawls employs are rather strong, and that the original position therefore offers less independent support to his conclusions than at first appears. The egalitarian liberalism which he develops and the conception of the good on which it depends are extremely persuasive, but the original position serves to model rather than to justify them. The contract approach allied with a non-liberal conception of the good would yield different results, and some conceptions of the good are incompatible with a contract approach to justice altogether. I believe that Rawls' conclusions can be more persuasively defended by direct moral arguments for liberty and equality, some of which he provides and some of which are indirectly represented in his present position. He remarks that it is worth

> noting from the outset that justice as fairness, like other contract views, consists of two parts: (1) an interpretation of the initial situation and of the problem of choice posed there, and (2) a set of principles which, it is argued, would be agreed to. One may accept the first part of the theory (or some variant thereof), but not the other, and conversely [p. 15].

He suggests that the principles are more likely to be rejected than their contractual basis, but I suspect the reverse. It seems to me likely that over the long term this book will achieve its permanent place in the literature of political theory because of the substantive doctrine that it develops so eloquently and persuasively. The plausibility of the results will no doubt be taken to confirm the validity of the method, but such inferences are not always correct. It is possible that the solution to the combinatorial problems of social choice can be reached by means of self-interested individual choice under carefully specified conditions of uncertainty, but the basis of such a solution has yet to be discovered.

This is already a famous and influential book, and inevitably for a certain time it will engage the attention of students of philosophy, politics, law, and economics. The longer life of a work and its broader impact on the habits of thought of reflective persons can never be predicted with certainty, but it is an interesting question. Although *A Theory of Justice* is for the most part very readable, it does not possess the literary distinction that has helped to make other important political works—those of Hobbes or Mill, for example—part of the common intellectual property of mankind. It does, however, possess another feature of great importance. Reading it is a powerful experience, because one is in direct contact at every point with a striking temperament and cast of mind. It is in that sense a very personal work, and the perceptions and attitudes one finds in it are vivid, intelligent, and appealing. The outlook expressed by this book is not characteristic of its age, for it is neither pessimistic nor alienated nor angry nor sentimental nor utopian. Instead it conveys something that today may seem incredible: a hopeful affirmation of human possibilities. Yet the hope has a basis, for Rawls possesses a deep sense of the multiple connections between social institutions and individual psychology. Without illusion he describes a pluralistic social order that will call forth the support of free men and evoke what is best in them. To have made such a vision precise, alive, and convincing is a memorable achievement.

2 The Original Position[1]

RONALD DWORKIN*

I

I trust that it is not necessary to describe John Rawls' famous idea of the original position in any great detail. It imagines a group of men and women who come together to form a social contract. Thus far it resembles the imaginary congresses of the classical social contract theories. The original position differs, however, from these theories in its description of the parties.

[1] I have benefited from discussion of a draft of this paper in a seminar in Oxford in the fall of 1972, and from discussions with H. L. A. Hart and Thomas Nagel. Nagel raises some of the issues discussed in the first part of this paper in his review of Rawls' book, 'Rawls on Justice', *Philosophical Review*, Vol. 81 (1973), this volume, p. 1.

* Professor of Jurisprudence and Fellow of University College, Oxford.

They are men and women with ordinary tastes, talents, ambitions, and convictions, but each is temporarily ignorant of these features of his own personality, and must agree upon a contract before his self-awareness returns.

Rawls tries to show that if these men and women are rational, and act only in their own self-interest, they will choose his two principles of justice. These provide, roughly, that every person must have the largest political liberty compatible with a like liberty for all, and that inequalities in power, wealth, income, and other resources must not exist except in so far as they work to the absolute benefit of the worst-off members of society. Many of Rawls' critics disagree that men and women in the original position would inevitably choose these two principles. The principles are conservative, and the critics believe they would be chosen only by men who were conservative by temperament, and not by men who were natural gamblers. I do not think this criticism is well-taken, but in this essay, at least, I mean to ignore the point. I am interested in a different issue.

Suppose that the critics are wrong, and that men and women in the original position would in fact choose Rawls' two principles as being in their own best interest. Rawls seems to think that that fact would provide an argument in favor of these two principles as a standard of justice against which to test actual political institutions. But it is not immediately plain why this should be so.

If a group contracted in advance that disputes amongst them would be settled in a particular way, the fact of that contract would be a powerful argument that such disputes should be settled in that way when they do arise. The contract would be an argument in itself, independent of the forces of the reasons that might have led different people to enter the contract. Ordinarily, for example, each of the parties supposes that a contract he signs is in his own interest; but if someone has made a mistake in calculating his self-interest, the fact that he did contract is a strong reason for the fairness of holding him nevertheless to the bargain.

Rawls does not suppose that any group ever entered into a social contract of the sort he describes. He argues only that if a group of rational men did find themselves in the predicament of the original position, they would contract for the two principles. His contract is hypothetical, and hypothetical contracts do not supply an independent argument for the fairness of enforcing their terms. A hypothetical contract is not simply a pale form of an actual contract; it is no contract at all.

If, for example, I am playing a game, it may be that I would have agreed to any number of ground rules if I had been asked in advance of play. It does not follow that these rules may be enforced against me if I have not, in fact, agreed to them. There must be reasons, of course, why I would have agreed if asked in advance, and these may also be reasons why it is fair to enforce these rules against me even if I have not agreed. But my hypothetical agreement does not count as a reason, independent of these other reasons, for enforcing the rules against me, as my actual agreement would have.

Suppose that you and I are playing poker and we find, in the middle of a hand, that the deck is one card short. You suggest that we throw the hand in, but I refuse because I know I am going to win and I want the money in the pot. You might say that I would certainly have agreed to that procedure had the possibility of the deck being short been raised in advance. But your point is not that I am somehow committed to throwing the hand in by an agreement I never made. Rather you use the device of a hypothetical agreement to make a point that might have been made without that device, which is that the solution recommended is so obviously fair and sensible that only someone with an immediate contrary interest could disagree. Your main argument is that your solution is fair and sensible, and the fact that I would have chosen it myself adds nothing of substance to that argument. If I am able to meet the main argument nothing remains, rising out of your claim that I would have agreed, to be answered or excused.

In some circumstances, moreover, the fact that I would have agreed does not even suggest an independent argument of this character. Everything depends on your reasons for supposing that I would have agreed. Suppose you say that I would have agreed, if you had brought up the point and insisted on your solution, because I very much wanted to play and would have given in rather than miss my chance. I might concede that I would have agreed for that reason, and then add that I am lucky that you did not raise the point. The fact that I would have agreed if you had insisted neither adds nor suggests any argument why I should agree now. The point is not that it would have been unfair of you to insist on your proposal as a condition of playing; indeed, it would not have been. If you had held out for your proposal, and

I had agreed, I could not say that my agreement was in any way nullified or called into question because of duress. But if I had not in fact agreed, the fact that I would have in itself means nothing.

I do not mean that it is never relevant, in deciding whether an act affecting someone is fair, that he would have consented if asked. If a doctor finds a man unconscious and bleeding, for example, it might be important for him to ask whether the man would consent to a transfusion if he were conscious. If there is every reason to think that he would, that fact is important in justifying the transfusion if the patient later, perhaps because he has undergone a religious conversion, condemns the doctor for having proceeded. But this sort of case is beside the present point, because the patient's hypothetical agreement shows that his will was inclined toward the decision at the time and in the circumstances that the decision was taken. He has lost nothing by not being consulted at the appropriate time, because he would have consented if he had been. The original position argument is very different. If we take it to argue for the fairness of applying the two principles we must take it to argue that because a man would have consented to certain principles if asked in advance, it is fair to apply those principles to him later, under different circumstances, when he does not consent.

But that is a bad argument. Suppose I did not know the value of my painting on Monday; if you had offered me $100 for it then I would have accepted. On Tuesday I discovered it was valuable. You cannot argue that it would be fair for the courts to make me sell it to you for $100 on Wednesday. It may be my good fortune that you did not ask me on Monday, but that does not justify coercion against me later.

We must therefore treat the argument from the original position as we treat your argument in the poker game; it must be a device for calling attention to some independent argument for the fairness of the two principles—an argument that does not rest on the false premise that a hypothetical contract has some pale binding force. What other argument is available? One might say that the original position shows that the two principles are in the best interests of every member of any political community, and that it is fair to govern in accordance with them for that reason. It is true that if the two principles could be shown to be in everyone's interest, that would be a sound argument for their fairness, but it is hard to see how the original position can be used to show that they are.

We must be careful to distinguish two senses in which something might be said to be in my interest. It is in my *antecedent* interest to make a bet on a horse that, all things considered, offers the best odds, even if, in the event, the horse loses. It is in my actual interest to bet on the horse that wins, even if the bet was, at the time I made it, a silly one. If the original position furnishes an argument that it is in everyone's interest to accept the two principles over other possible bases for a constitution, it must be an argument that uses the idea of antecedent and not actual interest. It is not in the actual best interest of everyone to choose the two principles, because when the veil of ignorance is lifted some will discover that they would have been better off if some other principle, like the principle of average utility, had been chosen.

A judgment of antecedent interest depends upon the circumstances under which the judgment is made, and, in particular, upon the knowledge available to the man making the judgment. It might be in my antecedent interest to bet on a certain horse at given odds before the starting gun, but not, at least at the same odds, after he has stumbled on the first turn. The fact, therefore, that a particular choice is in my interest at a particular time, under conditions of great uncertainty, is not a good argument for the fairness of enforcing that choice against me later under conditions of much greater knowledge. But that is what, on this interpretation, the original position argument suggests, because it seeks to justify the contemporary use of the two principles on the supposition that, under conditions very different from present conditions, it would be in the antecedent interest of everyone to agree to them. If I have bought a ticket on a longshot it might be in my antecedent interest, before the race, to sell the ticket to you for twice what I paid; it does not follow that it is fair to you to take it from me for that sum when the longshot is about to win.

Someone might now say that I have misunderstood the point of special conditions of uncertainty in the original position. The parties are made ignorant of their special resources and talents to prevent them from bargaining for principles that are inherently unfair because they favor some collection of resources and talents over others. If the man in the original position does not know his special interests, he cannot negotiate to favor them. In that case, it might be said, the uncertainty of the original position does not vitiate the argument from antecedent interest as I have suggested, but only limits the range within which self-interest might operate. The argument shows that the two principles are in everyone's interest once obviously unfair principles are removed from consideration by the device of uncertainty. Since the only additional knowledge contemporary men and women have over men and women in the original position is knowledge that they ought not to rely upon in choosing principles of justice, their antecedent interest is, so far as it is relevant, the same, and if that is so the original position argument does offer a good argument for applying the two principles to contemporary politics.

But surely this confuses the argument that Rawls makes with a different argument that he might have made. Suppose his men and women had full knowledge of their own talents and tastes, but had to reach agreement under conditions that ruled out, simply by stipulation, obviously unfair principles like those providing special advantage for named individuals. If Rawls could show that, once such obviously unfair principles had been set aside, it would be in the interest of everyone to settle for his two principles, that would indeed count as an argument for the two principles. My point—that the antecedent self-interest of men in the original position is different from that of contemporary men —would no longer hold because both groups of men would have the same knowledge about themselves, and be subject to the same moral restrictions against choosing obviously unfair principles.

Rawls' actual argument is quite different, however. The ignorance in which his men must choose affects their calculations of self-interest, and cannot be described merely as setting boundaries within which these calculations must be applied. Rawls supposes, for example, that his men would inevitably choose conservative principles because this would be the only rational choice, in their ignorance, for self-interested men to make. But some actual men, aware of their own talents, might well prefer less conservative principles that would allow them to take advantage of the resources they know they have. Someone who considers the original position an argument for the conservative principles, therefore, is faced with this choice. If less conservative principles, like principles that favor named individuals, are to be ruled out as obviously unfair, then the argument for the conservative principles is complete at the outset, on grounds of obvious fairness alone. In that case neither the original position nor any considerations of self-interest it is meant to demonstrate play any role in the argument. But if less conservative principles cannot be ruled out in advance as obviously unfair, then imposing ignorance on Rawls' men, so that they prefer the more conservative principles, cannot be explained simply as ruling out obvious unfair choices. And since this affects the antecedent self-interest of these men, the argument that the original position demonstrates the antecedent self-interest of actual men must therefore fail. This same dilemma can, of course, be constructed for each feature of the two principles.

I recognize that the argument thus far seems to ignore a distinctive feature of Rawls' methodology, which he describes as the technique of seeking a 'reflective equilibrium' between our ordinary, unreflective moral beliefs and some theoretical structure that might unify and justify these ordinary beliefs (pp. 48ff). It might now be said that the idea of an original position plays a part in this reflective equilibrium, which we will miss if we insist, as I have, on trying to find a more direct, one-way argument from the original position to the two principles of justice.

The technique of equilibrium does play an important role in Rawls' argument, and it is worth describing that technique briefly here. The technique assumes that Rawls' readers have a sense, which we draw upon in our daily life, that certain particular political arrangements or decisions like conventional trials, are just and others, like slavery, are unjust. It assumes, moreover, that we are each able to arrange these immediate intuitions or convictions in an order that designates some of them as more certain than others. Most people, for example, think that it is more plainly unjust for the state to execute innocent citizens of its own than to kill innocent foreign civilians in war. They might be prepared to abandon their position on foreign civilians in war, on the basis of some argument, but would be much more reluctant to abandon their view on executing innocent countrymen.

It is the task of moral philosophy, according to the technique of equilibrium, to provide a structure of principles that supports these immediate convictions about which we are more or less secure, with two goals in mind. First, this structure of principles must explain the convictions by showing the underlying assumptions they reflect; second, it must provide guidance in those cases about which we have either no convictions or weak or contradictory convictions. If we are unsure, for example, whether economic institutions that allow great disparity of wealth are unjust, we may turn to the principles that explain our confident convictions, and then apply these principles to that difficult issue.

But the process is not simply one of finding principles that accommodate our more-or-less settled judgments. These principles must support, and not merely account for, our judgments, and this means that the principles must have independent appeal to our moral sense. It might be, for example, that a cluster of familiar moral convictions could be shown to serve an undeserving policy—perhaps, that the standard judgments we make without reflection serve the purpose of maintaining one particular class in political power. But this discovery would not vouch for the principle of class egoism; on the contrary, it would discredit our ordinary judgments, unless some other principle of a more respectable sort could be found that also fits our intuitions, in which case it would be this principle and not the class-intent principle that our intuitions would recommend.

It might be that no coherent set of principles could be found that has independent appeal and that supports the full set of our immediate convictions; indeed it would be surprising if this were not often the case. If that does happen, we must compromise, giving way on both sides. We might relax, though we could not abandon, our initial sense of what might be an acceptable principle. We might come to accept, for example, after further reflection, some principle that seemed to us initially unattractive, perhaps the principle that men should sometimes be made to be free. We might accept this principle if we were satisfied that no less harsh principle could support the set of political convictions we were especially reluctant to abandon. On the other hand, we must also be ready to modify or adjust, or even to give up entirely, immediate convictions that cannot be accommodated by any principle that meets our relaxed standards; in adjusting these immediate convictions we will use our initial sense of which seem to us more and which less certain, though in principle no immediate conviction can be taken as immune from reinspection or abandonment if that should prove necessary. We can expect to proceed back and forth between our immediate judgments and the structure of explanatory principles in this way, tinkering first with one side and then the other, until we arrive at what Rawls calls the state of reflective equilibrium in which we are satisfied, or as much satisfied as we can reasonably expect.

It may well be that, at least for most of us, our ordinary political judgments stand in this relation of reflective equilibrium with Rawls' two principles of justice, or, at least, that they could be made to do so through the process of adjustment just described. It is neverthless unclear how the idea of the original position fits into this structure or, indeed, why it has any role to play at all. The original position is not among the ordinary political convictions that we find we have, and that we turn to reflective equilibrium to justify. If it has any role, it must be in the process of justification, because it takes its place in the body of theory we construct to bring our convictions into balance. But if the two principles of justice are themselves in reflective equilibrium with our convictions, it is unclear why we need the original position to supplement the two principles on the theoretical side of the balance. What can the idea contribute to a harmony already established?

We should consider the following answer. It is one of the conditions we impose on a theoretical principle, before we allow it to figure as a justification of our convictions, that the people the principle would govern would have accepted that principle, at least under certain conditions, if they had been asked, or at least that the principle can be shown to be in the antecedent interest of every such person. If this is so, then the original position plays an essential part in the process of justification through equilibrium. It is used to show that the two principles conform to this established standard of acceptability for political principles. At the same time, the fact that the two principles, which do conform to that standard, justify our ordinary convictions in reflective equilibrium, reinforces our faith in the standard and encourages us to apply it to other issues of political or moral philosophy.

This answer does not advance the case that the original position furnishes an argument for the two principles, however; it merely restates the ideas we have already considered and rejected. It is certainly not part of our established political traditions or ordinary moral understanding that principles are acceptable only if they would be chosen by men in the particular predicament of the original position. It is, of course, part of these traditions that principles are fair if they have in fact been chosen by those whom they govern, or if they can at least be shown to be in their antecedent common interest. But we have already seen that the original position device cannot be used to support either of these arguments in favor of applying the two principles to contemporary politics. If the original position is to play any role in a structure of principles and convictions in reflective equilibrium, it must be by virtue of assumptions we have not yet identified.

It is time to reconsider an earlier assumption. So far I have been treating the original position construction as if it were either the foundation of Rawls' argument or an ingredient in a reflective equilibrium established between our political intuitions and his two principles of justice. But, in fact, Rawls does not treat the original position that way. He describes the construction in these words:

> I have emphasized that this original position is purely hypothetical. It is natural to ask why, if this agreement is never actually entered into, we should take any interest in these principles, moral or otherwise. The answer is that the conditions embodied in the description of the original position are ones that we do in fact accept. Or if we do not, then perhaps we can be persuaded to do so by philosophical reflection. Each aspect of the contractual situation can be given supporting grounds.... On the other hand, this conception is also an intuitive notion that suggests its own elaboration, so that led on by it we are drawn to define more clearly the standpoint from which we can best interpret moral relationships. We need a conception that enables us to envision our objective from afar: the intuitive notion of the original position is to do this for us [pp. 21–2].

This description is taken from Rawls' first statement of the original position. It is recalled and repeated in the very last paragraph of the book (p. 587). It is plainly of capital importance, and it suggests that the original position, far from being the foundations of his argument, or an expository device for the technique of equilibrium, is one of the major substantive products of the theory as a whole. Its importance is reflected in another crucial passage. Rawls describes his moral theory as a type of psychology. He wants to characterize the structure of our (or, at least, one person's) capacity to make moral judgments of a certain sort, that is, judgments about justice. He thinks that the conditions embodied in the original position are the fundamental 'principles governing our moral powers, or, more specifically, our sense of justice' (p. 51). The original position is therefore a schematic representation of a particular mental process of at least some, and perhaps most, human beings, just as depth grammar, he suggests, is a schematic presentation of a different mental capacity.

All this suggests that the original position is an intermediate conclusion, a halfway point in a deeper theory that provides philosophical arguments for its conditions. In the next part of this essay I shall try to describe at least the main outlines of this deeper theory. I shall distinguish three features of the surface argument of the book—the technique of equilibrium, the social contract, and the original position itself—and try to discern which of various familiar philosophical principles or positions these represent.

First, however, I must say a further word about Rawls' exciting, if imprecise, idea that the principles of this deeper theory are constitutive of our moral capacity. That idea can be understood on different levels of profundity. It may mean, at its least profound, that the principles that support the original position as a device for reasoning about justice are so widely shared and so little questioned within a particular community, for whom the book is meant, that the community could not abandon these principles without fundamentally changing its patterns of reasoning and arguing about political morality. It may mean, at its most profound, that these principles are innate categories of morality common to all men, imprinted in their neural structure, so that man could not deny these principles short of abandoning the power to reason about morality at all.

I shall be guided, in what follows, by the less profound interpretation, though what I shall say, I think, is consistent with the more profound. I shall assume, then, that there is a group of men and women who find, on reading Rawls, that the original position does strike them as a proper 'intuitive notion' from which to think about problems of justice, and who would find it persuasive, if it could be demonstrated that the parties to the original position would in fact contract for the two principles he describes. I suppose, on the basis of experience and the literature, that this group contains a very large number of those who think about justice at all, and I find that I am a member myself. I want to discover the hidden assumptions that bend the inclinations of this group that way, and I shall do so by repeating the question with which I began. Why does Rawls' argument support his claim that his two principles are principles of justice? My answer is complex and it will take us, at times, far from his text, but not, I think, from its spirit.

II

A. *Equilibrium*

I shall start by considering the philosophical basis of the technique of equilibrium I just described. I must spend several pages in this way, but it is important to understand what substantive features of Rawls' deep theory are required by his method. This technique presupposes, as I said, a familiar fact about our moral lives. We all entertain beliefs about justice that we hold because they seem right, not because we have deduced or inferred them from other beliefs. We may believe in this way, for example, that slavery is unjust, and that the standard sort of trial is fair.

These different sorts of beliefs are, according to some philosophers, direct perceptions of some independent and objective moral facts. In the view of other philosophers they are simply subjective preferences, not unlike ordinary tastes, but dressed up in the language of justice to indicate how important they seem to us. In any event, when we argue with ourselves or each other about justice we use these accustomed beliefs—which we call 'intuitions' or 'convictions'—in roughly the way Rawls' equilibrium technique suggests. We test general theories about justice against our own intuitions, and we try to confound those who disagree with us by showing how their own intuitions embarrass their own theories.

Suppose we try to justify this process by setting out a philosophical position about the connection between moral theory and moral intuition. The technique of equilibrium supposes what might be called a 'coherence' theory of morality.[2] But we have a choice between two general models that define coherence and explain why it is required, and the choice between these is significant and consequential for our moral philosophy. I shall describe these two models, and then argue that the equilibrium technique makes sense on one but not the other.

I call the first a 'natural' model. It presupposes a philosophical position that can be summarized in this way. Theories of justice, like Rawls' two principles, describe an objective moral reality; they are not, that is, created by men of societies but are rather discovered by them, as they discover laws of physics. The main

[2] *See* Feinberg, 'Justice, Fairness and Rationality', *Yale Law Journal*, Vol. 81 (1972), pp. 1018–21.

instrument of this discovery is a moral faculty possessed by at least some men, which produces concrete intuitions of political morality in particular situations, like the intuition that slavery is wrong. Those intuitions are clues to the nature and existence of more abstract and fundamental moral principles, as physical observations are clues to the existence and nature of fundamental physical laws. Moral reasoning or philosophy is a process of reconstructing the fundamental principles by assembling concrete judgments in the right order, as a natural historian reconstructs the shape of the whole animal from the fragments of its bones that he has found.

The second model is quite different. It treats intuitions of justice not as clues to the existence of independent principles, but rather as stipulated features of a general theory to be constructed, as if a sculptor set himself to carve the animal that best fit a pile of bones he happened to find together. This 'constructive' model does not assume, as the natural model does, that principles of justice have some fixed, objective existence, so that descriptions of these principles must be true or false in some standard way. It does not assume that the animal it matches to the bones actually exists. It makes the different, and in some ways more complex, assumption that men and women have a responsibility to fit the particular judgments on which they act into a coherent program of action, or, at least, that officials who exercise power over other men have that sort of responsibility.

This second, constructive, model is not unfamiliar to lawyers. It is analogous to one model of common law adjudication. Suppose a judge is faced with a novel claim—for example, a claim for damages based on a legal right to privacy that courts have not heretofore recognized.[3] He must examine such precedents as seem in any way relevant to see whether any principles that are, as we might say, 'instinct' in these precedents bear upon the claimed right to privacy. We might treat this judge as being in the position of a man arguing from moral intuitions to a general moral theory. The particular precedents are analogous to intuitions; the judge tries to reach an accommodation between these precedents and a set of principles that might justify them and also justify further decisions that go beyond them. He does not suppose, however, that the precedents are glimpses into a

[3] I have here in mind the famous argument of Brandeis and Warren. *See* Brandeis and Warren, 'The Right to Privacy', *Harvard Law Review*, Vol. 4 (1890), which is a paradigm of argument in the constructive model.

moral reality, and therefore clues to objective principles he ends by declaring. He does not believe that the principles are 'instinct' in the precedents in that sense. Instead, in the spirit of the constructive model, he accepts these precedents as specifications for a principle that he must construct, out of a sense of responsibility for consistency with what has gone before.

I want to underline the important difference between the two models. Suppose that an official holds, with reasonable conviction, some intuition that cannot be reconciled with his other intuitions by any set of principles he can now fashion. He may think, for example, that it is unjust to punish an attempted murder as severely as a successful one, and yet be unable to reconcile that position with his sense that a man's guilt is properly assessed by considering only what he intended, and not what actually happened. Or he may think that a particular minority race, as such, is entitled to special protection, and be unable to reconcile that view with his view that distinctions based on race are inherently unfair to individuals. When an official is in this position the two models give him different advice.

The natural model supports a policy of following the troublesome intuition, and submerging the apparent contradiction, in the faith that a more sophisticated set of principles, which reconciles that intuition, does in fact exist though it has not yet been discovered. The official, according to this model, is in the position of the astronomer who has clear observational data that he is as yet unable to reconcile in any coherent account, for example, of the origin of the solar system. He continues to accept and employ his observational data, placing his faith in the idea that some reconciling explanation does exist though it has not been, and for all he knows may never be, discovered by men.

The natural model supports this policy because it is based on a philosophical position that encourages the analogy between moral intuitions and observational data. It makes perfect sense, on that assumption, to suppose that direct observations, made through a moral faculty, have outstripped the explanatory powers of those who observe. It also makes sense to suppose that some correct explanation, in the shape of principles of morality, does in fact exist in spite of this failure; if the direct observations are sound, some explanation must exist for why matters are as they have been observed to be in the moral universe, just as some explanation must exist for why matters are as they have been observed to be in the physical universe.

The constructive model, however, does not support the policy of submerging apparent inconsistency in the faith that reconciling principles must exist. On the contrary, it demands that decisions taken in the name of justice must never outstrip an official's ability to account for these decisions in a theory of justice, even when such a theory must compromise some of his intuitions. It demands that we act on principle rather than on faith. Its engine is a doctrine of responsibility that requires men to integrate their intuitions and subordinate some of these, when necessary, to that responsibility. It presupposes that articulated consistency, decisions in accordance with a program that can be made public and followed until changed, is essential to any conception of justice. An official in the position I describe, guided by this model, must give up his apparently inconsistent position; he must do so even if he hopes one day, by further reflection, to devise better principles that will allow all his initial convictions to stand as principles.[4]

The constructive model does not presuppose skepticism or relativism. On the contrary, it assumes that the men and women who reason within the model will each hold sincerely the convictions they bring to it, and that this sincerity will extend to criticizing as unjust political acts or systems that offend the most profound of these. The model does not deny, any more than it affirms, the objective standing of any of these convictions; it is therefore consistent with, though as a model of reasoning it does not require, the moral ontology that the natural model presupposes.

It does not require that ontology because its requirements are independent of it. The natural model insists on consistency with conviction, on the assumption that moral intuitions are accurate observations; the requirement of consistency with conviction as an independent requirement, flowing not from the assumption that these convictions are accurate reports, but from the different assumption that it is unfair for officials to act except on the basis of a general public theory that will constrain them to consistency, provide a public standard for testing or debating or predicting what they do, and not allow appeals to unique intuitions that

[4] The famous debate between Professor Wechsler, 'Toward Neutral Principles in Constitutional Law', *Harvard Law Review*, Vol. 73 (1959), and his critics may be illuminated by this distinction. Wechsler proposes a constructive model for constitutional adjudication, while those who favor a more tentative or intuitive approach to constitutional law are following the material model.

might mask prejudice or self-interest in particular cases. The constructive model requires coherence, then, for independent reasons of political morality; it takes convictions held with the requisite sincerity as given, and seeks to impose conditions on the acts that these intuitions might be said to warrant. If the constructive model is to constitute morality, in either of the senses I have distinguished, these independent reasons of political morality are at the heart of our political theories.

The two models, therefore, represent different standpoints from which theories of justice might be developed. The natural model, we might say, looks at intuitions from the personal standpoint of the individual who holds them, and who takes them to be discrete observations of moral reality. The constructive model looks at these intuitions from a more public standpoint; it is a model that someone might propose for the governance of a community each of whose members has strong convictions that differ, though not too greatly, from the convictions of others.

The constructive model is appealing, from this public standpoint, for an additional reason. It is well suited to group consideration of problems of justice, that is, to developing a theory that can be said to be the theory of a community rather than of particular individuals, and this is an enterprise that is important, for example, in adjudication. The range of initial convictions to be assessed can be expanded or contracted to accommodate the intuitions of a larger or smaller group, either by including all convictions held by any members, or by excluding those not held by all, as the particular calculation might warrant. This process would be self-destructive on the natural model, because every individual would believe that either false observations were being taken into account or accurate observations disregarded, and hence that the inference to objective morality was invalid. But on the constructive model that objection would be unavailable; the model, so applied, would be appropriate to identify the program of justice that best accommodates the community's common convictions, for example, with no claim to a description of an objective moral universe.

Which of these two models, then, better supports the technique of equilibrium? Some commentators seem to have assumed that the technique commits Rawls to the natural model.[5] But the alliance between that model and the equilibrium technique turns

[5] *See* e.g. Hare, 'Rawls' Theory of Justice—I', *Philosophical Quarterly*, Vol. 23 (1973); this volume, pp. 81 ff.

out to be only superficial; when we probe deeper we find that they are incompatible. In the first place, the natural model cannot explain one distinctive feature of the technique. It explains why our theory of justice must fit our intuitions about justice, but it does not explain why we are justified in amending these intuitions to make the fit more secure.

Rawls' notion of equilibrium, as I said earlier, is a two-way process; we move back and forth between adjustments to theory and adjustments to conviction until the best fit possible is achieved. If my settled convictions can otherwise be captured by, for example, a straightforward utilitarian theory of justice, that may be a reason, within the technique, for discarding my intuition that slavery would be wrong even if it advanced utility. But on the natural model this would be nothing short of cooking the evidence, as if a naturalist rubbed out the footprints that embarrassed his efforts to describe the animal that left them, or the astronomer just set aside the observations that his theory could not accommodate.

We must be careful not to lose this point in false sophistication about science. It is common to say—Rawls himself draws the comparison[6]—that scientists also adjust their evidence to achieve a smooth set of explanatory principles. But if this is true at all, their procedures are very different from those recommended by the technique of equilibrium. Consider, to take a familiar example, optical illusions or hallucinations. It is perfectly true that the scientist who sees water in the sand does not say that the pond was really there until he arrived at it, so that physics must be revised to provide for disappearing water; on the contrary, he uses the apparent disappearing as evidence of an illusion, that is, as evidence that, contrary to his observation, there was never any water there at all.

The scientist, of course, cannot leave the matter at that. He cannot dismiss mirages unless he supplements the laws of physics with laws of optics that explain them. It may be that he has, in some sense, a choice amongst competing sets of explanations of all his observations taken together. He may have a choice, for example, between either treating mirages as physical objects of a special sort and then amending the laws of physics to allow for disappearing objects of this sort, or treating mirages as optical illusions and then developing laws of optics to explain such illusions. He has a choice in the sense that his experience does

[6] Rawls draws attention to the distinction, p. 49.

not absolutely force either of these explanations upon him; the former is a possible choice, though it would require wholesale revision of both physics and common sense to carry it off.

This is, I take it, what is meant by philosophers like Quine who suppose that our concepts and our theories face our experience as a whole, so that we might react to recalcitrant or surprising experience by making different revisions at different places in our theoretical structures if we wish.[7] Regardless of whether this is an accurate picture of scientific reasoning, it is not a picture of the procedure of equilibrium, because this procedure argues not simply that alternative structures of principle are available to explain the same phenomena, but that some of the phenomena, in the form of moral convictions, may simply be ignored the better to serve some particular theory.

It is true that Rawls sometimes describes the procedure in a more innocent way. He suggests that if our tentative theories of justice do not fit some particular intuition, this should act as a warning light requiring us to reflect on whether the conviction is really one we hold (p. 48). If my convictions otherwise support a principle of utility, but I feel that slavery would be unjust even if utility were improved, I might think about slavery again, in a calmer way, and this time my intuitions might be different and consistent with that principle. In this case, the initial inconsistency is used as an occasion for reconsidering the intuition, but not as a reason for abandoning it.

Still, this need not happen. I might continue to receive the former intuition, no matter how firmly I steeled myself against it. In that case the procedure nevertheless authorizes me to set it aside if that is required to achieve the harmony of equilibrium. But if I do, I am not offering an alternative account of the evidence, but simply disregarding it. Someone else, whose intuitions are different, may say that mine are distorted, perhaps because of some childhood experience, or because I am insufficiently imaginative to think of hypothetical cases in which slavery might actually improve utility. He may say, that is, that my sensibilities are defective here, so that my intuitions are not genuine perceptions of moral reality, and may be set aside like the flawed reports of a color-blind man.

But I cannot accept that about myself, as an explanation for my own troublesome convictions, so long as I hold these convictions

[7] W. V. Quine, 'Two Dogmas of Empiricism', in *From a Logical Point of View* (2nd ed. rev. 1964).

and they seem to me sound, indistinguishable in their moral quality from my other convictions. I am in a different position from the color-blind man who need only come to understand that others' perceptions differ from his. If I believe that my intuitions are a direct report from some moral reality, I cannot accept that one particular intuition is false until I come to feel or sense that it is false. The bare fact that others disagree, if they do, may be an occasion for consulting my intuitions again, but if my convictions remain the same, the fact that others may explain them in a different way cannot be a reason for my abandoning them, instead of retaining them in the faith that a reconciliation of these with my other convictions does in fact exist.

Thus, the natural model does not offer a satisfactory explanation of the two-way feature of equilibrium. Even if it did, however, it would leave other features of that technique unexplained; it would leave unexplained, for example, the fact that the results of the technique, at least in Rawls' hands, are necessarily and profoundly practical. Rawlsian men and women in the original position seek to find principles that they and their successors will find it easy to understand and publicize and observe; principles otherwise appealing are to be rejected or adjusted because they are too complex or are otherwise impractical in this sense. But principles of justice selected in this spirit are compromises with infirmity, and are contingent in the sense that they will change as the general condition and education of people change. This seems inconsistent with the spirit, at least, of the natural model, according to which principles of justice are timeless features of some independent moral reality, to which imperfect men and women must attempt to conform as best they can.

The equilibrium technique, moreover, is designed to produce principles that are relative in at least two ways. First, it is designed to select the best theory of justice from a list of alternative theories that must not only be finite, but short enough to make comparisons among them feasible. This limitation is an important one; it leads Rawls himself to say that he has no doubt that an initial list of possible theories expanded well beyond the list he considers would contain a better theory of justice than his own two principles (p. 581). Second, it yields results that are relative to the area of initial agreement among those who jointly conduct the speculative experiments it recommends. It is designed, as Rawls says, to reconcile men who disagree by fixing on what is common ground among them (pp. 580–1). The test concededly will yield different results for different groups, and for the same group at different times, as the common ground of confident intuition shifts.

If the equilibrium technique were used within the natural model, the authority of its conclusions would be seriously compromised by both forms of relativism. If the equilibrium argument for Rawls' two principles, for example, shows only that a better case can be made for them than for any other principles on a restricted short list, and if Rawls himself is confident that further study would produce a better theory, then we have very little reason to suppose that these two principles are an accurate description of moral reality. It is hard to see, on the natural model, why they then should have any authority at all.

Indeed, the argument provides no very good ground for supposing even that the two principles are a better description of moral reality than other theories on the short list. Suppose we are asked to choose, among five theories of justice, the theory that best unites our convictions in reflective equilibrium, and we pick, from among these, the fifth. Let us assume that there is some sixth theory that we would have chosen had it appeared on the list. This sixth theory might be closer to, for example, the first on our original list than to the fifth, at least in the following sense: over a long term, a society following the first might reach more of the decisions that a society following the sixth would reach than would a society following the fifth.

Suppose, for example, that our original list included, as available theories of justice, classical utilitarianism and Rawls' two principles, but did not include average utilitarianism. We might have rejected classical utilitarianism on the ground that the production of pleasure for its own sake, unrelated to any increase in the welfare of particular human beings or other animals, makes little sense, and then chosen Rawls' two principles as the best of the theories left. We might nevertheless have chosen average utilitarianism as superior to the two principles, if it had been on the list, because average utilitarianism does not suppose that just any increase in the total quantity of pleasure is good. But classical utilitarianism, which we rejected, might be closer to average utilitarianism, which we would have chosen if we could have, than are the two principles which we did choose. It might be closer, in the sense described, because it would dictate more of the particular decisions that average utilitarianism would require, and thus be a better description of ultimate moral reality, than would the two principles. Of course, average utilitarianism might itself be rejected in a still larger list, and the choice we should then make might indicate that another member of the original list was better than either classical utilitarianism or the two principles.

The second sort of relativism would be equally damaging on the natural model, for reasons I have already explained. If the technique of equilibrium is used by a single person, and the intuitions allowed to count are just his and all of his, then the results may be authoritative for him. Others, whose intuitions differ, will not be able to accept his conclusions, at least in full, but he may do so himself. If, however, the technique is used in a more public way, for example, by fixing on what is common amongst the intuitions of a group, then the results will be those that no one can accept as authoritative, just as no one could accept as authoritative a scientific result reached by disregarding what he believed to be evidence at least as pertinent as the evidence used.

So the natural model turns out to be poor support for the equilibrium technique. None of the difficulties just mentioned count, however, if we assume the technique to be in the service of the constructive model. It is, within that model, a reason for rejecting even a powerful conviction that it cannot be reconciled with other convictions by a plausible and coherent set of principles; the conviction is rejected not as a false report, but simply as ineligible within a program that meets the demands of the model. Nor does either respect in which the technique is relative embarrass the constructive model. It is not an embarrassment that some theory not considered might have been deemed superior if it had been considered. The model requires officials or citizens to proceed on the best program they can now fashion, for reasons of consistency that do not presuppose, as the natural model does, that the theory chosen is in any final sense true. It does not undermine a particular theory that a different group, or a different society, with different culture and experience, would produce a different one. It may call into question whether any group is entitled to treat its moral intuitions as in any sense objective or transcendental, but not that a particular society, which does treat particular convictions in that way, is therefore required to follow them in a principled way.

I shall assume, therefore, at least tentatively, that Rawls' methodology presupposes the constructive model of reasoning from particular convictions to general theories of justice, and I shall use that assumption in my attempt to show the further postulates of moral theory that lie behind this theory of justice.

B. *The Contract*

I come, then, to the second of the three features of Rawls' methodology that I want to discuss, which is the use he makes of the old idea of a social contract. I distinguish, as does Rawls, the general idea that an imaginary contract is an appropriate device for reasoning about justice, from the more specific features of the original position, which count as a particular application of that general idea. Rawls thinks that all theories that can be seen to rest on a hypothetical social contract of some sort are related and are distinguished as a class from theories that cannot; he supposes, for example, that average utilitarianism, which can be seen as the product of a social contract on a particular interpretation, is more closely related to his own theory than either is to classical utilitarianism, which cannot be seen as the product of a contract on any interpretation (chapter 30). In the next section I shall consider the theoretical basis of the original position. In this section I want to consider the basis of the more general idea of the contract itself.

Rawls says that the contract is a powerful argument for his principles because it embodies philosophical principles that we accept, or would accept if we thought about them. We want to find out what these principles are, and we may put our problem this way. The two principles comprise a theory of justice that is built up from the hypothesis of a contract. But the contract cannot sensibly be taken as the fundamental premise or postulate of that theory, for the reasons I described in the first part of this article. It must be seen as a kind of halfway point in a larger argument, as itself the product of a deeper political theory that argues for the two principles *through* rather than *from* the contract. We must therefore try to identify the features of a deeper theory that would recommend the device of a contract as the engine of justice, rather than the other theoretical devices Rawls mentions, like the device of the impartial spectator (pp. 144 ff).

We shall find the answer, I think, if we attend to and refine the familiar distinction philosophers make between two types of moral theories, which they call teleological theories and deontological theories.[8] I shall argue that any deeper theory that would justify Rawls' use of the contract must be a particular form of deontological theory, a theory that takes the idea of rights so seriously as to make them fundamental in political morality. I shall try to show how such a theory would be distinguished, as a type, from other types of political theories, and why only such a theory could give the contract the role and prominence Rawls does.

I must begin this argument, however, by explaining how I shall use some familiar terms. (1) I shall say that some state of affairs is a *goal* within a particular political theory if it counts in favor of a political act, within that theory, that the act will advance or preserve that state of affairs, and counts against an act that will retard or threaten it. Goals may be relatively specific, like full employment or respect for authority, or relatively abstract, like improving the general welfare, advancing the power of a particular concept of human goodness or of the good life. (2) I shall say that an individual has a *right* to a particular political act, within a political theory, if the failure to provide that act, when he calls for it, would be unjustified within that theory even if the goals of the theory would, on the balance, be disserviced by that act. The strength of a particular right, within a particular theory, is a function of the degree of disservice to the goals of the theory, beyond a mere disservice on the whole, that is necessary to justify refusing an act called for under the right. In the popular political theory apparently prevailing in the United States, for example, individuals have rights to free public speech on political matters and to a certain minimum standard of living, but neither right is absolute and the former is much stronger than the latter. (3) I shall say that an individual has a *duty* to act in a particular way, within a political theory, if a political decision constraining such act is justified within that theory notwithstanding that no goal of the system would be served by that decision. A theory may provide, for example, that individuals have a duty to worship God, even though it does not stipulate any goal served by requiring them to do so.[9]

The three concepts I have described work in different ways,

[8] Rawls defines these terms at pp. 24–5 and 30.

[9] I do not count, as goals, the goal of respecting rights or enforcing duties. In this and other apparent ways my use of the terms I define is narrower than ordinary language permits.

but they all serve to justify or to condemn, at least *pro tanto*, particular political decisions. In each case, the justification provided by citing a goal, a right, or a duty is in principle complete, in the sense that nothing need be added to make the justification effective, if it is not undermined by some competing considerations. But, though such a justification is in this sense complete, it need not, within the theory, be ultimate. It remains open to ask why the particular goal, right, or duty is itself justified, and the theory may provide an answer by deploying a *more basic* goal, right, or duty that is served by accepting this less basic goal, right, or duty as a complete justification in particular cases.

A particular goal, for example, might be justified as contributing to a more basic goal; thus, full employment might be justified as contributing to greater average welfare. Or a goal might be justified as serving a more basic right or duty; a theory might argue, for example, that improving the gross national product, which is a goal, is necessary to enable the state to respect the rights of individuals to a decent minimum standard of living, or that improving the efficiency of the police process is necessary to enforce various individual duties not to sin. On the other hand, rights and duties may be justified on the ground that, by acting as a complete justification on particular occasions, they in fact serve more fundamental goals; the duty of individuals to drive carefully may be justified, for example, as serving the more basic goal of improving the general welfare. This form of justification does not, of course, suggest that the less basic right or duty itself justifies political decisions only when these decisions, considered one by one, advance the more basic goal. The point is rather the familiar one of rule utilitarianism, that treating the right or duty as a complete justification in particular cases, without reference to the more basic goal, will in fact advance the goal in the long run.

So goals can be justified by other goals or by rights or duties, and rights or duties can be justified by goals. Rights and duties can also be justified, of course, by other, more fundamental duties or rights. Your duty to respect my privacy, for example, may be justified by my right to privacy. I do not mean merely that rights and duties may be correlated, as opposite sides of the same coin. That may be so when, for example, a right and the corresponding duty are justified as serving a more fundamental goal, as when your right to property and my corresponding duty not to trespass are together justified by the more fundamental goal of socially efficient land use. In many cases, however, corresponding rights and duties are not correlative, but one is derivative from the other, and it makes a difference which is derivative from which. There is a difference between the idea that you have a duty not to lie to me because I have a right not to be lied to, and the idea that I have a right that you do not lie to me because you have a duty not to tell lies. In the first case I justify a duty by calling attention to a right; if I intend any further justification it is the right that I must justify, and I cannot do so by calling attention to the duty. In the second case it is the other way around. The difference is important because, as I shall shortly try to show, a theory that takes rights as fundamental is a theory of a different character from one that takes duties as fundamental.

Political theories will differ from one another, therefore, not simply in the particular goals, rights, and duties each sets out, but also in the way each connects the goals, rights, and duties it employs. In a well-formed theory some consistent set of these, internally ranked or weighted, will be taken as fundamental or ultimate within the theory. It seems reasonable to suppose that any particular theory will give ultimate pride of place to just one of these concepts; it will take some overriding goal, or some set of fundamental rights, or some set transcendent duties, as fundamental, and show other goals, rights, and duties as subordinate and derivative.[10]

We may therefore make a tentative initial classification of the political theories we might produce, on the constructive model, as deep theories that might contain a contract as an intermediate device. Such a theory might be *goal-based*, in which case it would take some goal, like improving the general welfare, as fundamental; it might be *right-based*, taking some right, like the right of all men to the greatest possible overall liberty, as fundamental; or it might be *duty-based*, taking some duty, like the duty to obey God's will as set forth in the Ten Commandments, as fundamental. It is easy to find examples of pure, or nearly pure, cases of each of these types of theory. Utilitarianism is, as my example suggested, a goal-based theory; Kant's categorical imperatives compose a duty-based theory; and Tom Paine's theory of revolution is right-based.

Theories within each of these types are likely to share certain very general characteristics. The types may be contrasted, for example, by comparing the attitudes they display toward individual choice and conduct. Goal-based theories are concerned

[10] But an 'intuitionist' theory, as Rawls uses that term, need not. *See* p. 34.

with the welfare of any particular individual only in so far as this contributes to some state of affairs stipulated as good quite apart from his choice of that state of affairs. This is plainly true of totalitarian goal-based theories, like fascism, that take the interest of a political organization as fundamental. It is also true of the various forms of utilitarianism, because, though they count up the impact of political decisions on distinct individuals, and are in this way concerned with individual welfare, they merge these impacts into overall totals or averages and take the improvement of these totals or averages as desirable quite apart from the decision of any individual that it is. It is also true of perfectionist theories, like Aristotle's, that impose upon individuals an ideal of excellence and take the goal of politics to be the culture of such excellence.

Right-based and duty-based theories, on the other hand, place the individual at the center, and take his decision or conduct as of fundamental importance. But the two types put the individual in a different light. Duty-based theories are concerned with the moral quality of his acts, because they suppose that it is wrong, without more, for an individual to fail to meet certain standards of behavior. Kant thought that it was wrong to tell a lie no matter how beneficial the consequences, not because hating this practice promoted some goal, but just because it was wrong. Right-based theories are, in contrast, concerned with the independence rather than the conformity of individual action. They presuppose and protect the value of individual thought and choice. Both types of theory make use of the idea of moral rules, codes of conduct to be followed, on individual occasions, without consulting self-interest. Duty-based theories treat such codes of conduct as of the essence, whether set by society to the individual or by the individual to himself. The man at their center is the man who must conform to such a code, or be punished or corrupted if he does not. Right-based theories, however, treat codes of conduct as instrumental, perhaps necessary to protect the rights of others, but having no essential value in themselves. The man at their center is the man who benefits from others' compliance, not the man who leads the life of virtue by complying himself.

We should, therefore, expect that the different types of theories would be associated with different metaphysical or political temperaments, and that one or another would be dominant in certain sorts of political economy. Goal-based theories, for example, seem especially compatible with homogeneous societies, or those at least temporarily united by an urgent, overriding goal, like self-defense or economic expansion. We should also expect that these differences between types of theory would find echoes in the legal systems of the communities they dominate. We should expect, for example, that a lawyer would approach the question of punishing moral offenses through the criminal law in a different way if his inchoate theory of justice were goal-, right- or duty-based. If his theory were goal-based he would consider the full effect of enforcing morality upon his overriding goal. If this goal were utilitarian, for example, he would entertain, though he might, in the end, reject, Lord Devlin's arguments that the secondary effects of punishing immorality may be beneficial.[11] If his theory were duty-based, on the other hand, he would see the point of the argument, commonly called retributive, that since immorality is wrong the state must punish it even if it harms no one. If his theory were right-based, however, he would reject the retributive argument, and judge the utilitarian argument against the background of his own assumption that individual rights must be served even at some cost to the general welfare.

All this is, of course, superficial and trivial as ideological sociology. My point is only to suggest that these differences in the character of a political theory are important quite apart from the details of position that might distinguish one theory from another of the same character. It is for this reason that the social contract is so important a feature of Rawls' methodology. It signals that his deep theory is a right-based theory, rather than a theory of either of the other two types.

The social contract provides every potential party with a veto: unless he agrees, no contract is formed. The importance, and even the existence, of this veto is obscured in the particular interpretation of the contract that constitutes the original position. Since no one knows anything about himself that would distinguish him from anyone else, he cannot rationally pursue any interest that is different. In these circumstances nothing turns on each man having a veto, or, indeed, on there being more than one potential party to the contract in the first place. But the original position is only one interpretation of the contract, and in any other interpretation in which the parties do have some knowledge with which to distinguish their situation or ambitions from those of others, the veto that the contract gives each party becomes crucial. The force of the veto each individual has de-

[11] *See* Dworkin, 'Lord Devlin and the Enforcement of Morals', *Yale Law Journal*, Vol. 75 (1966).

pends, of course, upon his knowledge, that is to say, the particular interpretation of the contract we in the end choose. But the fact that individuals should have any veto at all is in itself remarkable.

It can have no place in a purely goal-based theory, for example. I do not mean that the parties to a social contract could not settle on a particular social goal and make that goal henceforth the test of the justice of political decisions. I mean that no goal-based theory could make a contract the proper device for deciding upon a principle of justice in the first place; that is, the deep theory we are trying to find could not itself be goal-based.

The reason is straightforward. Suppose some particular overriding goal, like the goal of improving the average welfare in a community, or increasing the power and authority of a state, or creating a utopia according to a particular conception of the good, is taken as fundamental within a political theory. If any such goal is fundamental, then it authorizes such distribution of resources, rights, benefits, and burdens within the community as will best advance that goal, and condemns any other. The contract device, however, which supposes each individual to pursue his own interest and gives each a veto on the collective decision, applies a very different test to determine the optimum distribution. It is designed to produce the distribution that each individual deems in his own best interest, given his knowledge under whatever interpretation of the contract is specified, or at least to come as close to that distribution as he thinks he is likely to get. The contract, therefore, offers a very different test of optimum distribution than a direct application of the fundamental goal would dictate. There is no reason to suppose that a system of individual vetoes will produce a good solution to a problem in which the fairness of a distribution, considered apart from the contribution of the distribution to an overall goal, is meant to count for nothing.

It might be, of course, that a contract would produce the result that some fundamental goal dictates. Some critics, in fact, think that men in the original position, Rawls' most favored interpretation of the contract, would choose a theory of justice based on principles of average utility, that is, just the principles that a deep theory stipulating the fundamental goal of average utility would produce.[12] But if this is so, it is either because of

[12] John Mackie presented a forceful form of this argument to an Oxford seminar in the fall of 1972.

coincidence or because the interpretation of the contract has been chosen to produce this result; in either case the contract is supererogatory, because the final result is determined by the fundamental goal and the contract device adds nothing.

One counterargument is available. Suppose it appears that the fundamental goal will in fact be served only if the state is governed in accordance with principles that all men will see to be, in some sense, in their own interest. If the fundamental goal is the aggrandizement of the state, for example, it may be that this goal can be reached only if the population does not see that the government acts for this goal, but instead supposes that it acts according to principles shown to be in their individual interests through a contract device; only if they believe this will they work in the state's interest at all. We cannot ignore this devious, if unlikely, argument, but it does not support the use that Rawls makes of the contract. The argument depends upon a deception, like Sidgwick's famous argument that utilitarianism can best be served by keeping the public ignorant of that theory.[13] A theory that includes such a deception is ineligible on the constructivist model we are pursuing, because our aim, on that model, is to develop a theory that unites our convictions and can serve as a program for public action; publicity is as much a requirement of our deep theory as of the conception of justice that Rawls develops within it.

So a goal-based theory cannot support the contract, except as a useless and confusing appendage. Neither can a duty-based deep theory, for much the same reasons. A theory that takes some duty or duties to be fundamental offers no ground to suppose that just institutions are those seen to be in everyone's self-interest under some description. I do not deny, again, that the parties to the contract may decide to impose certain duties upon themselves and their successors, just as they may decide to adopt certain goals, in the exercise of their judgment of their own self-interest. Rawls describes the duties they would impose upon themselves under his most favored interpretation, the original position, and calls these natural duties (section 19). But this is very different from supposing that the deep theory, which makes this decision decisive of what these duties are, can itself be duty-based.

It is possible to argue, of course, as many philosophers have, that a man's self-interest lies in doing his duty under the moral

[13] H. Sidgwick, *The Methods of Ethics* (7th ed. 1907), 489 ff.

law, either because God will punish him otherwise, or because fulfilling his role in the natural order is his most satisfying activity, or, as Kant thought, because only in following rules he could consistently wish universal can he be free. But that says a man's duties define his self-interest, and not the other way round. It is an argument not for deciding upon a man's particular duties by letting him consult his own interest, but rather for his setting aside any calculations of self-interest except calculations of duty. It could not, therefore, support the role of a Rawlsian contract in a duty-based deep theory.

It is true that if a contract were a feature of a duty-based deep theory, an interpretation of the contract could be chosen that would dissolve the apparent conflict between self-interest and duty. It might be a feature of the contract situation, for example, that all parties accepted the idea just mentioned, that their self-interest lay in ascertaining and doing their duty. This contract would produce principles that accurately described their duties, at least if we add the supposition that they are proficient, for some reason, in discovering what their duties are. But then, once again, we have made the contract supererogatory, a march up the hill and then back down again. We would have done better simply to work out principles of justice from the duties the deep theory takes as fundamental.

The contract does, however, make sense in a right-based deep theory. Indeed, it seems a natural development of such a theory. The basic idea of a right-based theory is that distinct individuals have interests that they are entitled to protect if they so wish. It seems natural, in developing such a theory, to try to identify the institutions an individual would veto in the exercise of whatever rights are taken as fundamental. The contract is an excellent device for this purpose, for at least two reasons. First, it allows us to distinguish between a veto in the exercise of these rights and a veto for the sake of some interest that is not so protected, a distinction we can make by adopting an interpretation of the contract that reflects our sense of what these rights are. Second, it enforces the requirements of the constructive model of argument. The parties to the contract face a practical problem; they must devise a constitution from the options available to them, rather than postponing their decision to a day of later moral insight, and they must devise a program that is both practical and public in the sense I have described.

It seems fair to assume, then, that the deep theory behind the original position must be a right-based theory of some sort. There is another way to put the point, which I have avoided until now. It must be a theory that is based on the concept of rights that are *natural*, in the sense that they are not the product of any legislation, or convention, or hypothetical contract. I have avoided that phrase because it has, for many people, disqualifying metaphysical associations. They think that natural rights are supposed to be spectral attributes worn by primitive men like amulets, which they carry into civilization to ward off tyranny. Mr. Justice Black, for example, thought it was a sufficient refutation of a judicial philosophy he disliked simply to point out that it seemed to rely on this preposterous notion.[14]

But on the constructive model, at least, the assumption of natural rights is not a metaphysically ambitious one. It requires no more than the hypothesis that the best political program, within the sense of that model, is one that takes the protection of certain individual choices as fundamental, and not properly subordinated to any goal or duty or combination of these. This requires no ontology more dubious or controversial than any contrary choice of fundamental concepts would be and, in particular, no more than the hypothesis of a fundamental goal that underlies the various popular utilitarian theories would require. Nor is it disturbing that a Rawlsian deep theory makes these rights natural rather than legal or conventional. Plainly, any right-based theory must presume rights that are not simply the product of deliberate legislation or explicit social custom, but are independent grounds for judging legislation and custom. On the constructive model, the assumption that rights are in this sense natural is simply one assumption to be made and examined for its power to unite and explain our political convictions, one basic programatic decision to submit to this test of coherence and experience.

C. *The Original Position*

I said that the use of a social contract, in the way that Rawls uses it, presupposes a deep theory that assumes natural rights. I want now to describe, in somewhat more detail, how the device of a contract applies that assumption. It capitalizes on the idea, mentioned earlier, that some political arrangements might be said to

[14] Griswold v. Connecticut, 381 *U.S.* 479, 507 (1964) (dissenting opinion).

be in the antecedent interest of every individual even though they are not, in the event, in his actual interest.

Everyone whose consent is necessary to a contract has a veto over the terms of that contract, but the worth of that veto, to him, is limited by the fact that his judgment must be one of antecedent rather than actual self-interest. He must commit himself, and so abandon his veto, at a time when his knowledge is sufficient only to allow him to estimate the best odds, not to be certain of his bet. So the contract situation is in one way structurally like the situation in which an individual with specific political rights confronts political decisions that may disadvantage him. He has a limited, political right to veto these, a veto limited by the scope of the rights he has. The contract can be used as a model for the political situation by shaping the degree or character of a party's ignorance in the contractual situation so that this ignorance has the same force on his decision as the limited nature of his rights would have in the political situation.

This shaping of ignorance to suit the limited character of political rights is most efficiently done simply by narrowing the individual goals that the parties to the contract know they wish to pursue. If we take Hobbes' deep theory, for example, to propose that men have a fundamental natural right to life, so that it is wrong to take their lives, even for social goals otherwise proper, we should expect a contract situation of the sort he describes. Hobbes' men and women, in Rawls' phrase, have lexically ordered security of life over all other individual goals; the same situation would result if they were simply ignorant of any other goals they might have and unable to speculate about the chances that they have any particular one or set of these.

The ignorance of the parties in the original position might thus be seen as a kind of limiting case of the ignorance that can be found, in the form of a distorted or eccentric ranking of interests, in classical contract theories and that is natural to the contract device. The original position is a limiting case because Rawls' men are not simply ignorant of interests beyond a chosen few; they are ignorant of all the interests they have. It would be wrong to suppose that this makes them incapable of any judgments of self-interest. But the judgments they make must nevertheless be very abstract; they must allow for any combination of interests, without the benefit of any supposition that some of these are more likely than others.

The basic right of Rawls' deep theory, therefore, cannot be a right to any particular individual goal, like a right to security of life, or a right to lead a life according to a particular conception of the good. Such rights to individual goals may be produced by the deep theory as rights that men in the original position would stipulate as being in their best interest. But the original position cannot itself be justified on the assumption of such a right, because the parties to the contract do not know that they have any such interest or rank it lexically ahead of others.

So the basic right of Rawls' deep theory must be an abstract right, that is, not a right to any particular individual goal. There are two candidates, within the familiar concepts of political theory, for this role. The first is the right to liberty, and it may strike many readers as both plausible and comforting to assume that Rawls' entire structure is based on the assumption of a fundamental natural right to liberty—plausible because the two principles that compose his theory of justice gave liberty an important and dominant place, and comforting because the argument attempting to justify that place seems uncharacteristically incomplete.[15]

Nevertheless, the right to liberty cannot be taken as the fundamental right in Rawls' deep theory. Suppose we define general liberty as the overall minimum possible constraints, imposed by government or by other men, on what a man might want to do.[16] We must then distinguish this general liberty from particular liberties, that is, freedom from such constraints on particular acts thought specially important, like participation in politics. The parties to the original position certainly have, and know that they have, an interest in general liberty, because general liberty will, *pro tanto*, improve their power to achieve any particular goals they later discover themselves to have. But the qualification is important because they have no way of knowing that general liberty will in fact improve this power overall, and every reason to suspect that it will not. They know that they might have other interests, beyond general liberty, that can be protected only by political constraints on acts of others.

So if Rawlsian men must be supposed to have a right to liberty of some sort, which the contract situation is shaped to embody, it must be a right to particular liberties. Rawls does name a list of basic liberties, and it is these that his men do choose to protect

15 *See* Hart, 'Rawls on Liberty and Its Priority', *University of Chicago Law Review*, Vol. 40 (1973), this volume, p. 230.

16 Cf. Rawls' definition of liberty at p. 202.

through their lexically ordered first principle of justice (p. 61). But Rawls plainly casts this principle as the product of the contract rather than as a condition of it. He argues that the parties to the original position would select these basic liberties to protect the basic goods they decide to value, like self-respect, rather than taking these liberties as goals in themselves. Of course they might, in fact, value the activities protected as basic liberties for their own sake, rather than as a means to some other goal or interest. But they certainly do not know that they do.

The second familiar concept of political theory is even more abstract than liberty. This is equality, and in one way Rawlsian men and women cannot choose other than to protect it. The state of ignorance in the original position is so shaped that the antecedent interest of everyone must lie, as I said, in the same solution. The right of each man to be treated equally without regard to his person or character or tastes is enforced by the fact that no one else can secure a better position by virtue of being different in any such respect. In other contract situations, when ignorance is less complete, individuals who share the same goal may nevertheless have different antecedent interests. Even if two men value life above everything else, for example, the antecedent interest of the weaker might call for a state monopoly of force rather than some provision for private vengeance, but the antecedent interest of the stronger might not. Even if two men value political participation above all else, the knowledge that one's views are likely to be more unorthodox or unpopular than those of the other will suggest that his antecedent interests calls for different arrangements. In the original position no such discrimination of antecedent interests can be made.

It is true that, in two respects, the principles of justice that Rawls thinks men and women would choose in the original position may be said to fall short of an egalitarian ideal. First, they subordinate equality in material resources, when this is necessary, to liberty of political activity, by making demands of the first principle prior to those of the second. Second, they do not take account of relative deprivation, because they justify any inequality when those worse off are better off than they would be, in absolute terms, without that inequality.

Rawls makes plain that these inequalities are required, not by some competing notion of liberty or some overriding goal, but by a more basic sense of equality itself. He accepts a distinction between what he calls two conceptions of equality:

> Some writers have distinguished between equality as it is invoked in connection with the distribution of certain goods, some of which will almost certainly give higher status or prestige to those who are more favored, and equality as it applies to the respect which is owed to persons irrespective of their social position. Equality of the first kind is defined by the second principle of justice . . . . But equality of the second kind is fundamental [p. 511].

We may describe a right to equality of the second kind, which Rawls says is fundamental, in this way. We might say that individuals have a right to equal concern and respect in the design and administration of the political institutions that govern them. This is a highly abstract right. Someone might argue, for example, that it is satisfied by political arrangements that provide equal opportunity, for office and position on the basis of merit. Someone else might argue, to the contrary, that it is satisfied only by a system that guarantees absolute equality of income and status, without regard to merit. A third man might argue that equal concern and respect is provided by that system, whatever it is, that improves the average welfare of all citizens counting the welfare of each on the same scale. A fourth might argue, in the name of this fundamental equality, for the priority of liberty, and for the other apparent inequalities of Rawls' two principles.

The right to equal concern and respect, then, is more abstract than the standard conceptions of equality that distinguish different political theories. It permits arguments that this more basic right requires one or another of these conceptions as a derivative right or goal.

The original position may now be seen as a device for testing these competing arguments. It supposes, reasonably, that political arrangements that do not display equal concern and respect are those that are established and administered by powerful men and women who, whether they recognize it or not, have more concern and respect for members of a particular class, or people with particular talents or ideals, than they have for others. It relies on this supposition in shaping the ignorance of the parties to the contract. Men who do not know to which class they belong cannot design institutions, consciously or unconsciously, to favor their own class. Men who have no idea of their own conception of the good cannot act to favor those who hold one ideal over those who hold another. The original position is well designed to enforce the abstract right to equal concern and respect, which must be understood to be the fundamental concept of Rawls' deep theory.

If this is right, then Rawls must not use the original position to argue for this right in the same way that he uses it, for example, to argue for the rights to basic liberties embodied in the first principle. The text confirms that he does not. It is true that he once says that equality of respect is 'defined' by the first principle of justice (p. 511). But he does not mean, and in any case he does not argue, that the parties choose to be respected equally in order to advance some more basic right or goal. On the contrary, the right to equal respect is not, on his account, a product of the contract, but a condition of admission to the original position. This right, he says, is 'owed to human beings as moral persons', and follows from the moral personality that distinguishes humans from animals. It is possessed by all men who can give justice, and only such men can contract (chapter 77). This is one right, therefore, that does not emerge from the contract, but is assumed, as the fundamental right must be, in its design.

Rawls is well aware that his argument for equality stands on a different footing from his argument for the other rights within his theory:

> Now of course none of this is literally argument. I have not set out the premises from which this conclusion follows, as I have tried to do, albeit not very rigorously, with the choice of conceptions of justice in the original position. Nor have I tried to prove that the characterization of the parties must be used as the basis of equality. Rather this interpretation seems to be the natural completion of justice as fairness [p. 509].

It is the 'natural completion', that is to say, of the theory as a whole. It completes the theory by providing the fundamental assumption that charges the original position, and makes it an 'intuitive notion' for developing and testing theories of justice.

We may therefore say that justice as fairness rests on the assumption of a natural right of all men and women to equality of concern and respect, a right they possess not by virtue of birth or characteristic or merit or excellence but simply as human beings with the capacity to make plans and give justice. Many readers will not be surprised by this conclusion, and it is, as I have said, reasonably clear from the text. It is an important conclusion, nevertheless, because some forms of criticism of the theory, already standard, ignore it. I shall close this long essay with one example.

One form of criticism has been expressed to me by many colleagues and students, particularly lawyers. They point out that the particular political institutions and arrangements that Rawls says men in the original position would choose are merely idealized forms of those now in force in the United States. They are the institutions, that is, of liberal constitutional democracy. The critics conclude that the fundamental assumptions of Rawls' theory must, therefore, be the assumptions of classical liberalism, however they define these, and that the original position, which appears to animate the theory, must somehow be an embodiment of these assumptions. Justice as fairness therefore seems to them, in its entirety, a particularly subtle rationalization of the political *status quo*, which may safely be disregarded by those who want to offer a more radical critique of the liberal tradition.

If I am right, this point of view is foolish, and those who take it lose an opportunity, rare for them, to submit their own political views to some form of philosophical examination. Rawls' most basic assumption is not that men have a right to certain liberties that Locke or Mill thought important, but that they have a right to equal respect and concern in the design of political institutions. This assumption may be contested in many ways. It will be denied by those who believe that some goal, like utility or the triumph of a class or the flowering of some conception of how men should live, is more fundamental than any individual right, including the right to equality. But it cannot be denied in the name of any more radical concept of equality, because none exists.

Rawls does argue that this fundamental right to equality requires a liberal constitution, and supports an idealized form of present economic and social structures. He argues, for example, that men in the original position would protect the basic liberties in the interest of their right to equality, once a certain level of material comfort has been reached, because they would understand that a threat to self-respect, which the basic liberties protect, is then the most serious threat to equal respect. He also argues that these men would accept the second principle in preference to material equality because they would understand that sacrifice out of envy for another is a form of subordination to him. These arguments may, of course, be wrong. I have certainly said nothing in their defense here. But the critics of liberalism now have the responsibility to show that they are wrong. They cannot say that Rawls' basic assumptions and attitudes are too far from their own to allow a confrontation.

3 History and Reason in Rawls' Moral Theory

MILTON FISK*

I. *Contract Theory*

Traditional contract theorists worked with a distinction between the natural and the contingent. Returning to the position of the original social contract was viewed as a way to concentrate on the natural. What humans do in the contract position was held to be expressive of what they are by nature, rather than expressive of what they have become because of the contingent situation they are in. The social principles they adopt in this position, then, carry greater moral weight than any they might adopt in contingent circumstances.

This suggests a familiar dilemma for the traditional contract theorist. Which features of the human situation are to be left out in the attempt to isolate features of human nature? The first horn of the dilemma results from leaving out too much. Omitting all factors peculiar to a given epoch would imply that human nature involves only factors which can be shared by others, regardless of the social configurations they live through. Even if there were such trans-historical factors, it becomes a serious question whether they are sufficient to yield a set of principles on which a society can be built. The second horn of the dilemma results from including too much. Suppose the factors that express themselves in a set of social principles are internally related to a particular historical epoch. Then, treating these factors as if they defined a state of nature for humans means presenting historically specific factors as if they were independent of the accidents of history. This kind of fallacy of misplaced concreteness is just what C. B. Macpherson accuses Locke of when he says, 'Locke read back

* Professor of Philosophy, Indiana University.

into the state of nature, in a generalized form, the assumptions he made about differential rights and rationality in existing societies'.[1]

At first sight, John Rawls' contract theory seems to provide a way around this dilemma. He seems not to require any appeal to human nature at all. Rather, he is concerned to find what conditions are relevant in the determination of what is just. A person cannot appeal to his or her own wealth in arguing that a capital gains tax is unjust. The 'original position', which corresponds to the old state of nature, includes those conditions that 'we are ready upon due consideration to recognize as reasonable' (p. 21) limits on fair terms of social cooperation. So, just as in Locke's state of nature humans are free and equal under a law of reason, so too, in Rawls' original position, humans are free and rational beings concerned with furthering their own interests in a situation of equality. From the conditions making up the original position, and certain contingent assumptions, one deduces the principles of justice. Whether the original position really does capture what is relevant to justice depends, in the last analysis, on whether these principles agree with the judgments we make on the basis of our sense of justice, after that sense has been refined by the consideration of several alternative conceptions of justice (p. 48). This is the test of how correct it is for Rawls to leave out of the original position differences in intelligence and strength, class differences, an economic system, a state, and culture (p. 137). Our reflective sense of justice is the ultimate arbiter on admittance to the original position.

I wish to show, in the remainder of this section, that there is an analogue of the difficulty encountered in the older contractualist. Because of this difficulty it becomes necessary, as I show in the next section, for Rawls himself to appeal to the distinction between the humanly essential and the humanly accidental. This faces him with the original dilemma encountered by the traditional contractualists. Has he ways of coping with this predicament they lacked? My ultimate aim is to show that he does not.

If indeed Rawls' reflective reason is a genuinely 'critical' instrument, it must reveal whether any condition that is a candidate for the original position is or is not tied to a specific historical period. One possibility is that Rawls' reflective reason leaves out too much. Earlier contract theory was suspect because a sup-

[1] C. B. Macpherson, *The Political Theory of Possessive Individualism* (Oxford University Press: New York, 1962), p. 238.

posedly trans-historical human nature would prove too impoverished to imply social principles. Similarly, it might be suspected that Rawls' reflective reason, employed as a critical device, leaves so little substance to the original position that it fails to imply principles capable of shaping a social order. The other possibility is that Rawls' reflective reason leaves out too little. Because it implied relevant social principles, the state of nature in earlier contract theory appeared to be no more than the result of analyzing contemporary assumptions, not facts about human nature. Since Rawls derives currently relevant social principles from his original position, justified as it is by reflective reason, then perhaps his reflective reason is, in the end, an 'analytical', not a critical, device for determining the internal merits and demerits of views of justice. If so, then what Rawls calls the sense of justice in reflective equilibrium would be incapable of acting as a filter for historically contingent criteria of what is relevant to justice. In short, the conditions making up the original position would in no sense define our place in society from 'the perspective of eternity' (p. 587).

So, suspicions analogous to those surrounding traditional contract theory rise to the surface when Rawls sets forth his new approach. If, as a critical device, the reflective sense of justice eliminates from the original position whatever is not eternally relevant to considerations of justice, how does the original position have enough content to imply social principles? If, on the other hand the reflective sense of justice is only an analytically refined sense of justice, must it not allow into the original position historically bound conditions, and thus must not the moral weight of what is derived from that position be historically limited?

II. *Human Nature*

Let it be granted that Rawls successfully derives social principles from an original position that passes the test of reflective reason. To extract himself from the above dilemma, Rawls must still defend the neutrality of reflective reason. In other words, reflective reason must be shown to place current ideas about what is relevant to justice in historical perspective, not to assume them. It must be shown to be a critical device that, despite our suspicions above, permits construction of an original position from which social principles can be derived. Rawls does not take on such a defense in explicit terms, but he has an idea of what needs to be shown to give such a defense.

What needs to be shown, he suggests, is a conformity of the original position with human nature. When persons act on the principles adopted in the original position 'their actions do not depend upon social or natural contingencies ... By acting from these principles persons express their nature as free and equal rational beings ...' (p. 252). Suppose it can be shown that conditions making up the original position constitute our human nature. Then, reflective reason, in certifying that the original position contains just what is relevant to justice, proves that it is a device capable of reviewing historical and class differences and finding, nontheless, something significant about justice transcending these differences. It does not matter that reflective reason treats freedom and equality as conditions to be realized in choosing principles of justice unless reason has a basis for treating freedom and equality as aspects of human nature. The older contractualists prove themselves more perspicacious than Rawls by getting straight to the question of what humans are. Rawls only backs into this question after it becomes clear that reason needs a guarantee it can act as a neutral criterion in determining what is relevant to justice. The guarantee consists in the conformity between its conclusions and what human nature is. Rawls has no doubts there is such a conformity:

> But when we knowingly act on the principles of justice ... we deliberately assume the limitations of the original position. One reason for doing this ... is to give expression to one's nature [p. 253].

Rawls' view of humans is that they are by nature free and rational beings (p. 253). This was also Hobbes' and Locke's view. But Rawls appeals to Kant's authority to make more plausible the suggestion that the perspective of the original position is that of 'noumenal selves' (p. 255). One expresses human nature, so conceived, when one acts according to principles that would be adopted in the original position. Acting unjustly expresses not just human nature but contingencies of ability, income, and status. Once the original position is seen as natural, one understands how Rawls can freely translate talk about the moral arbitrariness (pp. 72, 75, 141) of features of our real life position into talk about what is contingent (pp. 19, 32, 136) relative to human nature. Since the original position constructed by reflective reason expresses human nature, the principles derived from it regain the unlimited moral weight they would otherwise lose if reflective reason were not historically neutral.

This defense of the neutrality of the reflective sense of justice depends on showing that the conception of humans as by nature free, equal, and rational itself has title to neutrality. This title, however, is defective in several ways. In the first place, this conception of human nature is supported only by showing that certain values characteristic of the institutions of liberal democracy are associated with it. Thus the arguments for it do not in any sense tend to establish its title to neutrality. In the second place, those values of liberal institutions with which this conception of human nature is associated are limited in that adherence to those institutions has enormous social costs. There are alternative conceptions of human nature, associated with values realized in alternative institutions, that reduce these costs, though they will certainly incur other costs. Thus the enormous social costs of the liberal institutions stand in the way of rejecting these alternative institutions out of hand. The existence of the associated alternative conceptions of human nature undercuts the title to neutrality of the liberal conception. It appears that Rawls' working conception of human nature does not supply the principles of justice he derives from the original position with moral weight outside the context of liberal democracy. To show that the title to neutrality is defective in these two ways, I shall consider the two aspects of human nature separately—freedom in Section III and equality in Section IV.

III. *Freedom*

Consider the freedom that humans are said to have by nature. This freedom includes the ability to choose a 'plan of life' in a way that puts no *antecedent* restrictions on the sorts of plan that one might choose (pp. 254, 409, 448). Of course, once the facts of my historical situation are clear to me, then the life plans it would be rational for me to choose are considerably restricted. But this is a restriction *consequent* upon bringing these 'arbitrary facts to light. Antecedently, I am not bound to choose a plan consistent with my taking an interest in the interests of another.

[To permit savings for the future, Rawls does, nonetheless, posit an interest in the next generation (pp. 128, 140, 292).] Taking an interest in another would be an arbitrary restriction on the life plans antecedently open to me. Natural freedom implies that taking an interest in the interests of another is not a natural but a contingent feature of humans who have it. Likewise, Rawls sees a sentiment of benevolence, such as that appealed to by Hume as the basis for justice,[2] as an admissible consequent, but not antecedent, restriction on the kinds of life plan one can choose. So, since freedom is natural, benevolence is accidental. In this way, Rawls realizes the equivalent of Locke's natural freedom according to which persons are able to act 'without asking leave or depending upon the will of any other man'.[3]

Locke advocated this conception of freedom since it provided a basis for saying that a 'market' society, characterized by few antecedent restrictions on the plans of economic agents, as opposed to a 'status' society, characterized by economic relations that are antecedently limited by custom, is a natural society. But in the 1970s, the motivation for adopting Locke's view of freedom is not to defend the market, and the political institutions that perpetuate it, against failing feudal institutions. Rather, it is to show that the current market economy and the current institutions of liberal democracy are not yet adequate reflections of principles of justice derived from freedom. Thus the Lockean conception of freedom is used by Rawls to prepare for a constructive criticism, and hence reform, of the market economy and liberal democratic institutions. These institutions are not being defended against feudal ones, but implicitly against the rising tide of belief that they cannot be reformed to rid them of their chronically oppressive features. The radical purpose of the older contractualists has given way to a newer, conservative contractualism. That an author like Daniel Bell should view Rawls as far from a conservative—because of the welfare measures proposed by Rawls and of the compatibility of Rawls' principles with market 'socialism'—merely indicates that the imagination of liberal critics does not extend beyond conserving the market and liberal democracy. To see that Rawls' contractualism serves such a purpose, rather than that of giving us tools for detecting what may be fundamentally wrong with current institutions, let us

[2] David Hume, *An Enquiry Concerning the Principles of Morals*, Sect. V, Part II.

[3] John Locke, *The Second Treatise of Government*, Sect. 4.

note some of the consequences that flow from his conception of freedom.

(i) *Freedom of Thought.* Natural freedom forms, for Rawls, part of the basis for the right to freedom of thought, interpreted in the traditional liberal way. Natural freedom implies that persons are self-interested, though not necessarily selfish. Saying that there are no antecedent restrictions on the choice of a life plan implies that the only antecedently identifiable basis for choice is the fact that one has unspecified interests. In addition, Rawls posits that, if, as a matter of fact, we do have moral, religious, and philosophical interests, that is, moral, religious, and philosophical beliefs, we will give special importance to them. Not only do we stand by them because we are self-interested, but we give them highest place among our interests. Since one may well have such beliefs and since these beliefs may well not be those of the majority, one will be unwilling to allow the suppression of minority beliefs. To avoid taking a chance with these specially important interests, persons in the original position will agree that there should be freedom of thought (p. 207). The grounds for freedom of thought are then, first, natural freedom, and second, the intensity of religious and moral interests.

Here beliefs are put on par with skills and savings as entities over which people have proprietary rights. One protects one's beliefs as the ruling class protects its wealth. But this liberal theory of freedom of thought ignores an important difference between interests in thoughts and interests in wealth. The thoughts one takes an interest in defending are one's own rarely in more than the sense that one is willing to defend them. They are not one's own mental productions from one's own raw experiential materials. Rather, these religious, moral, and philosophical thoughts are, by and large, inculcated by institutions. And these institutions are strengthened by people defending these very thoughts.

This creates a paradox. Natural freedom permits a mental and and social servitude by leading to liberal freedom of thought. If there are no antecedent restrictions that result from concern for others' interests, there will be no antecedent restriction on accepting any moral belief. This will be true however manipulative the process might be by which one is led to hold that belief. This will be true however antagonistic holding that belief might ultimately be to one's interests as a member of an actual social group. Should one not be careful or critical in accepting beliefs? Presumably one should; but groups that manipulate people's beliefs do so in part by manipulating the criteria for their acceptance. In short, liberal freedom of thought provides a cover for the hidden persuaders that aid oppressing groups.

Genuine freedom of thought, by contrast, presupposes antecedent restrictions on the choice of life plans. These restrictions flow from the fact that humans are members of groups. Moral and other beliefs that are propagated by oppressing groups have only a qualified claim to toleration. The qualification has two facets to it. First, an oppressed group has the right to hold beliefs that further the cause of its liberation. Second, the duty to tolerate beliefs that have been imposed on them and that thwart their liberation is only a conditional one. In what way conditional?

Consider, on the one hand, toleration of people who speak for an oppressing group. It is folly for an oppressed group to attempt to suppress beliefs harmful to it when the balance of power decidedly favors the oppressing group. It would risk losing the civil liberties that protect its own growth. The same consideration is not present when an oppressed group has the potential for throwing off its oppression. Being tolerant then might be decisive in stabilizing the oppression.

Consider, on the other hand, toleration of people within an oppressed group who speak for the continuation of oppression. Once again the duty of toleration is conditional. As consciousness rises within a group as to its characteristic interests, the toleration of voices speaking against those interests denies, in effect, the right of the group to overcome imposed consciousness. Thus there is a point at which blacks need not continue to debate with or provide media space for those of their members who might insist that blacks are suited only for unskilled jobs. Similarly, women at some point might take the same stance against other women who preach the eternality of contemporary sex roles. And workers would at some point rightly deny expression to those of their leaders who insist that their lot is optimum under the profit system.

The natural freedom, from which the liberal's right to freedom of thought comes, obscures an important fact. In a society where there is oppression between groups, there are antecedent restrictions on choice of life-plans. These result from one's group membership and from restrictions imposed by the institutions of the society as a whole. Once this fact is recognized, the right to toleration appears in a more conditional light than the advocate of natural freedom is able to place it. This right vanishes when the circumstances make the overcoming of imposed consciousness a necessity for advancing the liberation of the group.

(ii) *The State.* Natural freedom leads to the need for a coercive sovereign interpreted in the traditional liberal way. Self-interested individuals agree to the principles of justice in order to maximize the minimum satisfaction of their individual interests. But once those principles are agreed to, any party to the contract might suspect others' hope to get greater benefits for themselves by violating the compact of cooperation (p. 240). If one suspects that others are cheating, one is tempted to cheat oneself. To avoid the implied possibility of social breakdown, a coercive sovereign is needed who will not let the suspicion arise in the first place. We thus have a deduction of the liberal conception of the state. The state is an organ for guaranteeing that self-interested individuals do not let their self-interest destroy the compact of cooperation.

Two points are important about this view of the state. First, it is diversionary; the fundamental instability calling forth the state is concealed by it. Second, it makes concrete the result of an unwarranted simplifying abstraction; without the abstraction of group membership from human nature to get mutual disinterestedness, there would be no ground for saying that there must be a state in all circumstances.

The first point is supported by the general observation that the state more forcefully displays its coercive mechanism in precautions taken against conflicts arising from membership in antagonistic groups than in precautions taken against instances of freeloading by random individuals. This observation is explicable. It is these group conflicts, and not temptation to freeload, that are the roots of social instability. The criminal courts are a bulwark of social order against the lower classes. Armed police and militia are needed, not for directing traffic and putting sandbags in dikes, but to break strikes, generate respect for property, and quell riots. Yet, starting with the atomistic model of individuals afforded by the conception of natural freedom leads to overlooking these points and to locating the need for a coercive mechanism or sovereign in the isolated individual's tendency to cheat. The facts, however, indicate that the state winks at cheating when it is not associated with a threat to social stability expressed in the restiveness of a group. If random freeloading were the main problem, it could be dealt with in the way deviants in certain primitive societies are dealt with, through sanctions administered by peers rather than by the formal mechanism of a separate institution called the state. But such equilibrating mechanisms suppose strong social ties and not just self-interestedness. In sum, the doctrine that we must support the state because individuals might cheat diverts attention from the fact that the state controls conflicts that threaten those most advantaged by a given social order.

One way to try to avoid the second criticism of Rawls' view of the state is to restrict Rawls' theory to a situation in which there are no conflicting groups. If a classless society were achieved, Rawls' view of the origin of the state would come into its own. In the absence of group conflict, the temptation to cheat would replace inequality between groups as the major source of instability. This appeal will not help Rawls. In a classless society, as in one with classes, people are not mutually disinterested. If there are no classes, people extend their interests to the advancement of the classless society. Of course, if there are classes, their people will be concerned with their class interest. It is then illegitimate to draw conclusions about the need for a state from that mere abstraction, the self-interested person. In any society, class or otherwise, self-interest will be modified by interests derived from social memberships. Nor can one conclude that such modifications are too weak to counteract the destabilizing tendencies of self-interest. Thus Rawls' conclusion that a coercive sovereign is necessary, even in a classless society, is unwarranted. It is necessary only for an unrealizable case, a multiplicity of self-interested persons.

In view of these two points, one might imagine a contract agreement to form a state which differs at crucial points from Rawls' contract. Assume we are not concerned with classless societies, just as Rawls seems to do when he builds differently favored 'starting positions' into the basic structure of society (p. 96). Then, different groups will exist and there will be sources of conflict. Now this has a consequence Rawls rejects, namely, that the contracting parties recognize they are not simply self-interested, but rather, have unspecified interests in the aims of some unspecified group possibly involved in conflict. Any compact they agree to must reflect this fact, rather than, as Rawls would have it, the mere fact of their self-interestedness. Thus the need for a coercive sovereign or a state will be seen, even in the original position, primarily as a result of the need to preserve the social order in the face of the social conflict based on inequality. Self-interestedness will only play a secondary role in recognizing the need for a state.

But is such a contract based on group involvement at all feasible? I wish now to show that, on any other than the atomic conception of humans that is involved in the idea of natural freedom, contract theory must fail.

Assume that I know I am a member of a major group in a society, though I do not know which, and that these groups are locked in a struggle over factors controlling inequality. Why should I agree to a coercive mechanism to keep this struggle within the limits of social stability? I might turn out to be a member of a group that can protect its members' interests only by fundamentally altering the social structure. If so, a state would simply prevent my group from effectively engaging its natural enemies. Since more likely than not I will enter history as a member of a disadvantaged group, I would not willingly risk giving up the benefits of solidarity with that group by agreeing to social harmony. (The same point follows from maximin considerations without reliance on prior probabilities. For the contract calls for giving up the benefits of solidarity and thus accepting a level of welfare that is not the best of the worst such levels. The best level for the least advantaged is achieved only when the least advantaged are struggling in conflict with other groups for their welfare.) Thus parties to the contract would not reach agreement on an organ for stabilizing social arrangements. But without agreement on an organ for stabilization, the contract would be empty. For individuals conceived as members of groups, rather than as atomic, contractualism has no application.

Consequently, if human nature requires a conception of group interests, then a general social contract cannot be reached. The possibility remains of forming a contract with co-members of a group that is binding only for the group. But this possibility will be rejected in Section V. In the absence of group identifications, self-interested freeloading may itself become a sufficient threat to stability to warrant a state. By contrast, individuals who by nature belong to groups in conflict cannot agree to establish a state without running counter to their natures.

On the basis of (i) and (ii), we can evaluate Rawls' claim that the assumption of natural freedom or mutual disinterestedness is justified because it is a logically weak assumption, that is, because it does not posit social connections in order to get social obligation (p. 149). In contrast, the above assumption that humans have a group nature is logically stronger. There are, however, other factors affecting the 'strength' of the assumptions. The problem for Rawls is that natural freedom is so one-sided in its consequences. It eliminates many feasible social arrangements, ones that the group nature of humans allows. It eliminates a society without a state and it puts freedom of thought before the right to overcome imposed consciousness. In a sense, then, natural freedom is a very strong assumption. It excludes many social alternatives to liberal institutions.

(iii) *Community.* Rawls shows concern about the one-sidedness of these consequences in his discussion of 'social union' or community (p. 522). He says the contract doctrine does not pose what Hegel called the 'civil society', in which the interests of the individual are ultimate as an ideal. In civil society the aim of union is merely the 'protection of property and personal freedom'.[4] Yet Rawls does not acknowledge Hegel's cogent observation that, since contractualists do—and as I showed above must—start with the ultimacy of the interests of mutually disinterested individuals, they cannot advance beyond civil society as their ideal of union. To move beyond the ideal market to community, it is essential that Rawls' natural freedom, and hence his mutually disinterested individuals, be abandoned as a starting point.

Instead one must begin with individuals whose capacities—and their interests are capacities—are inexplicable apart from reference to the roles they would play in society. Since one's social role involves oneself with others in a coordinated activity, personal interests will not leave one indifferent to the interests of others. Mutual support is a condition of sustaining such activity. Thus mutual interestedness comes from the fact that any social structure involves social roles. Even in an original position we are mutually interested. It does not matter that there we cannot know precisely whose interests we take an interest in. After all, we are ignorant there of the specific nature of most of our self-interests. It is not through a fanciful sentiment of benevolence imposed on isolated individuals that we get mutual interestedness.

Rawls also places people in social roles once they enter

[4] Hegel, *Philosophy of Right*, Sect. 258.

actual society. So he too has mutually interested individuals, though not in the original position. Can the atomism of his original position provide just as adequate a framework for community as our social hypothesis? The reason it can not is clear from Rawls' discussion of moral psychology (cf. Sections 70–5). Rawls postulates a capacity of humans to have friendly feelings toward those with whom they engage in coordinated social activity. On the basis of this friendly feeling, he arrives at a feeling of justice, needed for social stability (p. 458). This leaves Rawls' concept of human nature in a shambles. Mutual disinterestedness and the awareness that one has fellow feeling toward unspecified fellows cannot be combined to form a coherent conception of human nature. There are only two ways to avoid this incoherence. Either fellow feeling is contingent, or, though necessary for phenomenal human nature, it is not part of the noumenal human of the original position. Neither way will work. If it is contingent, then the Rawlsian contract will not in general allow for more than civil society. If it is phenomenal but not necessary, then there are two human natures and, as in Kant, no basis for the claim that moral requirements for one should hold for the other.

Rawls cannot avoid our point. Community is possible only when, from the start, there are individual powers that have been 'recognized and organized' as social powers.[5] This social hypothesis is a necessary but by no means a sufficient condition for community, since in any actual society lacking true community individuals nonetheless play roles that give them social interests.

It is by no means easy to define the ideal of community satisfactorily. But the ideal surely suggests that many or all of the interests of others become, apart from the ulterior motive of self-interest, the interests of each individual. And it suggests with equal strength that many, but not all, human interests manifested outside community are transformed into drives for goals described simply as activities involving many individuals. This second aspect of community concerns many people cooperating to create *one* result, not just many people cooperating to create *similar* results, one of which pertains to each individual. Now Rawls thinks that a community exists when humans have shared final ends and value their common institutions and activities as

[5] Marx, 'Bruno Bauer, *Die Judenfrage*', in *Karl Marx: Early Writings*, trans. T. B. Bottomore (McGraw-Hill: New York, 1963).

good in themselves (p. 522). For him, those who accept his principles of justice:

(*a*) will have the shared end of bringing about and stabilizing institutions, and

(*b*) will prize such institutions as good in themselves rather than as burdensome ways of distributing private satisfactions.

Let us grant that (*a*) and (*b*) characterize a just society in Rawls' case. The question remains whether a society characterized by (*a*) and (*b*) is, as Rawls contends, a community or merely a civil society. I shall now show that on Rawls' interpretation of them (*a*) and (*b*) do not suffice for community.

To understand (*a*), let us ask why, for Rawls, there is such a shared end. Realizing the shared end would be a realization of our human nature, since the aim is the bringing about of just institutions and such institutions do express our nature. But Rawls supposes it is normal for people in a well-ordered society to cooperate 'to realize one another's nature' (p. 527). Yet what is this nature we cooperate to realize? Since we are naturally free, we will share the goal of realizing ourselves as self-interested. Does this promote community at all? Certainly it does not involve transforming self-interest into drives for common activity. Moreover, it is doubtful that it involves taking an interest in realizing the nature of others apart from the ulterior motive of self-interest. Rawls cannot avoid relying on this ulterior motive. As we observed, if he introduces a basic friendly feeling, he makes his conception of human nature incoherent. But even if we grant him this friendly feeling, the other requirement of community, interest in common activity, is not satisfied. The advance from the market to community is not accomplished simply by seeing the necessity to agree on market regulations as a method of maximizing the minimum likely satisfaction of self-interest.

Is it possible that (*b*) complements (*a*) in a way that gives rise to community? Probably not, once we see how Rawls argues that the institutions of justice are goods in themselves. In essence, we are told that such institutions are good because they interfere least with the realization of self-interest. We are told they are goods in themselves because, when we act to uphold them, our 'nature as moral persons is most fully realized' (p. 528). But he tells us it is our nature as moral persons to have an unrestricted antecedent choice and to be capable of reconciling conflicting self-interests on the basis of adopted principles (p. 505). Thus the institutions are 'good in themselves' because they interfere least with the realization of self-interest. Prizing such institutions as intrinsic goods implies neither an interest in others based on something other than self-interest nor a transformation of self-interests into drives for participating in common activity. The aim of these institutions is precisely that of Hegel's civil society and of the liberal institutions it contains. Thus on the basis of the atomistic human nature of Rawls' original position we derive, not surprisingly, atomistic conclusions that are not sufficient for community and seem incompatible with it.

One could easily go on to other examples than those in (i)–(iii), but the pattern is already clear. Rawls is forced to show that the reflective sense of justice is a neutral instrument for deciding what is relevant to determining justice. He must show it is a neutral instrument for deciding what is properly left out of the original position. His defense of this neutrality involved appealing to a conception of human nature which included freedom and equality. This was intended to protect him against the charge that the reflective sense of justice was only an analytical, and not a critical, device. That is, that it was only as good as the general social assumptions of the epoch in which it is exercised. However, we can now see that this retreat to the traditional fortress of contractualism, the view that human nature is free and equal, was to no avail. Deductions from this conception of human nature support a particular set of values—freedom of thought, the state as guarantor of honesty, and the protection of self-interest by society—associated with the liberal institutions of the capitalist epoch. Unfortunately, these values are unable to cope with genuine social concerns, like bondage due to imposed consciousness, class conflict that the state functions to dampen but not eliminate, and the absence of genuine community. The inability of the contractualist to meet these problems raises the question whether the values he defends are legitimate not only from all social but also from all temporal points of view. The supporting conception of human nature loses its claim to social and historical neutrality.

IV. *Equality and Universality*

We still must consider whether the second feature of human nature on Rawls' view, equality, has a claim to neutrality. I shall show that the conditions he requires for equality obtain only on a liberal presupposition that he does not attempt to support. Since the fate of equality and universality is tied together, I shall examine the former in the context of the latter.

The principles adopted in Rawls' original position are supposed to apply to everyone. In particular, there is to be no class relativity of rights (p. 132). This universality is understandable from both an external and an internal perspective. It is understandable from the perspective of social stability, which is an external perspective relative to the details of Rawls' view of human nature. The co-operation of members of all classes is needed to bring about this stability. Universality is the requirement any society puts on the public principles by which it hopes to stabilize itself, and to the extent that a theory of right is a theory in the service of social policy of stabilization, that theory must require universality of its principles. Rawls hints at such an external justification for the universality of principles of justice when he says that the justification is 'derived from the task of principles of right in adjusting the claims that persons make on their institutions and one another' (p. 131).

But he also sees the matter from a perspective internal to his view of human nature. It is with this internal perspective that I shall be concerned. We are all equal by nature; so the equality of the original position expresses our nature. It is important, though, to know what 'equality' means here. Rawls gives the notion of equality content through his notion of a moral person. He defines a 'moral person' as one who:

(*a*) chooses his or her own life plan, which plan then determines what his or her conception of the 'good' is, and
(*b*) can have a sense of justice, which is a skill in judging things to be just or unjust (p. 12).

Persons in the original position are then 'equal' since, on the one hand, they are moral persons and, on the other, it is only the respects in which they are moral persons that are admissible in the discussion of what principles to agree upon (p. 19).

How then does equality tie in with universality? Suppose principles are justified on the basis of considerations dealing exclusively with features common to all individuals of a certain kind. Then there is simply no basis for withholding the application of any of those principles from any member of that kind. Now since all persons are held by Rawls to be equal by nature, they have in common at least the two features of all moral persons. But among equals *only* these features are relevant in the justification of principles. Hence, any principles justified on this basis apply to all persons (pp. 132, 134). So, since persons are equal, principles applying to some apply to all.

I would like to make two responses to this argument. On the one hand, I shall show in (i) that it is doubtful that persons are equal in the above sense. It will suffice to show that it is doubtful persons have all of the features associated with what Rawls calls moral persons. Rawls' internal argument for the universality of principles of justice will then collapse. On the other hand, I shall show in (ii) that it is possible to modify the above conception of moral person so that it does indeed contain two features common to all persons. The modification is in the direction of a class conception of the moral person. But this modification cannot be used as a basis for universality. As noted in Section III, part (ii), this social conception of persons is incompatible with the possibility of mutual agreement on common principles in an original position.

(i) *The Sense of Justice.* Persons are not 'equal' since they do not have all of the features of a 'moral person'. The moral person is capable of a sense of justice. One assumption of the liberal view of equality tells us how this feature of a moral person is to be taken: People are naturally capable of taking the point of view of the whole (p. 475). This point of view enables one to judge the rightness of institutional arrangements for reconciling conflicting interests from a position above these interests. Without this point of view, the sense of justice would be mere partisanship. What is at stake here is whether conflict within society will make it impossible to take the broad view without special indoctrination. The idea that society is a collection of individuals who antecedently have self-interestedly made adjustments for survival and who in history have a sense of fellow feeling does tend to suggest that the broad view can be adopted naturally. All the individuals are cooperating to make the adjustments work.

But what if society, at least in the forms we have known and have had recorded for us, is constituted by an interrelation of roles? Moreover, suppose this interrelation of roles, when realized in various groups, sets these groups in conflict. The conflict stems from the greater or lesser remoteness from control over social production implied by the different roles. It is the conflict between landowner and peasant, union bureaucrat and the rank and file, welfare agent and the ghetto mother, board chairman and the worker subject to safety hazards. If indeed society has such a 'class structure', then a sense of justice that involves taking a view of the whole will not be natural for humans as social beings. Such a sense of justice implies, by definition, acceptance of the aim of avoiding open conflict within the society. But this is not the aim of those whose conditions will not be much more tolerable if conflict is avoided. In a class society, this sense of justice will not come about naturally. In many cases, it will result only from indoctrination by the forces in the society that can expect to benefit by reducing conflict and maintaining the power relations between the classes unchanged. A sense of justice, far from being neutral, would require the 'contingent', 'arbitrary' restriction of the normal interests of members of some classes by pressure from media, the courts, and the police.

Once we reject the idea that individuals are by nature mutually disinterested and once we recognize that our only workable concept of a person is of an individual that is a member of some social arrangement or other, then a sense of justice that puts one above the conflicts of groups must be regarded as, for many people, an artificially induced ability. Social arrangements of the forms we are familiar with and those allowed by what Rawls calls 'the basic structure of society' (p. 96) have group conflict as a built-in feature. The ability to have a universal sense of justice depends on the determination of the ruling class to maintain its hegemony by ideological pressure. But surely being a moral person should not be dependent on such a determination.

In a class society, it might well take manipulation to get people to take the point of view of the whole. But, one might reply, in a classless society, either the idealized market society of petty-capitalist producers, such as Kant's kingdom of ends, or the Marxist classless community, in which production is according to ability and reward according to need, taking the point of view of the whole would be quite natural. Does this exception show that taking the point of view of the whole is natural *in general*? Presumably, it might if we could agree that only in a classless society is human nature spontaneously manifested. But this point is no more likely than the claim that human nature evolves, having different content under different social conditions. Thus we cannot easily dismiss the conclusion reached above that the nature of the humans we find in class society is not the basis for the capacity to have a sense of justice.

Rawls tries to anticipate this criticism by his deduction of the sense of justice from individual psychology (pp. 473, 491). By this deduction, a sense of justice is acquired against the background of an already acquired sense of solidarity or fellow feeling within special groups within society, discussed above in Section III, part (iii). One sees oneself and the members of the groups with whom one has solidarity, on the basis of common roles, as benefiting from the just institutions of society. One's sense of solidarity and one's own self-interest are then the bases for supporting the just institutions from which these benefits come.

This argument overlooks a countervailing tendency produced by the sense of group solidarity. Because of group solidarity, it will be possible for members of a less advantaged group to interpret the benefits they receive from the overarching social arrangements for what they are. They are benefits designed to maintain social stability despite the exclusion of working groups from control over the means of survival. Rawls' understatement is perfectly accurate: 'To be sure, in any kind of well ordered society the strength of the sense of justice will not be the same in all social groups' (p. 500). For Rawls' psychological deduction to establish its conclusion, it must be recognized that more is needed than mere benefits distributed according to his difference principle. In addition, the artificial weapons of ideology will have to be mobilized to counteract the natural tendency for group solidarity to generate a partisan sense of justice, one that judges everything by whether it advances the group's development.

Equality, at least as interpreted by a sense of justice, has at best a doubtful title to being an aspect of human nature. Since the universality of principles of justice depends on everyone's being equal by nature, Rawls' justification for their universality is undermined.

(ii) *Beyond Liberal Equality.* A new interpretation of the notion of equality is needed if our discussion of Rawls is to take a more constructive turn. The new interpretation will require substituting a conception of moral person based on the realities of class society.

So far we have said nothing about (*a*) in Rawls' conception of a moral person. In so far as it suggests merely that persons have or can have 'a conception of their good', it seems innocuous enough. Rawls' mechanism for people achieving their conception of the good—they choose a plan of life—reflects his atomistic bent in psychology. To this one might counterpose the Aristotelian view that one's good is a function of one's character, which derives from many sources and is never more than partly a matter of choice.[6] But I shall accept (*a*) in so far as it claims moral persons have a conception of their good. I shall ignore its implication that it is literally a choice of a plan of life which determines what things become goods.

It is in regard to (*b*) that an important difference is introduced. In place of (*b*), put (*b′*):

(*b′*): A moral person is capable of a developed form of class consciousness which implies, among other things, a critique of the ideology of the society as a whole and of the special ideology, if any, imposed on the members of that person's class.

On our new interpretation of equality, then, the only factors relevant to choosing principles of justice are those implied by the fact that all parties to the decision are similar in that they are moral persons, that is, are similar in that (*a*) and (*b′*) apply to them.

It might seem as though this gives a justification of universality on class rather than atomistic grounds. In Section III, part (iii), however, we say that without the atomic conception of human nature, contract theory fails. In the original position, I will not know to which class I belong in history. But I know that there is a great likelihood that I will belong to a class society[7] and that I am capable of the class consciousness appropriate to my class. (In history I may of course be *déclassé* but that likelihood is too slim to be important.) Would I, then, agree to principles that bring stability to the entire society, as universal principles of justice must? If I were a member of a disadvantaged class, I would be able to recognize that such an arrangement is a direct attack on the tendency of my class to throw off oppression. I might be willing to agree, however, if I were assured that I would be a member of the ruling class. Though such an arrangement might slow down the natural tendency of a ruling class, it would at the same time be an essential mechanism for the preservation of its rule.

Of course, on Rawls' view that the individual is isolated by nature, the principles of justice will be seen primarily as

[6] Aristotle, *Ethics*, III, 4–5.

[7] Rawls assumes that any society will have differently favored 'starting places' (Sect. 16). The features given to these starting places taken together with the requirement of a 'coercive order' (Sect. 38) are the basis I have for claiming that for Rawls society is class society. Since such an identification seems to me unnecessary, I speak only about the great likelihood of entering a class society from the original position.

checking individual self-interest, not group tendencies. It is individual abuse of wealth and individual cheating, not the more or less coordinated efforts of group members to restrict liberties and maintain the maldistribution of income, that concern Rawls. It is easy enough for isolated individuals to accept the principles of justice for they are powerless anyway and thus do not generally have a historic mission—as a class does—which would be easier to accomplish without such a check on self-interest. Yet, for the individual as a class member, there can be no agreement on principles of justice except as expedients to survive an historical period during which the class could gather strength and prepare for the realization of its tendencies. This is true even though one does not know what one's social position in history might be. One could simply not risk tying one class's hands behind its back in the class struggle. Since one is capable of class consciousness, one would, from the unclouded perspective of the original position, see that settling for stability and relinquishing struggle is, for most classes, an invitation to increasing inequality. In history, it is the labor 'statesman' who trades off the interests of the working class for economic stability and industrial peace. It is the 'progressive' politician who, behind the guise of reform, turns struggle into apathy in order to stabilize the system that has provoked struggle. Consequently, the original position cannot yield agreement on principles of justice when the individuals in it are equal in our new class sense. There can be no society-wide contract, and *a fortiori* no contract with universal application.

V. *Relativizing Deontology*

I have attacked contract theory on the ground that it is tied to an unsupportable concept of human nature. Nonetheless, certain aspects of contract theory seem to me to point in the right direction. As developed by Rawls, contract theory affirms two theses that form the essence of what I shall call a social deontology. These theses are well worth maintaining. The first claims the inseparability of the right from social aims. The second affirms the primacy of the right in respect to the good. Rawls clearly adheres to both theses: 'The principles of justice... represent more or less definite social aims and restrictions... Once we realize a certain structure of institutions, we are at liberty to pursue our good within the limits which its arrangements allow' (p. 566). By contrast, the utilitarian attempts to define the right in terms of the good, as that which maximizes the good. The only question is whether it is possible to maintain the two theses once contract theory is abandoned. In short, if we cannot avoid utilitarianism or some other teleological position once we have abandoned the social contract, then these two useful themes of social deontology cannot be saved.

For Rawls, the first thesis is given concreteness through the goal of stability. Right is inseparable from the social aim of establishing a stable arrangement for resolving conflict among self-interested individuals. That aim is thought to be furthered by the principles of justice agreed to in the initial position (p. 4). For Rawls, the second thesis is given concreteness in terms of these principles of justice. These principles are prior to the good in the sense that what is morally good is within the limits of what is allowed by institutions formed on the basis of these principles (pp. 437–8).

However, one could define the social aim quite differently, yet still connect the right to it and subordinate the good to the right. The structure of the resulting theory would allow it to be called a social deontology even though it might not be possible to formulate it as a contract theory. I shall outline one such social deontology, but I need two assumptions. First, that the relevant societies, like present-day capitalist society, are class societies. Second, I assume that the human nature is in part determined by the classes people belong to. Thus, membership in a specific class is not, as Rawls supposes it is, a merely contingent feature of a human. This second assumption will, I think, eventually be justified by the fruitfulness of a class-centered concept of human nature in the social sciences. Rawls gives no convincing reason to reject such a concept in favor of his own.

By definition, a class society cannot endure without people filling specific roles that put them into conflict. The basic social aims with which the right is to be associated will be quite different from that of over-all stability. Indeed, the aim of stability need not be generally acknowledged in a situation of conflict. The basic social aims will quite naturally be the development of the full implications of being in those roles. If the role of productive workers is to be the basis for profit, by working longer than would be needed to produce what they need to live on, then one of their social aims will be to reduce this extra time and ultimately abolish the system that requires it. Since the roles are conflictive, what is right on the basis of the aim associated with one role need not cohere with what is right on the basis of the aim associated with another role. The workers' aim of reducing the extra time worked conflicts with the owners' aim of accumulating profits. The appropriate method for dealing with conflicting views of the right will not be to adopt the point of view of a reason that abstracts from the class structure of society. That would be abstract from something essential to the sorts of societies we are considering, and by our second assumption, from something essential to people in them. Such an abstraction would lead straight back to the abstraction of the isolated individual.

Conflicts of this sort over the right are not dissolved by reason (cf. p. 134), but are solved historically in the interaction of groups. Of course, this 'realism' about the right will be said to conflict with the reflective sense of justice. This conflict only mirrors the conflict in society between the centrifugal tendencies of class conflict and the centripetal forces of socialization. The realism is associated with the class struggle while the reflective sense of justice is part of the attempt to overcome that struggle by socialization.

The important thing is that the right, on this class relative view of it, is not defined on the basis of individual or class utility, in the manner of a teleological theory. Rather, what is right for members of a class is determined by the consequences for them of their role being what it is. The good for individuals with a given role in the society will be limited by what the principles of right for people in that role are. Thus the basic structure remains that of a social deontology.

The Locke-Kant-Rawls concept of human nature was seen, in Sections III and IV, to be a falsifying abstraction. Can we put in its place a conception that will avoid these relativist consequences? The problem of finding such a concept is this. Any concept of human nature sufficiently rich to generate a concept of right that will resolve conflicts arising on a class base will either stabilize the existing order or promote the forces tending to overthrow it. It will, thus, either perpetuate the ascendancy of the ruling class or promote a revolutionary class. In either case, it will conflict with members of a class realizing the tendencies they have. Are these tendencies really that important when the right is in question? On the above assumptions, they are important, so long as human nature has anything to do with the right. And how could they not be important? Even if we pretend to reject their importance, our resulting trans-class concept of human nature will resolve the conflict by benefiting one of the classes at the expense of the others. So, in effect, the tendencies of members of one class have been hypostatized as the tendencies of humanity generally. In sum, the alternative to a class-centered concept of human nature with relativist consequences is a supposedly trans-class concept with consequences that favor only one class. Either relativism or partisanship wrapped in absolutism! I shall now try to illustrate this point in regard both to what would now be stabilizing and revolutionary concepts of human nature.

The liberal view of human nature, if Rawls is right, would lead to his difference principle (p. 75) as an instrument for resolving conflicts regarding the distribution of goods. It follows from the principle that a certain level of inequality in distribution is just if lowering it would make a representative member of the least-advantaged group worse off (p. 78). Here we assume that the least-advantaged group is defined in terms of relative income and wealth, rather than by an economic role, as class would be defined (p. 98). Thus maximizing the advantages of the least advantaged need not be maximizing the advantages of part of all of an enduring class. The maximization may well destroy one lower class and substitute a different one for it.

It will of course be necessary to determine whether the benefits of the least advantaged have been maximized by reference to contrasting but specific economic structures. In a large industrial state, the income of the bottom quintile may be maximum under capitalism when it is no more than ten per cent of the total. But one cannot say this is a just maximum without first appealing to a contrasting system or systems. Under a bureaucratically controlled post-capitalist system, the income of the bottom quintile of the same population might be a larger sum when it is a maximum, whatever share of the total that sum happens to be. In this way, comparative maxima could be estimated in order to estimate relative justice under the difference principle. A distribution that yields an absolute maximum would require that there be an economic system more just than any other. Given the multiplicity of possibilities we already know, one hesitates to say there is such a unique system.

Despite the fact that the difference principle allows for these vast changes of economic system, it is nonetheless a principle that in another way tends to preserve privilege. The tendency of the working class is not merely to get a greater share of profits, but also to do away with the system of exploitation whereby the worker has to produce profits for an owner in order to live. So, too, a least-advantaged group will not simply struggle for its maximum share but will also struggle to get beyond the least-advantaged category. Yet the difference principle limits struggle to the former kind. Will people in the original position agree to limit it in this way? I think not. Any society they enter will have more than half its members with less than half the income and wealth. Their chances of entering this majority, least-advantaged group are better than one-half. Agreeing to the principle of difference limits their struggle to increasing the absolute size of this less than one-half share. They are, other things equal, debarred from attempting to unseat the privileged minority so that they might cease to be least advantaged. Such an attempt would be an expression of envy or what Nietzsche called the *ressentiment* of the slave (cf. Section 81). But whatever elitist philosophers call it, it is a struggle by the majority for the wherewithal to control their lives. There is, because of the need to struggle for such control, no basis for agreeing to a principle that denies to the least advantaged the right to struggle against being the least advantaged. In so far as people in the original position recognize they are likely to enter one of these classes, there is no basis for a compact involving the difference principle.

A revolutionary view of human nature, like Marx's, claims human nature is not being realized until human activity becomes an end rather than a means.[8] Such a view is no less biased in its class implications than the liberal view. This view of human nature corresponds to the nature of humans in the special role of working people in capitalist society. If one sells one's labor skills on the market, then indeed one's exercise of those skills has become a means rather than an end, a means of receiving the value of those skills' use for a designated time. Such a role in a capitalist society leads the humans that play it to feel they need to gain control over their activity. This need is ultimately expressed in the demand for social arrangements in which human spontaneity is possible. The struggles to shorten the working day and to increase wages as productivity increases are the immediate expressions within the capitalist context of this need.

[8] Marx, *Early Writings*, 'Alienated Labor'.

The liberal view of human nature reflects the tendency of a privileged minority to stabilize a social order in which its privileges are institutionalized. Similarly, the revolutionary view of human nature reflects the tendency of the lower classes to struggle to gain more control. These two conceptions belong to part of a historical antithesis. Neither can claim to be the basis of a system of right that is valid 'not only from all social but also from all temporal points of view'. In an original position defined in terms of the revolutionary conception of human nature, one could not expect all parties to agree to work toward the overthrow of the exploiting class, which, by exacting profits and controlling investment, decreases human spontaneity. For, on entering history, I might become a member of the exploiting class, and I would not want to risk committing myself to its overthrow. The revolutionary view of human nature is a useful conceptualization of the centrifugal implications of the social role of members of the working class. But, as Marx would himself insist, it would be uncritical to suppose that its validity extends beyond that role.[9]

Let us return to our relativized deontology and ask whether we can adjust contract theory to it by a corresponding relativization of contract theory. Instead of a *social* contract, why not a *class* contract? An original position for a non-ruling class would require eliminating the imposed consciousness that was instituted by the ruling class; indeed, the original position would be a situation in which class consciousness reaches its ultimate development. A charter for the class would result which would enable it to wage successful battle with its natural enemies. This extension of the notion of a contract reveals better than anything else could the fundamental weakness of contract theory in general.

Contract theory supposes that the task of arriving at moral principles is one of *reason*. By a rational idealization one constructs a concept of human nature and by a process of rational reflection one confirms the validity of this concept. The situation is not fundamentally altered if contract theory is relativized to classes. By a process of rational idealization one constructs a concept of human nature from what seems the obvious truth about the class. And similarly, by a process of rational reflection,

[9] A defense of a socially relative conception of human nature is given in the author's 'Human Nature and Social Change' presented at the first conference of the *Marxist Activist Philosophers*, September 1973.

it can be determined whether this concept is in fact what one would expect, given the role people in the class play. Moral principles arrived at on the basis of such rationally constructed and justified concepts of human nature will be distorted in the ways these concepts are themselves distorted. But why are the concepts distorted? Reason, as employed in the tasks of constructing and justifying these concepts, is analytic reason and hence it merely refines and abstracts from something it identifies as a consensus. When we are concerned with the consensus members of a class have their real identity, we must be aware of the surface assumptions within the consensus which must be broken through before the distortions can be eliminated. How can reason, as a device for refining the consensus view, be expected to avoid typical distortions, like the view that workers are incapable of controlling production, or that they will sacrifice improved conditions for higher wages? For breaking through these assumptions, attention must be paid to situations of struggle, which are the universal solvent for the surface assumptions making up a consensus. Struggles to maintain workers' control, in the few instances it has occurred, were defeated not from within by instability but from without by force. And struggles for improved conditions are defeated not because worker avarice gets control but because of the fear of decreased productivity.

To last, a social or class contract must have behind it an enduring conception of the nature of humans. Accordingly, it is appropriate that contract theory should insist on the methodology of analytic reason. If anything can develop enduringly valid concepts, it is the methodology of rational idealization and reflection. But, as I have argued, such concepts are enduringly valid only relative to those who wish to perpetuate the consensus on which they are based. If one aims not at perpetuation and stabilization, but at growth along lines indicated by the struggles one's class engages in, one will take a critical stance toward any conception that might be the basis for an enduring contract. In particular, one will not rely on analytic reason to construct and legitimate a conception that will then become the basis for an enduring agreement.

As the very idea of a contract suggests, the motivation for developing a contract theory is to support the forces of stabilization. It is not accidental, then, that contract theory mobilizes analytic reason, with its reputation for thoroughness, objectivity, and hence the permanence of its product, to provide the conception that is the basis for the contract.

But reason takes second place to history in a deontology that starts from the fact that, at least now, the social aspect of humans implies their involvement in class conflict. The development of their conflict parallels a development in the conception of the direction true class consciousness would point. Moreover, it yields changes in the humans involved in the conflict. Struggle is constantly shattering illusions and changing the participants. Struggle is thus antithetical to the idea of a contract. For an enduring agreement requires an enduring concept of human nature. And struggle will change this concept. The conclusion to all this is not that contract theory is restricted to situations of stability and compliance (p. 8). It is rather that because of the structural guarantees of conflict in the relevant sorts of societies, the situation of stability and compliance that must be possible for contract theory to be credible is excluded in advance.[10]

[10] My thanks to Norman Daniels, Alan Ritter, Frank Thompson, and George Wright for helping me to see that various criticisms in the earlier drafts applied to views I had mistakenly taken to be Rawls'.

QUESTIONS OF METHOD

4 Rawls' Theory of Justice[1]

R. M. HARE*

Any philosopher who writes on justice or on any other subject in moral philosophy is likely to propound, or to give evidence of, views on one or more of the following topics:

(1) *Philosophical methodology*—i.e., what philosophy is supposed to be doing and how it does it. Rawls expresses some views about this, which have determined the whole structure of his argument, and which therefore need careful inspection.

(2) *Ethical analysis*—i.e., the meanings of the moral words or the nature and logical properties of the moral concepts. Rawls says very little about these, and certainly does not treat them as fundamental to his enquiry (p. 51, l. 10).

(3) *Moral methodology*—i.e., how moral thinking ought to proceed, or how moral arguments or reasonings have to be conducted if they are to be cogent.

(4) *Normative moral questions*—i.e., what we ought or ought not to do, what is just or unjust, and so on.

I shall argue that, through misconceptions about (1), Rawls has not paid enough attention to (2), and that therefore he has

[1] Reprinted from *Philosophical Quarterly* 23, pp. 144, 241, April and July 1973. Revisions are limited to those consequential upon the combination of two articles into one, the addition of footnotes 6 and 9, and the omission of the last two paragraphs—an appraisal of the book which, though I still adhere to it, is more suited to a review than to this volume. I do not hope to explore in this paper all the convolutions of Rawls' book. I shall concentrate on what seems most important. I feel excused from discussing Rawls' treatment of liberty by my general agreement with Professor Hart's article, reprinted on p. 230. Of the many other people with whom I have discussed the book, and who have kept my courage up during two readings of it, I should like especially to thank Mr. Derek Parfit, who seems to me to see deeper and more clearly into these problems than any of us.

* Professor of Moral Philosophy, Corpus Christi College, Oxford.

lacked the equipment necessary to handle (3) effectively; so that what he says about (4), however popular it may prove, is unsupported by any firm arguments.

(1) Rawls states quite explicitly how he thinks moral philosophy should be done: 'There is a definite if limited class of facts against which conjectured principles can be checked, namely our considered judgments in reflective equilibrium' (p. 51). It is clear from the succeeding passage that Rawls does not conceive of moral philosophy as depending primarily on the analysis of concepts in order to establish their logical properties and thus the rules of valid moral argument. Rather, he thinks of a theory of justice as analogous to a theory in empirical science. It has to square with what he calls 'facts', just like, for example, physiological theories. But what are these facts? They are what people will *say* when they have been thinking carefully. This suggestion is reminiscent of Sir David Ross.[2] But sometimes (though not consistently) Rawls goes further than Ross. Usually he is more cautious, and appeals to the reflections of *bien pensants* generally, as Ross does (e.g. p. 18, l. 9; p. 19, l. 26). But at page 50 he says, 'For the purposes of this book, the views of the reader and the author are the only ones that count.' It does not make much practical difference which way he puts it; for if (as will certainly be the case) he finds a large number of readers who can share with him a cosy unanimity in their considered judgments, he and they will think that they adequately represent 'people generally', and congratulate themselves on having attained the truth.[3] This is how phrases like 'reasonable and generally acceptable' (p. 45) are often used by philosophers in lieu of argument.

Rawls, in short, is here advocating a kind of subjectivism, in the narrowest and most old-fashioned sense. He is making the answer to the question 'Am I right in what I say about moral questions?' depend on the answer to the question 'Do you, the reader, and I agree in what we say?' This must be his view, if the considered judgments of author and reader are to occupy the place in his theory which is occupied in an empirical science by the facts of observation. Yet at page 516, line 15 he claims objectivity for his principles.

[2] Cf. *The Right and the Good*, pp. 40 ff.

[3] *See* page 104, lines 3–14 for a 'considered judgement' with which many of us would agree, but which differs from the views of most writers of other periods than the present and is not argued for.

It might be thought that such a criticism can be made only by one who has rejected (as Rawls has apparently accepted) the arguments of Professor Quine and others about the analytic-synthetic distinction and the way in which science confronts the world. But this is not so. Even Quine would hardly say that scientific theories as a whole are to be tested by seeing what people say when they have thought about them (it would have been a good thing for medieval flat-earthers if they could be); but that is what Rawls is proposing for moral principles.

In order not to be unfair to Rawls, it must be granted that *any* enquirer, in ethics as in any other subject, and whether he be a descriptivist or a prescriptivist, is looking for an answer to his questions which he can accept. I have myself implied this in my *Freedom and Reason*, page 73 and elsewhere. The element of subjectivism enters only when a philosopher claims that he can 'check' his theory against his and other people's views, so that a disagreement between the theory and the views tells against the theory. To speak like this (as Rawls does constantly throughout the book) is to make the *truth* of the theory *depend on* agreement with people's opinions. I have myself been so often falsely accused of this sort of subjectivism that it is depressing to find a self-styled objectivist falling as deeply into it as Rawls does—depressing, because it makes one feel that this essentially simple distinction will never be understood: the distinction between the view that thinking something can make it so (which is in general false) and the view that if we are to say something sincerely, we must be able to accept it (which is a tautology).

Intuitionism is nearly always a form of disguised subjectivism. Rawls does not call himself an intuitionist; but he certainly is one in the usual sense. He says, 'There is no reason to suppose that we can avoid all appeals to intuition, of whatever kind, or that we should try to. The practical aim is to reach a reasonably reliable agreement in judgment in order to provide a common conception of justice' (p. 44; cf. p. 124, l. 38). It is clear that he is here referring mainly to moral intuitions; perhaps if he appealed only to linguistic intuitions it would be all right. He reserves the name 'intuitionist' for those (including no doubt Ross) who advocate a *plurality* of moral principles, each established by intuition, and not related to one another in an ordered structure, but only weighed relatively to each other (also by intuition) when they conflict. The right name for this kind of intuitionism would be 'pluralistic intuitionism'. Rawls' theory is more systematic than this, but no more firmly grounded. There can also be another, non-pluralistic kind of intuitionist—one who intuits the validity of a single principle or ordered system of them, or of a single method, and erects his entire structure of moral thought on this. Sidgwick might come into this category—though if he were living today, it is unlikely that he would find it necessary to rely on moral intuition.

'Monistic intuitionism' would be a good description of this kind of view. It might apply to Rawls, did it not suggest falsely that he relies only on one great big intuition, and only at one point in his argument. Unfortunately he relies on scores of them. From page 18, line 9 to page 20, line 9, I have counted in two pages thirty expressions implying a reliance on intuitions: such expressions as 'I assume that there is a broad measure of agreement that'; 'commonly shared presumptions'; acceptable principles'; 'it seems reasonable to suppose'; 'is arrived at in a natural way'; 'match our considered convictions of justice or extend them in an acceptable way'; 'which we can affirm on reflection'; 'we are confident'; 'we think'; and so on. If, as I have done, the reader will underline the places in the book where crucial moves in the argument depend on such appeals, he may find himself recalling Plato's remark: 'If a man starts from something he knows not, and the end and the middle of his argument are tangled together out of what he knows not, how can such a mere consensus ever turn into knowledge?' (*Rep.* 533 c). Since the theoretical structure is tailored at every point to fit Rawls' intuitions, it is hardly surprising that its normative consequences fit them too—if they did not, he would alter the theory (p. 19, ll. 26 ff; cf. p. 141, l. 23); and the fact that Rawls is a fairly typical man of his times and society, and will therefore have many adherents, does not make this a good way of doing philosophy.

Rawls' answer to this objection (p. 581, l. 9) is that *any* justification of principles must proceed from some consensus. It is true that any justification which consists of a 'linear inference'[4] must so proceed; but Rawls' justification is not of this type. Why should it not *end* in consensus as a result of argument? There may have to be a prior consensus on matters of fact, including facts about the interests of the parties (though these themselves may conflict); and on matters of logic, established by analysis. But not on substantial moral questions, as Rawls seems to require. A review is not the place for an exposition of my own views of how

[4] *See* my *Freedom and Reason*, pp. 87–8.

moral argument can succeed in reaching normative conclusions with only facts, *singular* prescriptions and logic to go on; all that I wish to say here is that the matter will never be clarified unless these ingredients are kept meticulously distinct and the logic carefully attended to.[5]

(2) I shall mention only in passing Rawls' views about the meanings of the moral words or the natures, analyses and logical properties of the moral concepts. It would be wrong to take up space on something which Rawls evidently thinks of little importance for his argument. He wishes to 'leave questions of meaning and definition aside and to get on with the task of developing a substantive theory of justice' (p. 579). There is in fact a vast hole in his 600-page book which should be occupied by a thorough account of the meanings of these words, which is the only thing than can establish the logical rules that govern moral argument. If we do not have such an account, we shall never be able to distinguish between what we have to avoid saying if we are not to contradict ourselves or commit other *logical* errors, and what we have to avoid saying if we are to agree with Rawls and his coterie.[6] So far as he does say anything about the meanings of the moral words, it is mostly derivative from recent descriptivist views, my arguments against which it would be tedious to rehearse. I found this reliance surprising, in view of the fact that what he says about justice, at any rate, clearly commits him to some form of prescriptivism: the principles of justice determine how we *are* to behave, not how we are to *describe* certain kinds of behaviour (p. 61, l. 7; p. 145, ll. 12, 14, 33; p. 149, l. 16; p. 351, l. 15). My quarrels with Rawls' main theory do not depend at all on the fact that I am a prescriptivist.

There are significant passages in which Rawls compares moral philosophy with mathematics (p. 51, l. 23) and linguistics (p. 47, l. 5; p. 49, l. 8). The analogy with these sciences is vitiated by the fact that they do not yield substantial conclusions, as moral philosophy is supposed, on Rawls' view, to do, and in some sense

[5] *See* my *Freedom and Reason*, esp. chapters 6, 7, 10 and 11, and the paper cited in footnote 10, p. 89. A bibliography of my writings up to 1972 appears in my *Practical Inferences*. For my latest shot at giving an account of the relation between the meanings of the moral words and the rules of moral argument, see the paper 'Wrongness and Harm' in my *Essays on the Moral Concepts*.

[6] *See* the paper 'The Argument from Received Opinion' in my *Essays on Philosophical Method*, which might have been written with Rawls' book in mind, although in fact at the time I had not had the opportunity of reading it.

clearly should. It is quite all right to test a linguistic theory (a grammar) against what people actually say when they are speaking carefully; people's *linguistic* 'intuitions' are indeed, in the end, authoritative for what is correct in their language. The kind of interplay between theory and data that occurs in all sciences can occur here, and it is perfectly proper for the data to be the utterances of native speakers. But the only 'moral' theories that can be checked against people's actual moral judgments are anthropological theories about what, in general, people *think* one ought to do, not moral principles about what one ought to do. That these latter can be so checked is not, indeed, what Rawls is suggesting in this passage; but do not the whole drift of his argument, and the passage quoted above (p. 51), suggest it?

The case of mathematics is more controversial. Rawls seems to imply that if we had a 'moral system' analogous to the systems of logic and mathematics, then we could use such a system to elucidate the meanings of moral judgments, instead of the other way about, as I have suggested. There is no objection, so far as I can see, to such a claim in mathematics and logic, provided that we realize that the concepts used in the formal systems may be different from (perhaps more useful for certain purposes than) our natural ones. Such a procedure is all right in logic and mathematics, since the construction of artificial models can often illuminate the logic and meaning of our ordinary speech; but whichever way the illumination goes (why not both ways?) it can work only if the system in question is purely formal. If what Rawls calls 'the substantive content of the moral conceptions' (p. 52) is part of the system, then what will be revealed by it are not the meanings of moral judgments but the moral opinions of those who adhere to the system. And when he proposes (p. 111, l. 6) to *replace* our concept of right by the concept of being in accordance with the principles that would be acknowledged in the original position, he is in effect seeking to foist on us not a new meaning for a word, but a substantial set of moral views; for he thinks that he has tailored the original position so as to yield principles which fit his own considered judgments.

(3) Rawls' moral methodology takes the form of a picture or parable—and one which is even more difficult than most to interpret with any confidence. We are to imagine a set of people gathered together (hypothetically, not actually) to agree upon a set of 'principles of justice' to govern their conduct.[7] The 'prin-

[7] It is tempting to say 'their *subsequent* conduct'; but the tenses in Rawls'

ciples of justice' are those principles to which these 'people in the original position' (POPs) would agree for the conduct of all of them as 'people in ordinary life' (POLs,) if, when making the agreement, they were subject to certain conditions.

It is obviously these conditions which determine the substance of the theory (indeed they *are* its substance, the rest being mere dramatization, useful for expository purposes, but also potentially misleading). Rawls' theory belongs to a class of theories which we may call 'hypothetical choice theories'—i.e., theories which say that the right answer to some question is the answer that a person or set of people *would* choose if subject to certain conditions. The best-known example of such a theory is the 'ideal observer' theory of ethics, about which Rawls says something, and which we shall find instructive to compare with his own. The important thing to notice about all such theories is that *what* this hypothetical person would choose, if it is determinate at all (which many such theories fail to make it) has to be determined by the conditions to which he is subject. If the conditions, once made explicit, do not deductively determine the choice, then the choice remains indeterminate, except in so far as it is covertly conditioned by the prejudices or intuitions of the philosopher whose theory it is. Thus intuition can enter at two points (and in Rawls' case enters at both; cf. p. 121, ll. 7–15). It enters in the choice of the conditions to which the chooser is to be subject; and it enters to determine what he will choose in cases where the conditions, as made explicit, do not determine this (see below).

account are one of its most baffling features. On the one hand, these 'people in the original position' (POPs) are to make a 'contract'; this and terms like 'original position' and 'initial situation' (p. 20, l. 18), seem to indicate that this conclave is temporally prior to the time at which these same people are to enter the world as we know it, become 'people in ordinary life' (POLs) and carry out their contract. But on the other hand Rawls seems to speak commonly in the present tense (e.g. p. 520, ll. 27 ff.), as if they were somehow simultaneously POPs and POLs. Not surprisingly this, and other obscurities, make it often difficult, and sometimes (to me at any rate) impossible, to determine whether some particular remark is intended to refer to POPs or POLs: Who, for example, are 'they' in page 206, line 5? And in page 127 line 25, is it the POLs who are being said to be mutually disinterested, as the passage seems to imply, and as is suggested by the reference to 'circumstances of justice' on page 128 line 5 (which seems usually, though not on page 130, lines 1–5, to mean circumstances of POLs, not POPs)? But if so, how are page 129 lines 14–18 or page 148 lines 2 ff. consistent? Again, do POLs lack envy, or only POPs? (see pp. 151, ll. 22–4; 143, l. 4; sect. 80–1). A review as long as Rawls' book itself could be spent on such questions of interpretation.

The more important of the conditions to which Rawls' POPs are subject are the following (sect. 22–5):

(1) They know certain facts about the world and the society in which POLs live, but have others concealed from them by a 'veil of ignorance'. It is obviously going to be crucial *which* facts they are allowed to know, and which they are not.

(2) They are motivated in certain ways, especially in being selfish or mutually disinterested, and also in lacking envy and in being unwilling to use the principle of insufficient reason. They are also 'rational' (i.e., take the most effective means to given ends) (p. 14, l. 5).

(3) They are subject to 'the formal constraints of the concept of right'. Rawls explicitly says that he does not 'claim that these conditions follow from the concept of right, much less from the meaning of morality' (p. 130, l. 16). Instead, he as usual says that it 'seems reasonable' to impose them (p. 130, l. 14). He does not tell us what he would say to somebody to whom they did not 'seem reasonable'.

(4) There are also certain important procedural stipulations, such as that the POPs should all agree unanimously in their choice of principles. Later in the book, the procedure is very much elaborated, and takes the form of a series of stages in which the 'veil of ignorance' is progressively lifted; but I shall ignore this complication here.

In comparing Rawls' theory with other theories, it is most important to notice the roles played by these groups of conditions. If I may be allowed to mention my own theory, I would myself place almost the whole emphasis on (3), and would at the same time aim to establish the 'constraints' on the basis of a study of the logical properties of the moral words.[8] This still seems to me the most rigorous and secure procedure, because it enables us to say that *if* this is how we are using the words (*if* this is what we mean by them), then we shall be debarred from saying so-and-so on pain of self-contradiction; and this gives moral arguments a cutting edge which in Rawls they lack. In a similar way, Achilles should have answered the Tortoise by saying, 'If you mean by "If" what we all mean, you have to accept *modus ponens*; for this is the rule that gives its meaning to "if".' It is of course in dispute

[8] *See* note 5 above. I shall be returning to this question of how logic constrains moral argument in an article in a new volume in the series *Contemporary British Philosophy*, ed. H. D. Lewis, forthcoming.

how much we can do by this method; but I think, and have tried elsewhere to show, that we can do much more than Rawls allows.

The 'ideal observer' theory (in a typical form) differs from Rawls' theory in the following respects. Under (1), it allows the principle-chooser to know everything; there is no 'veil of ignorance'. On the other hand, under (2), he is differently motivated; instead of being concerned with his own interest only, he is impartially benevolent. Now it is possible to show that on a certain simple and natural 'rational contractor' theory of the Rawls type (though not, it is fairly safe to say, on Rawls' own version of this type of theory) these two changes exactly cancel one another, so that the normative consequences of the 'ideal observer' and 'rational contractor' theories would be identical. To see this, let us remember that the main object of these conditions is to secure impartiality. This is secured in the case of the rational contractor theory by not allowing the POPs to know what are to be their individual roles as POLs in the society in which the contract has to be observed; they therefore cannot choose the principles to suit their own selfish interests, although they are selfishly motivated. It is secured in the case of the ideal observer theory by express stipulation; he is required to be impartially benevolent. It looks, therefore, as if *these* versions of the two theories are, as I have said elsewhere,[9] practically equivalent.

We must next ask *how much* the POPs have to be ignorant of, in order to secure impartiality. It must be noticed that much of the work is already done by the 'formal constraint' that the principles have to be 'general'.[10] Rawls himself says that the formal constraints rule out egoism (p. 136, l. 13); it might therefore be asked what there is left for the 'veil of ignorance' to do, since to abandon egoism (and for the same formal reasons the pursuit of the interests of any other particular person or set of them) is *eo ipso* to become impartial. I do not think that this objection sticks; for a POP, if he had full knowledge of his own role as a POL, might adopt principles which were formally 'general' or universal but were rigged to suit his own interest. Rawls, however, thinks (wrongly)[11] that such rigged principles

[9] 'Rules of War and Moral Reasoning', *Philosophy and Public Affairs* 1, 1971/2, p. 166.

[10] Rawls' word; I have commented on his use, and given reasons for preferring the word 'universal', which *he* uses for something else, in my paper 'Principles', *P.A.S.* 73 (1972/3), p. 2.

[11] I have hinted why in 'Principles' (*op. cit.*), p. 4. For my own answer to the 'rigging' difficulty, see my *Freedom and Reason*, p. 107.

can be ruled out on the formal ground of lack of 'generality', and so is open to the objection *ad hominem*. That is to say, *he* has left nothing for the veil of ignorance to do as regards impartiality.

Be that as it may, however, we need to be clear how thick a veil of ignorance is required to achieve impartiality. To be frugal: all that the POPs need to be ignorant of are their roles as individuals in the world of POLs. That is to say, it would be possible to secure impartiality while allowing the POPs to know the entire history of the world—not only the general conditions governing it, but the actual course of history, and indeed the alternative courses of history which would be the result of different actions by individuals in it, and in particular to know that there would be in the world individuals *a*, *b*, . . . *n* who would be affected in specific ways by these actions—*provided* that each of the POPs did not know which individual he was (i.e., whether he was *a* or *b*, etc.). Impartiality would be secured even by this very economical veil, because if a POP does not know whether he is *a* or *b*, he has, however selfish, no motive for choosing his principles so as to suit the interests of *a* rather than those of *b* when these interests are in conflict.

A superficial reading of Rawls' rather ambiguous language at page 37, line 4, page 2, line 2 and page 198, line 20 might lead one to suppose that this 'economical veil' is what he has in mind. But this cannot be right, in view of page 200, line 7 and other passages. We need to ask, therefore, why Rawls is not content with it, if it suffices to secure impartiality. The answer might be that he is unclear as between two things: (1) the POPs' not knowing which of them is going to be *a* and which *b*; (2) their knowing this, but not knowing how *a* and *b* are going to fare. Much of his language could bear either interpretation. And page 141, line 25 seems to imply that Rawls thinks that the 'economical veil' would allow the POPs to use threats against each other based on the power which as individual POLs they would have; but this is obviously not so if they do not know *which* individuals they are going to be, however many particular facts about individuals they may know.

Nevertheless, sooner than accuse Rawls of a mere muddle, let us look for other explanations. One is, that he wants, not merely to secure impartiality, but to avoid an interpretation which would have normative consequences which he is committed to abjuring. With the 'economical veil', the rational contractor theory is practically equivalent in its normative consequences to the ideal observer theory and to my own theory (see above and below), and these normative consequences are of a utilitarian sort. Therefore Rawls may have reasoned that, since an 'economical veil' would make him into a utilitarian, he had better buy a more expensive one. We can, indeed, easily sympathize with the predicament of one who, having been working for the best part of his career on the construction of 'a viable alternative to the utilitarian tradition' (p. 150), discovered that the type of theory he had embraced, in its simplest and most natural form, led direct to a kind of utilitarianism. It must in fairness be said, however, that Rawls does not regard this motive as disreputable; for he is not against tailoring his theory to suit the conclusions he wants to reach (see above, and page 141, where he says, 'We want to define the original position so that we get the desired solution'). I shall be examining in the second part of this review the question of whether Rawls' thicker veil *does* help him to avoid utilitarianism; it is fairly clear from section 28 that he *thinks* it does.

A further motive for the thicker veil is a desire for simplicity both in the reasoning and in the principles resulting from it (p. 140, l. 31; p. 142, l. 8; but cf. p. 141, l. 22). By letting the POPs know only the general facts about the world in which the POLs live, and also by other devices (e.g., p. 95, l. 14; p. 96, l. 6; p. 98, l. 28), Rawls effectively prevents them from going into much detail about the facts. This means that his principles can and must be simple; but at the same time it raises the question of whether they can be adequate to the complexities of the actual world. Rawls is, in fact, faced with a dilemma. If he sticks to the 'economical veil', then there will be no difficulty of principle in doing justice even in highly specific and unusual cases in the actual world; but this will involve very complex calculations, in advance, on the part of his POPs. On the other hand, if in order to avoid these complex calculations, he limits the POPs' knowledge to 'general' facts about the world, he is in danger of having his POPs choose principles which may, in particular cases, result in flagrant injustice, because the facts of these cases are peculiar.

This is merely the analogue, in Rawls' system, of the dilemma which afflicts utilitarians, and which I have tried to solve in two articles already referred to.[12] The solution lies in distinguishing between two levels of moral thinking, in one of which (for use 'in a cool hour') we are allowed to go into all the details, and in the other of which (for normal use under conditions of ignorance of

[12] 'Principles' (*op. cit.*), pp. 7 ff.; *Philosophy and Public Affairs* 1 (1971/2), p. 166.

the future, stress and temptation, and in moral education and self-education) we stick to firm and simple principles which are most likely in general to lead to right action—they are not, however, to be confused with 'rules of thumb', a term whose undiscriminating use has misled many. The first kind of thinking (let us call it 'level-2') is used in order to select the principles to be adhered to in the second kind ('level-1'), choosing those principles which are best for situations likely to be actually encountered. If this kind of solution were applied to Rawls' system, he would allow his POPs to know everything but their individual roles as POLs (the 'economical veil'); but since their task would be to choose the best *level-1* principles for the thinking of POLs. they could still, since these principles have to be simple and observed only in general, attend only to the general facts about the POL society and the general run of cases. The contract would then not be a contract to act universally in certain ways, but rather a contract to employ certain firm principles in the moral education of POLs themselves and their children, and to uphold such principles as the norm in their society. For unusual cases, and for those in which the principles conflicted, the POLs would be allowed (in Aristotelian fashion)[18] to do a bit of POP-thinking for themselves.

Rawls does not adopt this solution, although he shows some awareness of the distinction between level-1 and level-2 thinking on page 28, line 19. Ross's different but related distinction between '*prima facie* duties' and 'duties all things considered' is referred to and indeed used on page 340, line 15. On the whole Rawls' principles are treated as unbreakable ones for universal observance (e.g. p. 115, l. 36); but they are supposed to have the simplicity which in fact only level-1 principles can, or need, have (p. 132, l. 17). Other passages which *might* be relevant to this question are page 157, line 32; page 159, lines 16 ff; page 161, line 17; page 304, line 13; page 337, line 11; page 340, line 28; page 341, line 14; page 454, line 6; but I have been unable to divine exactly what Rawls' view is.

He has tried to get over the difficulty of conflicts between principles and unusual cases in two ways. The first is by means of a rigid 'lexical' ordering of his principles (which could be guaranteed in unusual cases to yield absurd results); the second is by his 'four-stage sequence' (pp. 195 ff.), whereby the 'veil of ignorance' is progressively lifted, and at each 'lift' the knowledge

[18] *Eth. Nic.* 1137 b 24 ff. (cf. Rawls, p. 19, l. 9; p. 138, l. 20).

of extra facts is absorbed and the principles expanded to deal with them. The sequence ends with the complete disappearance of the evil. Since Rawls can say this, he cannot have any objection on grounds of practicability to unrestricted knowledge from the start, and his reasons for forbidding it must be theoretical ones.

The four-stage sequence would only work if at each later stage the principles inherited from the stage before *determined*, in the light of the new information, what further principles were to be adopted. At least this is so, if the method is required to be rigorous; Rawls can perhaps escape this requirement by using intuitions all down the line. But if the principles chosen in the original position do *determine*, in conjunction with each new batch of facts, all the additional principles that are to be adopted at each stage, then the moral law is likely to turn out to be an ass. Some victim of the application of one of these lower-order principles may be found complaining that if the POPs had only *known* about him and his situation, likes and dislikes, then they would have complicated their principles a little to allow them to do justice to him (perhaps he does not give a fig for the 'priority of liberty'; or perhaps his preferences do not coincide with the POPs' ranking of the 'primary goods'). If it were rigidly applied, Rawls' system would be like a constitution having a legislature in which reading of the newspapers was forbidden, and law-courts without any judicial discretion. But of course he does *not* apply it rigidly.

I will conclude this part of the review by showing why the ideal observer theory, the rational contractor theory and my own theory must on certain interpretations of each of them yield the same results. As pointed out above, the 'economical veil' version of the rational contractor theory secures impartiality between the individuals in society. The ideal observer theory includes impartiality as an express stipulation. My own theory secures impartiality by a combination of the requirement that moral judgments be universalizable and the requirement to prescribe for hypothetical reversed-role situations as if they were actual (I am not sure whether the second is an independent condition or not). So, as regards impartiality, the theories are on all fours. Next, some degree of benevolence is required by all three theories; the ideal observer is expressly required to be impartially benevolent; my universal prescriber, since he has to treat everybody as one and nobody as more than one, and since one of the persons included in 'everybody' is himself, to whom he is benevolent, has to be positively and equally benevolent to everybody; the rational contractor, although he is selfish, does not know which individual POL it is whose interests he should favour (since he does not know which is himself) and so his selfish or partial benevolence has the same results as impartial benevolence. For the same kind of reason the ideal observer and the universal prescriber, though they have additional knowledge (viz., knowledge of their own individual roles, if any, in the situations for which they are prescribing) are prevented by the previous requirements from using it for selfish ends.

Rawls himself says that the ideal observer theory leads to utilitarianism (p. 185, l. 24); and—at least if it takes a certain form, if it involves what Rawls calls 'sympathetic identification' with all affected parties—this seems plausible. In stating this form of the theory, he echoes some phrases of my own[14] and later treats my theory and the ideal observer theory as equivalent.[15] So, then, the rational contractor theory, in the version I have been discussing (which is not Rawls') should also lead to utilitarianism. Rawls is aware of this possibility (p. 121, l. 33). He even seems to imply on page 149, line 3 that *his own* theory is practically equivalent to the ideal observer theory; but this is not his usual view. I shall shortly take up the difficult task of deciding whether Rawls, by the departures he makes from this simple version of the rational contractor theory, succeeds in establishing a non-utilitarian conclusion.

Of the three theories that I have just shown to be practically equivalent, it is largely a matter of taste which one adopts. Philosophers will differ in the use they like to make of dramatizations of their theories, and in the particular scenarios chosen. Such dramatizations do not help the argument, though they may help to expound it; Rawls himself seems to agree (p. 138, l. 31). For myself, I think such devices useful (though they can also mislead), and I had much greater hopes of Rawls' enterprise than have in fact been realized. For, knowing that the simplest and most natural version of the rational contractor theory was practically equivalent to my own position, I was optimistic enough that Rawls' elaborate exploration of the normative consequences of such a theory might illuminate those of my own, and thus enable me in the most favourable outcome) simply to plug in to his results. Such good luck, however, seldom befalls philosophers; and in fact Rawls' constant appeal to intuition instead of argument, and his tailoring of his theory to suit his anti-utilitarian pre-

[14] Page 186, ll. 30, 34; *Freedom and Reason*, p. 123.
[15] See *Freedom and Reason*, p. 94 n.

conceptions, have deprived it of the value which it could have had as a tracing of the normative consequences of views about the logic of the moral concepts.

It is interesting that in his peroration (p. 587) Rawls as good as drops into an ideal-observer way of speaking. I myself should be happy to use any of these images (including C. I. Lewis's 'all lives *seriatim*' picture).[16] But the work needs to be done on the logic of the argument, which has to be shown to be valid by the procedures of philosophical logic, involving the analysis of concepts, natural, or if need be artificial. Without this, a 'theory of justice' is nothing but a suggestive picture.

We must now examine Rawls' normative views about justice, and in particular the question of whether he succeeds in establishing them by argument. Since my own argument will be for the most part destructive (though I shall give some hints as to how I think the job might be done better), I must start by making it clear that I am not criticizing the project of bringing philosophical argument to bear on practical questions, but only Rawls' attempted execution of it. It is my hope, as it is his, that philosophers can so clarify these matters that we shall be able to argue more cogently than we do at present about what is just or unjust. I differ from him in thinking that unless the philosophers who attempt this base themselves on a thorough understanding of the concepts used and their logical properties, which is their proper and peculiar philosophical contribution, they run the risk of doing no more for the topic of justice than journalists and politicians.

I have argued that the simplest form of the rational contractor theory would have the same normative consequences as a certain version of the ideal observer theory and as my own theory, and that these consequences would be of a utilitarian sort. I suggested that that was why Rawls himself does not adopt this simple version. It remains, therefore, to ask whether his more complex version (if we can ever determine exactly what it is) can bring him to the non-utilitarian conclusions that he so earnestly desires.

The crucial questions are:

(i) Who are to be included among the parties in the original position?

(ii) What are they to be allowed to know?

[16] *Analysis of Knowledge and Valuation*, p. 547; Rawls page 189, line 12; cf. my *Freedom and Reason*, p. 199, where through ignorance I failed to acknowledge Lewis's use of this picture.

(iii) How, given these restrictions on membership and knowledge, will the POPs set about making their choice? Under this heading come questions about their use of the 'principle of insufficient reason', and of their motivation (especially their aversion to risk).

We have to ask whether the answers to these three questions suffice to determine, without appeal to Rawls' own preconceptions or intuitions, what principles of justice the POPs would choose.

The principles which Rawls says they would choose are summed up by him in their 'final statement' on page 302, thus:

> *First Principle.* Each person is to have an equal right to the most extensive total system of equal basic liberties compatible with a similar system of liberty for all.
> *Second Principle.* Social and economic inequalities are to be arranged so that they are both: (*a*) to the greatest benefit of the least advantaged, consistent with the just savings principle, and (*b*) attached to offices and positions open to all under conditions of fair equality of opportunity.

He also has what he calls 'priority rules' determining the relative priority that we are to give to these principles and their parts, and to the principles of justice over other principles. I shall not have space to deal with these. The most important of them is that which gives priority to liberty over the other principles. All the principles of justice are said to rest on what he calls the same 'General Conception', which runs thus:

> All social primary goods—liberty and opportunity, income and wealth, and the bases of self-respect—are to be distributed equally unless an unequal distribution of any or all of these goods is to the advantage of the least favoured [p. 303].

Thus Rawls founds his theory of justice (rightly, as I think) on an account (though I think it is the wrong account) of distributive justice. This in turn is founded on a view about procedural justice in the selection of principles.

(i) *The Membership of the POP Committee*

It might be thought that the simplest and safest membership-rule for the assembly of POPs would be to include everybody who might be affected by their choice of principles—i.e. 'everybody who could live at some time' (p. 139). This simple rule Rawls rejects, on the ground that 'to conceive of the original position [thus] is to stretch fantasy too far; the conception would cease to be a natural guide to intuition' (p. 139). In a book which relies so much on fantasy, this is a surprisingly weak ground. The membership is restricted in several ways:

(*a*) Animals are excluded, thus neatly removing them from the direct protection of principles of justice.

The discussion of this topic is full of expressions like 'presumably' (pp. 505, 506), 'the natural answer seems to be (p. 505) and 'seems necessary to match our considered judgments' (p. 509). Rawls himself says, very frankly, 'Now of course none of this is literally argument' (p. 509); and it would certainly not convince a determined vegetarian. It is of course difficult to include animals among the POPs (they could not make speeches in the assembly); but that is an unfortunate effect of the dramatic scenario—which might perhaps be overcome if we could suppose that the POPs do not know whether they may not be, or be going to become, animals.

(*b*) By adopting the 'present time of entry interpretation' (p. 140, l. 11—an opaque phrase which I have nowhere found explained) he seems to exclude all but the members of just one generation in the world's history.

He thus lays up for himself, as he believes, troubles about justice between generations. These troubles are quite gratuitous, and could easily have been solved by allowing all generations to join the assembly. Rawls himself attempts to get over the difficulty by an obscure and contrived assumption that the POPs are to be 'thought of as representing continuing lines of claims, as being so to speak deputies for a kind of everlasting moral agent or institution' (p. 128). 'Everlasting' is toned down, two lines further on, to 'over at least two generations'. It might be thought that there was an important difference between two generations and eternity, and that, for securing impartiality between all generations, only eternity would do. But in fact it is strictly unnecessary to make any such assumption to secure impartiality if, as Rawls also rightly stipulates, the POPs, though they all belong to one generation, are not to know which this is (p. 137, l. 14; p. 288, l. 9). For in that case they will not favour a particular generation, any more than they will favour particular individuals, not knowing who they themselves are or when they are born (cf. p. 137, l. 23).

It is not the restriction of membership to one generation, by itself, that gets Rawls into trouble at this point, but rather this, combined with another feature of his theory to which we must now attend. He writes as if the POPs were not prescribing universally (or, as he would put it, 'generally') in choosing their principles of justice, but only prescribing for their *own* behaviour (and possibly also for that of *subsequent* generations—p. 13, l. 4). From this it follows that (in default of the *ad hoc* restriction which he imposes) they can happily say 'Let our generation, whichever it is, consume all the world's resources and leave none for succeeding generations.' If, on the contrary, they were prescribing universally for all men at whatever time, and did not know at what time *they* were to be in the world, they could not happily universalize this prescription; for they would then be prescribing equally for their own predecessors. Thus Rawls has (characteristically, and as a consequence of his contempt for such logical tools) failed to avail himself of one of the 'formal constraints of the concept of right' to which he himself has earlier drawn attention (p. 131, l. 14). If the POPs do not know to what generation they all belong, and are prescribing universally for the conduct of all generations, they will have (if they are rational) to adopt principles of justice which maintain impartiality between the interests of all generations. We can say that they are either prescribing for the past as well as the present and future, or choosing the principles by which they want society to be governed in the future, and hope that it has been governed in the past. I have heard rumours that Rawls himself is now attracted by this manoeuvre. Alternatively, we might take a hint from Kant via Mr. Richards,[17] and suppose that, when they make their choice, they are in a noumenal atemporal green room, and do not know at what point they are to come upon the stage of time as POLs. Any difficulties which attend these modifications to the scene arise from the creakings of the stage-machinery and not from the logic of the argument, which could be set out in universal-prescriptive

[17] D. A. J. Richards, *A Theory of Reasons for Action* (Oxford, 1970), pp. 88–9, 310.

terms without any such machinery. That the POPs cannot *affect* the past (p. 292, l. 5) is strictly irrelevant; Rawls thinks it relevant only because he takes his machinery too seriously.

(*c*) The principles of justice are to be chosen from the 'perspective' or 'point of view' of 'representative men in all relevant social positions' (p. 96, l. 1; cf. p. 64, l. 9).

This is not strictly a membership-restriction, but it has a similar effect. Those who are on the committee are compelled to choose from the *point of view* of representatives of these rather gross classes (in defining which, for example, we are not allowed to differentiate between large and small farmers (p. 96, l. 4); this perhaps illustrates how prone Rawls is to iron out material differences between cases in the interests of 'a coherent and manageable theory' (p. 96, l. 7)).

Does any of these membership-restrictions enable Rawls to avoid utilitarianism? It seems not; what they can do is, not to establish or refute any particular principle of distributive justice, but only to confine or limit the class of those protected by it. They will do this if the POPs *know* that the membership is so limited (as, perhaps, in the case of the disfranchisement of animals); if they do *not* know, then the membership-restriction can make no difference at all, since each POP will have to envisage the possibility of his being *any* person out of the set *a*, *b* . . .*n*, even though the person who is going to be, say, *b*, or *m*, is not actually present at the meeting. It does not matter if some person is not present, provided that nobody knows that he is not; and for that reason a committee is strictly unnecessary; one POP would do, provided that he did not know which POL he was going to be (cf. p. 139, l. 23).

However, there is one membership-restriction stipulated by Rawls that does seem to make a big difference:

(*d*) Only people who actually do or will exist are allowed on the committee.

In a passage which I quoted earlier, it has been implicitly laid down, on rather slender grounds, that (merely) *possible* people, as opposed to actual people, are to be blackballed (p. 139, l. 4); later it is explicitly stated that the POPs 'know that they already hold a place in some particular society' (p. 166, l. 12)—though of course they do not know what place in which society. This means that in Rawls' system the interests of possible people are simply not going to be taken account of. This would seem to be crucial for questions about population policy and abortion, for example.

The person that the foetus *would* turn into if not aborted, and the people who *will* be born if contraception is not practised, get no say if they are not actual people—i.e., if it is actually the case that abortion and contraception *are* practised. This would seem to have the curious consequence that, simply by performing an abortion, I can make sure that my act does not contravene Rawlsian justice, because I shall thereby have disfranchised the abortee. Since POPs have to have reached 'the age of reason' (p. 146, l. 15), it looks as if the same would apply to infanticide (since merely possible adults are excluded). But page 248, lines 37 ff. and page 509, lines 19 ff. seem to support a different and more orthodox view (if we can swallow the assumption that the veil of ignorance could conceal from one of the contracting parties the fact that he is a babe in arms).

It is only this membership-restriction that enables Rawls to pronounce so easily that the average utility principle, which bids us maximize average utility, is superior to the classical utility principle, which bids us maximize total utility (p. 166, l. 3). To understand this, consider a possible person *P* whose birth would have lowered average utility but raised total utility, because his own happiness would have been less than the previous average, but more than the combined losses suffered by the others owing to his arrival. If a POP might, for all he knows, be *P*, he will find the classical principle more attractive; but if he knows that he cannot be *P*, he will prefer the average principle. This is because the classical principle would require population policies which debarred him from existence. By excluding *P* from the committee, and allowing this to be known, Rawls makes sure that it will disregard *P*'s interests, and thus brings it about that, from the POP's point of view, the average utility principle is a stronger candidate against which to pit his own principles of justice than the classical utility principle. If the exclusion is unjust, he is perhaps choosing the weaker and not the stronger opponent; but I do not think that in fact this gives him much advantage. In any case, it would not seem that this membership-restriction helps him much in his fight with utilitarianism in general.

We may note here an embarrassing consequence of the inclusion of possible people among the POPs, if Rawls' own normative principles are adopted, and if it is assumed that to have any life at all is better than not to be born. The unborn will then be his least advantaged class; and so his difference or maximum principle (see below) will require him to say that before anything is done for the rest of us, we ought to secure the birth of all these possible people.[18] This would lead us to a duty of procreation on a vast scale; we could stop only when the earth would support no more people above the starvation level. But Rawls could reasonably escape this consequence by rejecting the assumption that any life is better than none; he does not need for this purpose to disfranchise merely possible people. The classical principle also has been thought by some to require a very expansive population policy; but this too will depend on what weight is put upon quality of life as opposed to mere life.

(ii) *The Thickness of the Veil*

We saw above that Rawls does not adopt the simplest and most economical version of the 'veil of ignorance' to which his POPs are to be subject—namely that which deprives them only of knowledge of which *individual* each of them is to be. He prefers to say that they are not to know what *properties* of various sorts they are to have, and that they know only the 'general' facts about society.[19] That is, besides, or instead of, being ignorant of what individual each of them is to be, the POPs are ignorant of everything that cannot be described in general terms. So, for example, they are allowed to understand 'political affairs and the principles of economic theory' (p. 137) and 'the laws of moral psychology' (p. 138). There is one exception to the ban on particular knowledge: the POPs are allowed to know the 'particular fact' that the POLs' society is subject to the 'circumstances of justice'—a phrase under which Rawls covers such facts as that human beings are vulnerable to attack and that natural resources are limited (p. 127, ll. 3, 6), as are also people's powers of reasoning, memory and attention (p. 127, l. 27).

I have suggested that section 28 indicates Rawls thinks that his thick veil of ignorance helps him to avoid utilitarianism. It

[18] I owe this point, and much else, to Mr. Parfit.

[19] It will perhaps not be necessary to enquire precisely what he means by 'general' (let alone such minutiae as whether he means the same by 'general' on page 197, line 9 as he does on page 137, line 1, or whether, rather, as seems necessary in order to preserve consistency, what are called 'general facts' on page 197, line 9 are what have been called 'particular circumstances' on page 137, line 11).

is not easy to see why he thinks this; but the key to the understanding of the reasoning seems to lie in the relation between ignorance of particular facts and refusal to use the 'principle of insufficient reason' (IR, see below). Rawls says that the POPs 'discount estimates of likelihood that are not based on a knowledge of the particular facts' (p. 173; cf. p. 155. ll. 25 ff.). This seems to imply that if they *did* have knowledge of particular facts (even if they did not know their individual places in the world constituted by those facts), they would be able to work out relative frequencies of sorts of events, and thus 'the objective probabilities' of occurrences in the POL society (cf. p. 168, l. 15; p. 171, l. 31). Thus they would have no need to use IR, but could base their predictions of how each individual POP-turned-POL was likely to fare, given the adoption of any one set of principles of justice, on 'objective probabilities'. And this, he may have thought, would lead to utilitarianism (see below). I am far from certain that this *is* Rawls' argument. But if it is, it is fallacious. For even if the POPs had knowledge of objective probabilities of sorts of occurrences in the world, they still might not have knowledge of the particular facts' (p. 173; cf. p. 155, l. 25 ff.). This vidual POL, and so might not know how likely it was that any individual POP-turned-POL would fare well, or ill. For *this*, they might have to rely on IR. In other words, if the sortition which results in this POP being this POL is not subject to objective probabilities, no amount of objective probability in the distribution of welfare among the POLs will help any individual POP to know how likely it is that *he* will get a certain degree of welfare, unless he is allowed to use IR. So, since the knowledge of particular facts would not by itself make the use of IR unnecessary, Rawls does not gain anything by refusing to allow knowledge of these particular facts. The upshot is that *everything* in his argument for the rejection of utilitarianism depends on his refusal to allow the POPs to use IR. To this, therefore, we must now turn.

(iii) *Insufficient Reason and Aversion to Risk*

The principle of insufficient reason (IR) requires us to assign equal probabilities to two or more outcomes when we have no reason to suppose that the probability of one is greater than that of another. In some forms it leads to paradox: if we are drawing blindfold from an urn containing black, red and yellow balls in unknown proportions, it gives us different values for the probability of drawing a black ball according as we state the alternatives, which by IR will be equiprobable, as black, red and yellow or as black and coloured. If, therefore, the situation of the POPs were such that they had no basis for listing the members of the set of outcomes uniquely, then it would be hard to quarrel with Rawls' insistence that rational POPs would not use IR. However (to use a Rawlish phrase) it seems reasonable to suppose that a POP, knowing that he will be one of n individuals, all in some respects faring differently, will say that there are, correspondingly, n different outcomes, and will, if he uses IR, conclude that the probability of getting any one of them is $1/n$. At any rate, we have at least as much reason to suppose this as Rawls has to suppose anything else.

He admits (p. 165) that 'if the parties are viewed as rational individuals who have no aversion to risk and who follow the principle of insufficient reason in computing likelihoods . . ., then the idea of the initial situation leads naturally to the average [utility] principle'. It is interesting that in this passage he seems to imply that it is possible to have no aversion to risk and to use IR, without thereby ceasing to be rational; if so, it cannot be the rationality of the POPs that he relies on. But his more considered view seems to be that it is not rational to use IR (cf. p. 172, l. 30).

It is not clear whether he thinks that the POPs have an aversion to risk. On page 172 he says, 'The essential thing is not to allow the principles chosen to depend on special attitudes towards risk . . . What must be shown is that choosing *as if* one had such an aversion is rational given the unique features of that situation irrespective of any special attitudes towards risk' (my italics). This seems inconsistent with page 169, lines 15 ff., where he implies that the POPs *actually have*, and ought to have, a reluctance to take great risks, on the inadequate ground that their descendants may reproach them if they do (might not their descendants be just as likely to say 'Nothing venture, nothing win'?). It also seems inconsistent with page 165, line 35, quoted above, if that implies that one who was not averse to risk and used IR would not be, on that score, irrational. In page 172 he says 'the parties do not know whether or not they have a characteristic aversion to taking chances'; and the reader is not to know either.[20]

[20] Nor whether the referent of 'they' is the POPs or the POLs. For if it is

It looks, at any rate, as if the principal weight is being placed on the rejection of IR. I am sure that reviewers more competent than me in the theory of probability will have discussed this topic at length. But the important thing to notice is that the answer to the question of whether it would be rational for the POPs, in default of objective probabilities, to use IR *does not matter*. Suppose we grant that it would not. The important point would then be: Why are they denied knowledge of objective probabilities? Rawls says that each of his assumptions about the POPs 'should by itself be natural and plausible' (p. 18), but this particular feature seems to be quite arbitrary. It is only there because it may help to lead by arguments which Rawls finds acceptable to conclusions which he finds acceptable (cf. again the remark on p. 141, 'We want to define the original position so that we get the desired solution'). It is obvious that, if Rawls had wished to reach utilitarian conclusions, he could have so arranged it that the POPs were going to have their place as POLs assigned to them by means of a well-conducted mechanical or electronic lottery of the usual kind. In that case, the POPs, knowing this, *would* have known the objective probabilities of their getting any particular POL-role, and would have known that they were equal, and utilitarianism would have resulted. So the very most that Rawls may have done towards setting up a non-utilitarian theory of justice is to show that it is possible, if one desires, so to rig the assumptions of the theory that it does not lead straight to a utilitarian conclusion.

We next have to examine the conclusions that Rawls thinks it does lead to. He thinks that the POPs would, in choosing their principles, 'maximin'—that is to say, choose the course which has the best worst outcome. They will seek to maximize the welfare of the least-advantaged representative members of their society. It is important to distinguish this strategy from another, which I will call the 'insurance' strategy; for Rawls uses arguments in favour of maximining which are really only arguments in favour of insuring against utter calamity (cf. p. 156, l. 30; p. 163, l. 18; p. 176, l. 23). We insure our houses against fire because we think that a certain outcome, namely having one's house burnt down and having no money to buy another, is so

the POPs, Rawls is saying that a veil of ignorance can conceal from someone an aversion which he presently has (a rather steep assumption); but if the POLs, how is this condition relevant to the present motivation of the POPs as regards risk-taking?

calamitous that we should rule it out. This is not at all the same strategy as maximining. If the POL society were going to be affluent enough to provide a more than just acceptable standard of living for even the least advantaged, the insurance strategy would allow the POPs to purchase a very great gain for the more advantaged at the cost of a small loss for the least advantaged; but the maximin strategy would forbid this. Maximiners would end up refusing to let the man with, say, $50,000 a year have any more if the man with the minimum income of $40,000 received in consequence a dollar or two less; but the follower of the insurance strategy would by that time have lost interest.

POPs following an insurance strategy would fix a minimum and frame their principles of justice to secure that. They would not have to know whether the minimum was feasible; all they would have to do would be to say that below it, the interests of the more advantaged were always to be sacrificed to those of the less (as in wartime rationing). Rawls has not actually given us anything to determine what minimum the POPs would fix (might they not differ on this?). But neither has he given us anything to determine the minimum which he himself requires for his argument. It is what they 'can be sure of by following the maximum rule' (p. 154); but how do they know what this amounts to? And if they do not know, how can they tell that anything less is 'intolerable' (p. 156, l. 24)?

The POP game is in effect played by imagining ourselves in the original position and then choosing principles of justice. Rawls' POPs come to the decisions that they come to simply because they are replicas of Rawls himself with what altruism he has removed and a veil of ignorance clapped over his head. It is not surprising, therefore, that they reach conclusions which he can accept. If I myself play this game, I import into the original position *my* prejudices and inclinations, which in some respects are different from Rawls'. I have some inclination to insure against the worst calamities, in so far as this is possible.[21]

[21] It is a difficult question (too difficult for discussion here) to what extent an insurance strategy on the part of the POPs is compatible with utilitarianism. Utilitarian POPs could insure against calamities if the premium were such as to maximize their expectation of utility; and this is what many *POLs* try to do. To be willing to pay a high premium may simply be an indication that one attaches a high negative value to the calamity in question, and less value to the possible affluence one is sacrificing. The diminishing marginal utility of affluence is relevant here. So, on the other hand, is the sheer impossibility of insuring against some of the worst calamities (for example, that of

But I have no inclination to maximin, once the acceptable minimum is assured; after that point I feel inclined to take chances in the hope of maximizing my expectation of welfare, as I do in actual life (for example, I do not entirely refrain from investing my own cash because I might lose it). And in certain cases I do not feel inclined to maximin even in very reduced circumstances. If, when I was a prisoner of war, a benevolent and trustworthy Japanese officer had said that he would play poker with me and, if I won enough, allow me to buy myself a ticket home through neutral territory with a safe conduct, then I should have accepted the invitation, in order to give myself a chance, however small, of freedom (the priority of liberty!) rather than forgo this chance and husband my money to buy smokes with as I languished on the Burma railway.

Thus the maximin strategy does not appeal to me as *in general* a good one for choices under uncertainty. Even Rawls does not go so far as to claim that. He states three features of situations which give 'plausibility' to the maximin strategy (p. 154, l. 11). The first is the ignorance of objective probabilities; but we have seen that the imposition of this condition is entirely arbitrary. The second is that the chooser has to have 'a conception of the good such that he cares very little, if anything, for what he might gain above the minimum stipend' (p. 154); but this condition is clearly inapplicable, for the POPs 'do not know their conception of the good' (p. 155).[22] The third feature is that some outcomes are 'intolerable' (p. 156); but this justifies only insurance, not maximining. It looks, therefore, as if Rawls has not succeeded in making his choice of strategy even 'plausible'. But in spite of this he says 'the original position has been defined so that it is a situation in which the maximin rule applies' (p. 155). We can only say 'Amen'.

I do not claim to have shown that the maximin strategy is a bad one for POPs, only that Rawls has given no good reason for holding that it is a good one. The truth is that it is a wide

being a person whose temperament simply prevents him being happy). A utilitarian POP might well achieve the results of an insurance strategy without the strategy, if he assigned a high acceptance-utility to a level-one (*prima facie*) principle enjoining *compassion*—a sentiment which is perhaps better able than 'justice as fairness' to make us look after the unfortunate.

[22] They therefore make do with assumptions about 'primary goods' (which seem in this passage, though not always, all to have monetary values). The fate of a man who was made miserable because he lacked something which he valued very much, but which was not on Rawls' list of primary goods, is therefore not even insured against.

open question how the POPs would choose; he has reduced the information available to them and about them so much that it is hard to say *what* they would choose, unless his own intuitions supply the lack. Rawls, however, has one recourse, and that is that the results of his theory have to tally with his 'considered judgments'. But they do not tally with mine. A maximin strategy would (and in Rawls does) yield principles of justice according to which it would always be just to impose any loss, however great, upon a better-off group in order to bring a gain, however small, to the least advantaged group, however affluent the latter's starting point. If intuitions are to be used, this is surely counterintuitive; at least, not many of us are as egalitarian as that.

It is to Rawls' credit that he does not avail himself of some well-worn but fallacious polemical arguments against utilitarianism. It is true that on page 156, he insinuates, without stating, that utilitarianism could justify 'if not slavery or serfdom, at least serious infractions of liberty for the sake of greater social benefits'. But he very fairly admits on page 26 that 'certain common precepts of justice, particularly those which concern the protection of liberties and rights' *can* be accounted for in a utilitarian system as 'those precepts which experience shows should be strictly respected and departed from only under exceptional circumstances if the sum of advantages is to be maximized'. It would be unfair on our part to expect Rawls to explore more fully the possibilities of showing the place of common notions of justice in a utilitarian system—that is not his enterprise. But he has not shown that there are no such possibilities; and until this has been shown, philosophers would do well to go on looking for them. It may be that they could be found, without anything like so much intuitive scaffolding as Rawls needs for his own system; it may be that the world *is* so constituted that to fail to inculcate and strenuously pursue principles of justice fairly closely related to *some* of the commonly accepted ones will result in a diminution of utility. I am indeed inclined to think that this is so; but obviously the question calls for further investigation.

5 Rawls and Intuitionism

JOEL FEINBERG*

Rawls conceives of his own theory of justice as an alternative both to utilitarianism and its major rival in recent literature, the theory Rawls calls 'intuitionism in a broad sense'. The latter is the theory that there is 'a plurality of first principles which may conflict to give contrary directives in particular types of cases', and that there is 'no explicit method, no priority rules, for weighing these principles against one another . . .' (p. 34).[1] That intuitionism, so defined, has its own severe difficulties, I do not wish to deny; nor can I deny that its strongest support is a kind of argument from the default of its rivals. What I shall suggest here, however, is that intuitionism is a more plausible view than Rawls allows, especially when it is construed as a theory about what we ought (on balance) to do, as opposed to a theory about the character of just institutions (Rawls' main concern). I also argue, in the second part of this essay, that Rawls' theory, in one of its most important parts, is itself clearly intuitionistic, for it provides no method for weighing distinct principles of justice, and the 'natural duties' they impose, in certain contexts of conflict in the 'nonideal world'.

I

I find some confusion at the start in understanding precisely what question Rawls' contractarian theory is intended to answer. In particular, I am not sure whether he is proposing an ultimate standard for determining when actions and social arrangements are *right*, all things considered, or whether he is proposing a criterion for answering what I take to be the somewhat narrower

* Professor of Philosophy, Rockefeller University.

[1] Rawls distinguishes this theory from 'intuitionism in the traditional sense' which contains certain characteristic epistemological doctrines ['for example those concerning the self evidence and necessity of moral principles' (p. 34)]. The 'intuitionism' referred to here, on the contrary, is epistemologically neutral.

question of when acts and arrangements are *just*. If there are considerations other than, or in addition to, justice that can enter into a decision about what is right, a demonstration that a policy or practice is just does not necessarily settle the question of 'on balance rightness' with finality. Conducibility to liberty, general health and happiness, perfection of character, social progress, and economic growth, I should think, are also considerations that have weight, even if we admit that considerations of justice are those that should have the most weight. It follows that a practice can be right even though to some extent unjust, and that we can sometimes be *justified*, all things considered, in treating some persons to some extent unjustly. Because of these unhappy truths, some philosophical writers[2] have resorted to an invented technical term 'justicize' meaning 'to show something to be just', in contrast with the word 'justify' which means 'to show something to be right, all things considered'. Rawls, however, apparently rejects the line of thought reflected in this usage. It seems to be his view that if one set of social arrangements is more just than another, it is decisively better than the other, and it would be wrong not to choose it if we can. The other considerations that have a bearing on rightness are relevant only when we choose between arrangements that are equally just.

Rawls' leading rivals, the utilitarians, have also tended over the centuries to obscure the distinction (if there is one) between justice and on-balance rightness. In their writings about justice it is not always clear whether they are proposing a utilitarian analysis of the concept of justice itself, or applying a utilitarian standard of right conduct to specific social problems often at the expense of the dictates of justice. The following are the main possible views about the relations between justice and utility:

(1) Individual justice and social utility, properly conceived, can *never* conflict. That is because 'justice is a name for certain moral requirements, which, regarded collectively, stand higher in the scale of social utility, and are therefore of more paramount obligation, than any others . . .'[3]

[2] William Frankena, 'The Concept of Social Justice', in *Social Justice*, R. Brandt ed. (Englewood Cliffs, N.J.: Prentice Hall, 1962), and A. D. Woozley, 'Injustice', *American Philosophical Quarterly Monograph* 7 (Oxford: Blackwell, 1973).

[3] J. S. Mill, 'Utilitarianism', in *Utilitarianism, Liberty and Representative Government* (Everyman's ed., 1910), p. 59.

(2) Individual justice and social utility do sometimes conflict, and when they do, so much the worse for individual justice.

(3) Individual justice and social utility do sometimes conflict, and when they do, so much the worse for social utility.

(4) Individual justice and social utility do sometimes conflict, but it is impossible in advance to say that one must always have a stronger claim than the other. These opposing irreconcilable claims can only be 'balanced' against each other in the concrete circumstances of their conflict.

Views (1) and (2) are both utilitarian, but they are answers to different questions. The first *analyzes* justice in terms of utility and yields the arbitrary and implausible result that the two can never conflict. The second view, characteristic of Stalinists and other wicked Machiavellians, is more realistic, but 'realistic' in the pejorative sense of that term. The third position, which accords with the ancient and honorable tradition of '*fiat iustitia, et pereat mundus*', is probably that of Rawls. What I am uncertain about is whether Rawls' view is to be understood as a direct rival to utilitarian view (1) only or to utilitarian view (2) as well. Most of Rawls' arguments seem to be directed against (1). Many of these show that justice as analyzed by utilitarians endorses arrangements as 'just' which are plainly contrary to our natural shared convictions of justice. I find Rawls entirely convincing in these arguments. Utilitarian view (2), however, is wholly consistent with a Rawlsian *analysis* of justice, and argues with some plausibility that utility, while different from justice, is nevertheless superior to it. I do not find the arguments in Rawls' book against this second kind of utilitarianism quite as convincing as those by which Rawls destroys the first kind. I am not myself tempted to the second utilitarian view, however, because it seems to me an extreme exaggeration, and therefore a caricature, of the insight of its proponents. Surely, we would not judge it to be right on balance to choose a slightly more 'useful' social practice in preference to a vastly more just one. Individual justice must provide *some* limit to the claims of the general welfare, even though the latter may be considerable.

I am inclined, therefore, to reject utilitarianism as either a misconstruction of the relevance of social utility or an exaggeration of its importance, and to consider the fourth view distinguished above to be the principal rival to Rawls' theory. Rawls calls theories of this kind 'intuitionistic', because they deny that a plurality of moral principles can be reduced to one ultimate and superior one, and they also deny that 'priority rules' can be formulated which assign weights to conflicting considerations so that we can determine in particular cases which has priority. Hence, irreducibly distinct principles must be 'weighed' or 'balanced' in particular cases of conflict by 'intuition' (see pp. 34–41).

Rawls' own view (*if* we interpret him to be talking about on-balance rightness, or justification, and not only 'justicization') is that there are principles other than justice that have some force, but in respect to them, justice has absolute weight, always deserving priority in cases of conflict. Rawls argues against his less absolutistic opponent in a characteristically undogmatic way. He expresses his respect for intuitionism as a plausible rival, and he concedes that he cannot prove it false. His own theory, however, is urged to be superior to it in at least one way: it has no priority problem. Rawls not only has a rigid priority rule giving justice absolute weight over other moral considerations; he also proposes a rigid ordering scheme among the various parts of his principles of justice themselves. If Rawls' priority rules are defensible, he claims, then his theory is clearly superior to intuitionism:

> The assignment of weights is an essential and not a minor part of a conception of justice. If we cannot explain how these weights are to be determined by reasonable ethical criteria, the means of rational discussion have come to an end. An intuitionist conception of justice [rightness?] is, one might say, but half a conception [p. 41].

This is fair enough. If the sort of 'explanation' Rawls seeks is in principle achievable, then the theory that supplies it carries the day. But there is also the possibility that rigid priority rules are, in the very nature of the case, impossible to formulate. Other things being equal, simplicity is preferable to complexity, but a distorting simplicity is worse than none. Existentialist writers have spoken of a 'tragic surdity' among equally stringent ethical principles. If they and others are right, the conflict between justice and other social principles may pointedly illustrate Aristotle's remark that we should 'look for precision in each class of things just so far as the nature of the subject admits . . .'[4]

[4] Aristotle, *Nicomachean Ethics*, bk. I: chap. 3, 1094b 25.

Even if precise priority rules between equally valid moral principles are impossible, the results may not be as unacceptable as Rawls suggests, nor indeed contrary to the expectations of sensitive persons experienced in the dilemmas of practical judgment and decision. Rational discussion may indeed come to an end before final agreement is possible, but it might still carry us a good part of the way toward that elusive goal in discussions of social questions as in all other practical contexts. A movie critic follows reason in judging a film by its dramatic tension and flow, its theme or moral, its acting, its photography and music. These components are relevant to its final evaluation in a way in which its length, or the previous experience of its director, is not, and reason can demonstrate such judgments of relevance to the satisfaction of any reasonable person. A skilled critic, moreover, can adduce nearly conclusive evidence for the positive and negative judgments he makes about the relevant elements of the film. Still, in the end, he may have to tote up the film's overall score and *balance* it against incommensurable excellences and defects of other films, if he is to make a judgment of comparative worth. There is no cut and dried formula for assigning numbers, or for deciding whether a well-acted poor story rates higher or lower than a poorly acted good story. At best, 'priority rules' in such contexts are useful rules of thumb, telling us that in general one kind of element should be weighed more than another kind, but not *how much* more, nor even that it *invariably* weighs more. All of this is well enough understood in cinema criticism. A plurality of standards puts a limit on 'the means of rational discussion' short of necessary and universal agreement, but the limit is not all that binding, and it gives no support to the view that film criticism is 'all a matter of taste'. The same I think is true of moral criticism and judgment. We may not be able to say whether the duty to keep promises or the duty to tell the truth must *always* take precedence in the cases where they cannot both be honored. Nevertheless we can have rational grounds for honoring both duties in the multitudinous cases where they don't conflict, and for arranging our lives in such a way that the occasions for conflict among acknowledged duties are kept to a minimum. Even in actual cases of conflict, the concrete context may make it perfectly obvious which conflicting duty takes precedence in the circumstances. The burden is on Rawls, then, to establish his proposed rigid ordering of justice and social utility in the teeth of skeptical intuitionists (like this writer) and we need not be surprised, much less distressed, if he has failed.

Part of the test of a proposed priority rule in moral philosophy is whether it conforms with our considered confident judgments in specific cases, actual and hypothetical. The intuitive force of Rawls' position is well captured in an example from William James:

> If the hypothesis were offered us of a world in which Messrs. Fourier's and Bellamy's and Morris's utopias should be all outdone, and millions kept permanently happy on the one simple condition that a certain lost soul on the far-off edge of things should lead a life of lonely torture, what except a specifical and independent sort of emotion can it be which would make us immediately feel, even though an impulse arose within us to clutch at the happiness so offered, how hideous a thing would be its enjoyment when deliberately accepted as the fruit of such a bargain.[5]

The 'emotion' to which James refers is the very sense of justice whose deliverances Rawls venerates. Even those of us who are so hardened against its voice that we would make the deal with James' devil would be troubled ever after by the thought that the great net gain in human happiness so achieved was made *unfairly* at the expense of the innocent victim.

Most of us, I feel confident, would join James and Rawls in declining the devil's offer. No conceivable gain in social well-being could possibly balance the cosmic injustice done the lost soul. Even utilitarians might be forced by this example to interpret 'social utility' in a broad enough way to escape its thrust. If every human soul, as such, has infinite value, they might argue, then moral mathematics will not dictate the sacrifice of one for the sake of the others. The laws of ordinary arithmetic do not apply to infinite magnitudes. The value of one infinite soul would be precisely equal to that of two billion other infinite souls taken together, and therefore the suffering of the sacrificed one would have no less negative value than the perfect happiness of the benefited souls had positive value. But this way of interpreting the metaphor of 'infinite value' will not completely save

[5] W. James, 'The Moral Philosopher and the Moral Life', in *Essays in Pragmatism* (Hafner, 1948), p. 68. This example is not cited or quoted in Rawls' book.

the utilitarian from embarrassment, if he is forced to admit that on his principles the decision whether to accept the offer of James' devil is a matter of moral indifference, a 'six of one and half dozen of the other'.

When we consider examples of a different kind, however, Rawls' absolutism loses plausibility. Suppose that our option is not whether to promote social *gains* unfairly at the expense of sacrificed individuals, but rather whether to prevent social *harms* —widespread suffering, misery, poverty, and ignorance—at the expense of sacrificed individual or class interests. Such options have in fact been faced over and over again by political leaders since the time of the first industrial revolution, and the case for ruthless means to the end of industrial growth, power, and prosperity has often been made with disconcerting persuasivness. Even our own unrivalled affluence has been said to be an historical vindication of the atrocious conditions imposed on the working masses in the earlier periods of industrial capitalism. But the case was put most forcefully by the defenders of Stalin's ruthlessness in transforming a backward agricultural nation into the world's second industrial power. In Koestler's novel *Darkness at Noon*, the commissar Ivanov 'justifies' the murders of millions of kulaks to the prisoner Rubashov:

> 'For a man with your past,' Ivanov went on, 'this sudden revulsion against experimenting is rather naive. Every year several million people are killed quite pointlessly by epidemics and other natural catastrophes. And we should shrink from sacrificing a few hundred thousand for the most promising experiment in history? Not to mention the legions of those who die of undernourishment and tuberculosis in coal and quicksilver mines, rice-fields, and cotton plantations. No one takes any notice of them; nobody asks why or what for; but if here we shoot a few hundred thousand objectively harmful people, the humanitarians all over the world foam at the mouth. Yes, we liquidated the parasitic part of the peasantry and let it die of starvation. It was a surgical operation which had to be done once and for all; but in the good old days before the Revolution just as many died in any dry year—only senselessly and pointlessly. The victims of the Yellow River floods in China amount sometimes to hundreds of thousands. Nature is generous in her senseless experiments on mankind. Why should mankind not have the rights to experiment on itself?'

He paused: Rubashov did not answer. He went on: 'Have you ever read brochures of an anti-vivisectionist society? They are shattering and heartbreaking; when one reads how some poor cur which has had its liver cut out, whines and licks his tormentor's hands, one is just as nauseated as you were tonight. But if these people had their say, we would have no serums against cholera, typhoid, or diphtheria . . .'[6]

A sophisticated utilitarian philosopher, of course, need not agree with the commissar Ivanov. He could (and probably would) argue that Stalin's 'utilitarianism' was very naive indeed, and that the cruel murders done in its name were done in deplorable hastiness and ignorance of actual causal connections, and through egregious miscalculations of probability that made them self-defeating and unnecessary. Valuing the general welfare does not commit one to ineffective and uneconomical means to achieve it, means which spew misery and distrust as side effects. In short, there is no utilitarian justification of idle or uncertain *experiments* with high initial costs in suffering. We do not sacrifice human beings, even volunteers, on the *mere chance* that we may find the cure for cancer, or the like.

Wholly utilitarian policies, however, are not necessarily naive and uninformed, and in any case as philosophers we need only to resort to plausible hypothetical examples of a sophisticated kind. Suppose then that political leaders *know* what means are necessary to avert or eliminate a horrible widespread evil, and that these means require that the interests of innocent persons be sacrificed for the sake of the remainder, as in the many examples in criminal law books of 'necessity' or 'forced choice of the lesser evil'. Two variants of the problem can be distinguished. In the first, those who are sacrificed would be lost in any case, so that the choice is between saving some or saving none at all.[7] In that kind of situation, there is no clear conflict between justice and the greater good. In the other variant, however, the interests of some persons are sacrificed for the sake of the interests of others, leading to a vast net improvement in

[6] A. Koestler, *Darkness at Noon* (1941), pp. 161–2.

[7] *See* Stephen's example in his discussion of Regina v. Dudley and Stephens, 14 *Q.B.D.* 273: 'Several men are roped together on the Alps. They slip and the weight of the whole party is thrown on one, who cuts the rope in order to save himself. Here the question is not whether some shall die but whether one shall live.' J. Stephen, *Digest of the Criminal Law*, n. 4 at 24 (4th ed., 1887).

social well-being, in circumstances that do permit an alternative in which the minority interests would not be sacrificed and social misery not remedied. A case in point would be the option of a dictator in a backward country to sacrifice the liberties of many of his subjects in order to accelerate the economic development held necessary for the elimination of widespread misery in the present and future generations. It is by no means evident to me that the morally right policy in such circumstances is *always* to 'Let justice be done though the heavens fall and the masses perish.' Nor is it self-evident that the correct policy must always be to sacrifice some for the sake of the rest. An intuitionist is a philosopher who cannot commit himself in advance to a single moral principle that can be a trump card in all cases of this kind. There are, after all, different degrees of injustice, different amounts and types of evils, and different probabilities in our projections of consequences. Many of these factors are mutually incommensurable. To be sure, there are easy cases too, and even well-supported general maxims, such as that directing us to take individual justice more seriously, by and large, than social utility, especially given our propensity to miscalculate and overestimate the latter. Still, there are close cases too, and few sensitive persons will be satisfied with a theory that would represent even the difficult problems as simple by declaring in advance that one type of conflicting consideration must always triumph over the other.

So much for Rawls' view of the relation between justice and other social values. His main concern, however, is not with that question at all but with the analysis of the concept of justice and the derivation of its basic normative principles. In the performance of that task, Rawls' treatise, while not without its difficulties, is unexcelled in the philosophical literature. His main rival in this enterprise is utilitarian theory (1) above, and against it Rawls wins a decisive victory.

II

There is a qualification that Rawls makes which is essential to an understanding of his theory. His book is an essay in what he calls 'ideal theory' (as opposed to 'non-ideal theory') or more specifically, 'strict-compliance theory' (as opposed to 'partial-compliance theory'). He presumes that his original choosers are to select principles that will regulate a 'well-ordered society', that is, a society in which everyone always acts justly, all laws are just, and all citizens always comply with them. (It is apparently also a society in which no one ever acts negligently and there are no automobile collisions, since Rawls consigns questions about compensatory justice to non-ideal theory.) Rawls admits that the really pressing and important problems about justice belong to partial-compliance theory (e.g., questions in the theory of criminal justice, in tort law, in the theory of civil disobedience, and justice between nations) but 'assumes' that the question of ideal theory is more fundamental since its answer will provide direction to our inquiries in non-ideal theory, and will be 'the only basis for the systematic grasp of these more pressing problems'.[8]

One of the ways the ideal theory helps with the real-life problems of the non-ideal world, according to Rawls, is by mediating our 'natural duty' to promote just institutions. In so far as our actual institutions depart from Rawls' basic principles of justice, we have a duty, he says, to work toward their reform. But in our actual imperfect world things are rarely that straightforward. For example, Sidgwick's paradox of 'conservative justice' confronts us at every turn. Every reform of an imperfect practice or institution is likely to be unfair to someone or other. To change the rules in the middle of the game, even when those rules were not altogether fair, will disappoint the honest expectations of those whose prior commitments and life plans were made in genuine reliance on the continuance of the old rules. The propriety of changing the rules in a given case depends upon (*inter alia*) the degree of unfairness of the old rules and the extent and degree of the reliance placed upon them. Very often, when we consider reform, we must weigh quite legitimate incompatible claims against each other in circumstances such that whichever judgment is reached it will be unfair to someone or other. Rawls admits that intuitive balancing is unavoidable in dealing with problems of non-ideal theory, but I find very little acknowledgment (if any) that justice can be in *both* pans of the balance beam when claims are weighed. By and large, however, Rawls, in talking about non-ideal theory, makes large concessions to the skeptical intuitionist who insists on the necessity of claim-

[8] Hugo Bedau, review of Carl Cohen, *Civil Disobedience: Conscience, Tactics, and the Law*, *Yale Law Journal*, LXIX, 7 (6 April 1972): 179–86, p. 185.

balancing. His sensitive treatment of the duty to obey an unjust law and its limits (pp. 350–5) is a good example of this. So is his grudging admission (on p. 303) that in the more 'extreme and tangled instances of non-ideal theory' there will be a point where his rigid priority rules designed for ideal theory will fail, and there may be 'no satisfactory answer at all'.

Rawls' chapter devoted to civil disobedience and conscientious refusal is the one place in the book where political ties are analyzed, and so the one place where 'social-contract theory', in a strict and traditional sense, might come into play. It is also worth discussing it here since it is the one place where Rawls' theory most closely contains intuitionistic elements.

The problem of civil disobedience is primarily a problem in individual ethics. To ask under what conditions, if any, an individual citizen is morally justified in engaging in a 'public non-violent, conscientious yet political act contrary to law' (p. 63) is to ask a question very much like those about when an individual is morally justified in telling lies, breaking promises, inflicting pain, or otherwise acting contrary to normally binding moral rules. That is because there is normally a presumption against disobeying the law in a just, or near-just, society—not an unconditional moral prohibition, but a kind of standing case that must be overridden in a given instance by sufficient reasons. [Rawls does *not* hold the discredited view, effectively attacked by Hugo Bedau, that there is such a presumption in favor of obedience to '*any* law, however instituted and enforced, whatever its provisions, and no matter what would be the consequences of universal unswerving compliance with it'. His whole discussion of civil disobedience assumes the special context of a near-just society with 'legitimately established democratic authority' (p. 63).] To solve the problem, it is not sufficient to have a set of principles for determining the justice of the basic structure of society; rather we need supplementary principles to guide the individual conscience in a society already assumed to have more or less just institutions.

Rawls derives his principles for individuals in the same rationalistic way he derived his principles of social justice. Once again, the 'contractarian method' is employed, and we must ask ourselves which principles of right conduct would be chosen unanimously by the rational and self-interested parties in the original position *after* they have chosen the principles of social justice. These form a relatively untidy miscellany, and as Rawls enumerates and clarifies them, the reader is naturally of Hobbes' 'Laws of Nature or Dictates of Reason'. The main distinction Rawls draws among them is between those which impose 'natural duties' and those which impose 'obligations'. Obligations arise from voluntary acts, e.g., express or tacit promises, or accepting benefits; their content derives in part from the specifications of institutional rules; they are owed to definite individuals, namely, those 'cooperating together to maintain the arrangement in question' (p. 113). On the other hand, such natural duties as the duty not to be cruel and the duty to help others in need 'apply to us without regard to our voluntary acts . . . have no necessary connection with institutions or social practices . . . and hold between persons irrespective of their institutional relationships' (pp. 114–15). The principles imposing natural duties are irreducibly diverse, but all obligations ultimately are derived from a single principle which Rawls calls the 'principle of fairness'. This expresses the requirement that an individual 'do his part as defined by the rules of an institution when . . . (1) the institution is just . . . and (2) one has voluntarily accepted the benefits of the arrangement or taken advantage of its opportunities (pp. 111–12).

In so far as the presumption in favor of obedience to law is grounded in the principle of fairness in Rawls' philosophy, his theory of political obligation falls squarely within the (or a) social-contract tradition. (Indeed, that interpretation of his theory is reminiscent of Socrates in his jail cell.) In fact, however, Rawls does not use the principle of fairness to provide much support for the presumption in favor of obedience. 'There is . . . no political obligation strictly speaking,' he says, 'for citizens generally' (p. 114). Those members of society whose 'equal liberties' are worth very little because of economic deprivation, social discrimination, and exclusion from powerful offices (even under just and enlightened rules) and those whose 'consent' to the governing institutional rules has been coerced by a kind of 'exhortation' are free of any genuine *obligation* to obey the law even in a society whose *institutions* (as opposed to policies and practices) are just. Society is not a 'mutually advantageous venture' (p. 343) for these citizens, and they do not 'voluntarily' restrict their liberties under law in it. The principle of fairness, then, does not establish even the presumption of an obligation of obedience for them: 'only the more favored members of society are likely to have a clear political obligation as opposed to a political duty' (p. 376). Since the principles of natural duty 'do not presuppose any act of consent, express or tacit, or indeed any voluntary act, in order to apply' (p. 115), even this part of Rawls' system is not a 'social-contract theory' except in a watered-down and untraditional sense.

The principle of natural duty that *does* account for the general presumption in favor of obedience in a just society is the principle imposing what Rawls calls the 'duty to uphold justice'. That principle which Rawls argues would be acknowledged in the original position and is in that sense 'derived from reason', requires individuals to 'support and comply with' already existing just institutions and help bring about new just arrangements (p. 115). It is *this* principle that binds people generally to their political institutions, and it is a 'contractarian principle' only in the sense that it is derived by Rawls' so-called contractarian method.

Under what conditions can the presumption in favor of obedience be overridden? The problem of justifying civil disobedience, as Rawls conceives it, is a problem for individual choice, and 'the difficulty is one of a conflict of [natural] duties': 'At what point does the duty to comply with laws enacted by a legislative majority (or executive acts supported by such a majority) cease to be binding in view of . . . the duty to oppose injustice?' (p. 363). (The latter, presumably, is another of the 'natural duties'.) If I understand Rawls correctly, the 'intuitionism' that he rejected in his account of social principles is reintroduced here in his discussion of conflicting individual duties (though Rawls denies that he resorts to intuitionism even here). Intuitionism, as I understand Rawls' use of the term, is the view that there are no rigid priority rules assigning weights to normative principles that can conflict. Yet in his discussion of the conflict of duties that makes the problem of civil disobedience difficult, he cautions us that 'Precise principles that straightway decide actual cases are clearly out of the question' (p. 364). He modestly claims for his own discussion only that 'it identifies the relevant considerations and helps us to assign them their correct weights in the more important instances', thus 'clearing our vision' generally (p. 364).

Rawls then lists a number of conditions whose satisfaction usually or generally (he calls them mere 'presumptions') makes civil disobedience reasonable. First, it should be limited to the protest of wrongs that are 'instances of substantial and clear injustice', in effect to 'serious infringements of the principle of equal liberty' (p. 373), that is to say, to denials of the basic political rights of citizenship. (It is worth noting in passing that this condition is not satisfied by civilly disobedient protests against cutting down sycamore trees to widen a city road, against busing pupils, over-severe marijuana laws, failure to install a traffic light at an intersection that is unsafe for children, or excessive air pollution. The weight Rawls assigns to the presumption for obedience is not easily outbalanced.) Second, civil disobedience is justified as, but only as, a last resort after legal means of redress have failed. Third, the case for civil disobedience weakens in proportion to the extent to which others have recently resorted to it or have as good a case for resorting to it, for 'there is a limit to which civil disobedience can be engaged in without leading to a breakdown in the respect for law and the constitution' (p. 374).

The question that divides both utilitarian and contractarian theories into 'act' and 'rule' varieties is the following: In choosing and justifying our actions, when may we appeal *directly* to an ethical first principle and when (if ever) must our appeal stop at some subordinate rule, itself justified by an ethical first principle? The act-utilitarian permits (indeed requires) each of us always to do the act that promises to produce the greatest gain in net utility. He admits, of course, that often (even usually) we can best maximise utility by conforming to rules and regulations that summarize the experience of many generations that acts of certain kinds tend to have bad consequences. He might even agree with G. E. Moore that some moral rules should be taken as absolutely binding, but only because the chances of a given murder, say, being optimific are always less than the chances that our predictions of optimificity are mistaken. Still he will hold, at least in principle, that we ought to violate any moral or legal rule whenever doing so will produce consequences that are better on the whole than the consequences of our obeying it. On the other hand, the philosopher who holds the view suggested by Rawls' 'Two Concepts of Rules', which I have elsewhere called 'Actual-Rule Utilitarianism',[9] will interpret our duties much more strictly. He may admit that some 'moral rules' are mere 'summaries', or rules of thumb, to be violated whenever the expected consequences of doing so are better than those of conformity (most 'rules' of sexual ethics can be interpreted that way), but he will insist that valid legal rules and legal-like rules

[9] John Rawls, 'The Forms and Limits of Utilitarianism', *Philosophical Review*, LXXVI, 3 (July 1973): 368–81, p. 378.

governing such practices as promising and punishing cannot rightly be broken merely to achieve a small gain in utility, but only to avoid a disastrous *loss* in utility. And he will support his strict legalism, paradoxically, by a kind of appeal of his own to utility. He will point out that it is conducive to social utility to have some rules (e.g., those pertaining to promising) that deprive persons, under certain conditions, of the right to appeal directly to the principle of utility in deciding what to do.

At first sight, it is not easy to see how a similar act-rule division would apply to contractarian theories, partly because it is difficult to say what corresponds to an 'ethical first principle' in Rawls' system. It is implausible, I think, to take Rawls' statement of the contractarian method itself to fill the same role in his theory as the principle of utility does in utilitarianism. The principle that moral principles are correct if and only if they would be chosen by the parties in the original position is not itself an ultimate moral principle so much as a test of truth for proposed ultimate principles. The principles of social justice, though several in number, do have sufficient cohesion to play the role of a single ultimate principle in virtue of the strict priority rules that govern their application, but they are principles for the design of institutions and practices, not principles for individual actions. The most plausible candidate for a rival to the principle of utility as a standard of right conduct is the whole collection of principles assigning 'natural duties' and obligations. However, these are *not* ordered by rigid priority rules and, thus, they lack the unity of their utilitarian counterpart. Still, the principle that imposes the 'duty to uphold justice', directing us to obey the rules of established just practices and institutions, is very stringent and fundamental among these. So, for the sake of simplifying this discussion, we could consider *it* to be 'the ethical first principle' for individuals in Rawls' system, at least for actions that fall within the ambit of already established near-just institutions. But we would have to remember that there are other natural duties that can conflict with the 'duty to uphold justice' even for clearly rule-governed conduct. Some of these might very well have included the word 'justice' in *their* names; for surely the duties not to harm the innocent, not to disappoint reasonable expectations, not to assign arbitrarily heavy burdens, etc. have as much to do with justice as the duty to uphold just institutions has.

Now, one way of interpreting Rawls' rule-contractarianism is as follows. Normally we have the discretion, morally speaking, to appeal directly to the principles of natural duty in deciding what would be the right thing to do. Some actions, for example, are seen to be ineligible for our choice since they would violate the natural duty not to inflict harm upon the innocent. But when we are to act in our role as citizens in a fairly functioning democracy, when obedience to law is at issue, or when we occupy a special office such as juryman in a just institution, then we forfeit our right to appeal directly to the (other) first principles of natural duty, and the duty to uphold just institutions will normally trump. Thus, when the evidence establishes beyond a reasonable doubt that the defendant committed the crime with which he is charged, then we must find him guilty even though he is a morally innocent and admirable person charged under an odious valid law. The example is an instance of 'quasi-pure procedural justice', but it has elements of imperfect procedural justice too, since the fair procedures of a just institution, fairly followed, lead to a result which is unjust by a criterion that is independent of the institution itself. In a society whose basic structure is itself unjust or in an otherwise just society where a law has been created without proper regard to constitutionally specified procedures, the juryman's normal duty might be canceled. But the example in question is a case where there is a duty to perform an act (voting 'guilty') that will have an unjust result. As such it is exactly parallel to the case under rule-utilitarianism where a person has a duty to perform an action with less than optimific consequences simply because he promised to perform that act and thus forfeited his right to appeal directly to considerations of utility in deciding what to do.

But suppose now that you are on a jury and the evidence establishes beyond a reasonable doubt that the ten-year-old defendant did steal turnips as charged and thereby committed a capital felony under duly established law. In this case the duty to uphold just institutions would have you commit not merely an unfortunate but routine injustice; rather it would have you become a party to a monstrous perversion of natural justice, a result so disastrously severe that the normally trumping effect of the duty to uphold just institutions would be nullified in this case. This example is exactly parallel to the case under rule-utilitarianism where a person must deliberately break his promise not because the net consequences of so doing are likely to be somewhat better on the whole than the consequences of keeping the promise (rule-utilitarianism would not permit that) but rather because breaking the promise is necessary to prevent some *severe* harm to third parties. The rules of promising themselves, having utilitarian grounding, would permit *that* kind of breach.

Violating one's oath as a juryman is an example more like conscientious refusal than like civil disobedience, but the principle involved is much the same. In both cases the natural duty to uphold just institutions conflicts with what can be called 'the natural duty to oppose unjust laws, policies, and actions' (the latter a summary of all other natural duties that could well include the term 'justice' in their names). When the conflict is close, our natural duty on balance will be to try somehow to support just institutions and oppose injustice both, and civil disobedience, as Rawls conceives it, is a way of doing both these things at one stroke, since it is a way of 'expressing disobedience to law within the limits of fidelity to law, although it is at the outer edge thereof' (p. 366). In a nearly just society where the sense of justice is deeply entrenched, justified civil disobedience actually functions as 'a final device to maintain the stability of a just constitution' (p. 384). This is a welcome and ingenious idea, but one wonders whether an ideally just constitution will itself make some reference to civil disobedience and the conditions of its permissibility. If not, why not? If so, in what sense is civil disobedience 'illegal'?

Rawls' theory of civil disobedience may well be the nearest thing we have yet to an adequate account of these subtle matters, but, for the reasons given above, I think that Rawls overestimates the role that contract and 'pure procedural justice' play in it and in his theory of justice generally, and underestimates the extent of his own intuitionism.

6 Non-Neutral Principles

GERALD DWORKIN*

Concerning the demand that Professor Richard Herrnstein be fired, I asked a demonstration leader the following question:

Q: You mean Harvard should say, 'This group of professors supports the ruling class and should be fired and this group of

* Visiting Fellow, Battelle Seattle Research Center.

professors supports the oppressed workers and should not be fired?'

A: Precisely.

ERGO
September 22, 1971

Now, to be sure, an intolerant man will say that he acts in good faith and that he does not ask anything for himself that he denies to others. His view, let us suppose, is that he is acting on the principle that God is to be obeyed and the truth accepted by all. The principle is perfectly general and by acting on it he is not making an exception in his own case. As he sees the matter, he is following the correct principle which others reject.

John Rawls,
A Theory of Justice, p. 217

It is by now widely accepted that those who act claiming moral justification for their conduct must be prepared to accept as legitimate certain universalizations of their action. There must be a consistency in conduct, a refusal to make special pleas in one's own behalf or to consider oneself an exception to general principles. One way of testing this consistency is to ask of one proposing to engage in a certain course of action, 'What if everyone did that?' Another way of bringing out the same point is to present a clear case of some action that would not be considered justified by the agent and ask him to specify a relevant difference in the two cases. Thus those who defend the civil disobedience of Martin Luther King are asked to specify a relevant difference between his actions and those of George Wallace. Those who attack the use of political criteria in academic hiring by the Regents of the University of California are asked how they can defend the use of political criteria by students when they urge that students have a part in the hiring and retention of faculty. Those who defend the right to publish attacks on our current economic and political system but would refuse their opponents 'after the revolution' the same right are accused of inconsistency.

There are obviously a number of ways of defending oneself against such charges of inconsistency as against charges of unprincipled behavior in general.[1] I want to focus in this essay on one particular way of meeting the accusation of inconsistency. Let me illustrate by giving examples of the defense in each of the above cases. In the case of civil disobedience the defense is that the laws which King broke were unjust while those Wallace violated were just. In the case of political criteria for hiring it is that those urged by the Regents were reactionary and racist while those used by students would be progressive and anti-racist. In the case of censorship 'after the revolution' the defense is that the opponents of the current regime tell the truth and have the interest of all at heart while the opponents of the post-revolutionary regime distort the truth and have only the interests of their narrow class in mind. This pattern of argument, while perhaps made notorious most recently by Marcuse in his discussion of repressive tolerance, occurs in widely different areas of social and political life.[2] In each case we have persons proposing to act in ways that restrict the liberty of others or take more liberties than others on the basis of moral and political beliefs, who are then faced with the challenge of inconsistency in so far as they are not prepared to let others act as they do and meet the challenge by providing a principle which they are prepared to let anyone act on. It is characteristic of the principles appealed to that their application to particular cases is a matter of controversy for the parties whose conduct is supposed to be regulated by the principle in question. I shall call such principles non-neutral principles.[3]

It is important to realize that the controversy in question here is not one concerning the correctness or incorrectness, rightness or wrongness, of the principle, but one concerning whether or not the controversial predicate in question applies to the particular case. Thus a principle which states that killing of redheaded

[1] Consider the German tariff law which was designed to exlude Dutch and Russian cattle but allow entry of Swiss cattle. It gave an extra low duty rate to 'large dapple mountain cattle or brown cattle reared at a spot 300 meters above sea level and which have at least one month's grazing at a spot at least 800 meters above sea level'.

[2] 'Repressive Tolerance' in Robert P. Wolff, Barrington Moore, Jr., and Herbert Marcuse, *A Critique of Pure Tolerance* (Beacon Press, 1965).

[3] Those who are acquainted with the jurisprudential controversy concerning principled rules for judicial decision-making should be warned that my discussion while relevant to that issue is not directly connected with it. In view of the possible confusion I searched for an alternative terminology but could not find anything which was not also misleading (pseudo-universalizable) principles or too cute (disagreeable principles).

people is justified is neutral in my sense—since one can tell which people are redheaded and which are not—while one which states that killing in self-defense is legitimate is non-neutral since parties will often differ as to when a case is one of self-defense. It is clear that the distinction is one that lies on a continuum rather than a sharp one, but for the purpose of my argument we can consider cases which lie at the extremes of the spectrum. Thus a principle which states that civil disobedience is not justified if the law in question has been found unconstitutional by the Supreme Court is neutral in spite of the fact that there is some slippage as to which laws have the status in question. Suppose the vote is 4 to 4 or because a judge disqualifies himself from the decision, thus upholding a lower court ruling. Or suppose the vote is 5 to 4 and one of the majority judges ought to have disqualified himself. Still, relative to a principle which maintains that civil disobedience is not justified if the law in question is immoral or contrary to the wishes of God, the former principle is neutral.

The question I wish to discuss in this essay is whether there is anything that can be said, in general, by way of criticism of all such defenses. By virtue of their character of non-neutrality do they suffer from an inherent defect which rules them out, as a class, as illegitimate?

The first attack I would like to dispose of is the claim that people who appeal to such principles are hypocritical or inconsistent in some straightforward logical way. Whether or not one is inconsistent depends on whether one is trying to make an exception for oneself to a principle which one believes ought to be binding on all alike. Therefore, to assert of someone that he is acting inconsistently depends on a correct assessment of what principle is being appealed to. It is, of course, tempting to suppose that the principles one accepts for oneself are so obviously the correct ones that we must assume that others must really be acting in accordance with them—even when they claim they are not. But there seems to be no reason to suppose that persons acting on non-neutral principles are being insincere. It is not hypocritical to urge tolerance for one's political party under a capitalist regime while admitting that one will not be tolerant to opposition parties once the present regime is replaced. One may think that position wrong, even wicked, but that is because one rejects what, in many cases, is a perfectly sincerely held principle (tolerance for those who advance the interests of mankind or the working class) in favor of a different principle (say, tolerance for all as long as they confine themselves to political processes).

A similar charge of inconsistency is often levelled in the form of the rhetorical question, 'Wouldn't you be the first to complain if the treatment you are advocating for others were inflicted on you?' The answer is usually 'Yes', at which point the accusation of hypocrisy emerges again. For a person's right to complain of certain treatment is limited to violations of principles which he has accepted as binding for all. But the obvious reply to this objection is that in accordance with one and the same principle you are not justified in doing to us what we are justified in doing to you. Hence we have a right to complain and you do not. Again, this may not be true but the mere fact that it is being claimed that some persons have a right to act in ways which others do not have a right to act doesn't demonstrate inconsistency. To suppose that those who appeal to non-neutral principles are making some kind of logical mistake is incorrect.

Nor will it do to say that what is wrong with such principles is that they are particularistic, for that just amounts to saying that not everybody is to be allowed to act in the same way, and the question at issue is: what, if anything, is wrong with that? We don't object to the criminal law on the grounds that it states conditions under which it is right to limit the liberties of some persons (criminals) while at the same time denying that it would be right to limit the liberties of others (law-abiding citizens).

One way of attacking such principles is by invoking an extreme form of moral skepticism; by, in effect, denying that it makes sense to think of the principles in question as being applied either correctly or incorrectly. Consider the following dialogue.

A. I believe that those who engaged in civil disobedience by violating the Fugitive Slave laws were completely justified. They were unjust laws and provided certain other conditions are met, e.g., the disobedients have exhausted legal channels of redress, their actions are non-violent, etc., they had the right to break such laws.

B. What about Governor Wallace? Do you think he was entitled to violate the law and prevent black children from attending integrated schools? He claimed quite sincerely, let us suppose, that the laws he was opposing also violated fundamental rights of liberty and association and were, therefore, unjust.

A. But I did not say that one was justified in breaking the law if one believed it to be unjust. Only if it was unjust. Governor Wallace was quite wrong in thinking the law he was opposing was unjust.
B. But you are making a distinction between believing a law to be unjust or thinking a law unjust and the law being unjust. There is no such distinction to be made.
A. Of course there is a distinction. There is all the difference in the world between thinking X is good or right or fair and X being good or right or just or fair. People may have thought slavery was just or good. But they were wrong. Ethics makes no sense unless there is a difference between thinking one is right and being right.

In this argument A has the better of it. If the only objection to his mode of argument relies on a radical moral skepticism such as B proposes, then the cure is worse than the disease.
Another try.

B. You say Governor Wallace was wrong. How are you going to convince him of that? He would say the same thing about your claim and he will not accept your arguments any more than you would accept his.
A. All I can do is try to convince him by rational argument that he is mistaken and I am correct. It may be that some people are so prejudiced or ignorant or irrational that they will not be able to be convinced of the incorrectness of their application of the principle. Nevertheless, all I am required to do is show that I am acting on a principle which I am prepared to have everybody act on, and I am indeed doing that.
B. But even if he is not applying the principle correctly—I shall assume with you that there is a distinction between applying a principle correctly and mistakenly applying it—he thinks he is applying it correctly and hence will act in ways which will have bad consequences. You can't generalize over the class of people who are correctly applying principles. You have to consider the possibility of mistakes. And once you consider the possibility of mistakes, you will see that great harm will come from adopting non-neutral principles.

It is this consequential argument that I want to spend some time considering. I want to argue that this question of mistakes raises a very deep problem for a rule-utilitarian which affects the basic question of how he formulates his position.

I

I shall want to consider two versions of the appeal to consequences. One might be called the hypothetical universalization version and the other the ideal moral code version.

The first takes the following form. In deciding whether to act in a certain way one must consider the consequences of others acting in a similar fashion. So, the man who proposes to suppress false views must ask himself: what would be the consequences of everyone in a similar situation acting in that manner? But the question immediately arises, 'What situation?' For is his situation to be considered one in which he believes that the view he is suppressing is false, or is it to be considered one in which he believes correctly that the view he is suppressing is false. The defender of non-neutral principles wishes to generalize over the latter class. Thus, he says, I am perfectly prepared to accept the consequences of everyone who believes correctly that the view he is suppressing is false suppressing it. I agree that the consequences of everyone suppressing views which they believe to be false would be bad. But they are not doing what I propose to do.

Now it might seem that this view is *ad hoc* and indeed perverse, but there is actually a perfectly general principle which, when applied to other cases, is generally recognized as perfectly legitimate. In determining the general character or type of action in question, one looks to see whether the action can be specified in a more particular and relevant way. If the action is of type X and also of type Y and Y is a subset of X which has consequences which it is more relevant for moral decision making to consider, then we should consider the consequences of everyone doing Y and not X. Thus if a certain action is a lie, then rather than considering the consequences of everyone saying certain words to another we should consider the consequences of saying certain words believed to be false with the intention of deceiving. Of course the subset must be a morally relevant one. The set of lies told on Tuesday is a proper subset of the set of all lies, but the property of being told on a Tuesday isn't relevant to the moral assessment.

Now to return to the defender of censorship. He believes his action belongs to two types: (1) suppressing a view believed to be false and (2) suppressing a view believed correctly to be false.

It seems clear that the latter is a subset of the former and it also seems that the property of being correct in one's beliefs is a morally relevant property to consider.

It might be objected that while the principle of finding a narrower relevant action-type over which to generalize is usually corect, in this case it isn't because considered as guides to action the two types are equivalent. Equivalent not in the sense that the results of applying the two general rules are the same. Clearly they are not. In the one case one gets only false views suppressed; in the other one gets (probably) some true views suppressed as well. But equivalent in the sense that from the standpoint of the agent they do not enjoin different courses of action. After all, it might be said, a person can only act on the basis of the way things appear to him. If he believes that a certain view is false and finds that on reviewing all his evidence there is an overwhelming case, he can't go on to ask the further question: Is the view, in fact, false? But, even considered merely as guides to action, the two rules dictate, at least sometimes, different courses of conduct. Suppose, for example, I wanted for some reason to suppress the view that snow is white—a view which I now believed to be true. Then if I acted in accordance with the rule, 'Suppress views believed to be false', I could adopt the strategy of getting myself to believe that snow is white is false. I could, for example, pay someone to hypnotize me into having a false belief; but if I act in accordance with the rule, 'Suppress views which are (really) false', then there is nothing I can set about doing in view of the physical laws governing the color of snow to bring the view under my rule.

It seems then that the defender of non-neutral principles can hold his own against consequential objections. For by generalizing in the way indicated, he can accept the consequences of everyone acting as he proposes to act.

It might be thought that the ideal moral code version of rule-utilitarianism is superior to non-neutral principles based on their possible misuse. But, in fact, I think it fares no better.

Consider Brandt's definition of a right action.[4] He says that an action is right if it conforms with that learnable set of rules, the adoption of which by everyone would maximize intrinsic value. But, as our previous argument would lead one to suppose, there is an ambiguity here in the idea of adoption. It can either

[4] R. B. Brandt, 'Some Merits of One Form of Rule Utilitarianism', *University of Colorado Studies* (Boulder, 1967), pp. 58–9.

mean that set of rules which people try to act in accordance with and succeed or that set of rules which they try to act in accordance with and we leave the question open as to whether and how often they succeed or not. Now, clearly, which interpretation one accepts will affect the set of consequences that one has to estimate and may lead to different views as to whether a rule produces good consequences or not. If everyone tried to act in accordance with the rule, 'Suppress false views', and succeeded, it might be the case—we shall assume for the sake of argument—that, on balance, good consequences resulted. Whereas if everyone tried to act in accordance with the rule but only succeeded, say, half the time, the other half succeeding only in suppressing true views, that might have, on balance, very bad consequences.

Intuitively, it would seem that a utilitarian should opt for the weaker notion of adoption. He is interested in the way things are likely to work out and shouldn't proceed on what he believes to be the false assumption that people are infallible. But if he opts for this trying model, I don't think he can meet the basic act-utilitarian challenge which is why one should care about the hypothetical consequences of everyone adopting a rule. What one should judge are the particular consequences of what one is proposing to do—including, of course, its causal influences on other people's behaviour. But concern for the consequences of some hypothetical adoption is mere 'rule worship'.

If one adopts the everyone trying and succeeding interpretation of 'adoption', the following reply to the act-utilitarian is available. Because I want to consider the problem of misuse and hence disutility rather than utility, let me consider the formulation in terms of an action being wrong rather than right. An action is wrong if it conforms to a rule, the adoption of which by everyone would have bad consequences. This is then to be glossed in terms of everyone succeeding in acting in accordance with the rule. Now since we want to avoid the bad consequences, we need to insure that the action in question is not performed—at least by most people. But the responsibility for insuring this falls on everyone alike; unless, of course, there is a more particular rule which picks out a relevant sub-class of the population, in which case we should be considering the adoption of a different rule. But if the responsibility falls on everyone equally, then if one proposes to act in accordance with the rule anyway, one is failing to shoulder one's proper share of the responsibility for insuring the action is not generally performed. One is free-riding by getting benefits without assuming the corresponding sacrifices.

If one adopts the trying interpretation, this argument is not available. For in the kind of cases we have been considering the bad consequences result from those instances of people trying but failing to act in accordance with the rule. Therefore the burden does not fall on all the members of the population equally. It is the failures not the successes which produce the bad results. Hence if I have reason to suppose that I will succeed, then I should go ahead and act in accordance with the rule even though I am aware that everyone trying to act in accordance with the rule has bad consequences. Hence the act-utilitarian challenge cannot be met.

My argument, then, is in the form of a dilemma. Either the rule-utilitarian adopts the trying and succeeding interpretation, in which case he has no argument from bad consequences against non-neutral principles (since he can't consider the harmful consequences of trying but failing to act in accordance with them), or he adopts the trying interpretation, in which case—given various plausible empirical assumptions—he has an argument against their use, but he has no reason to be a rule-utilitarian. As for the act-utilitarian, he can hardly object to the non-neutrality of principles since the only principles he has—as opposed to rules of thumb—is a paradigm case of such a principle: Do that act which of the alternatives open to you produces the greatest balance of good over bad consequences. All of this indicates that the use of non-neutral principles raises some crucial problems for a 'generalizing' utilitarian—to adopt a term which covers both of the versions I have considered.

There is, however, a more decisive objection to this whole way of attacking non-neutral principles. Namely, the question of misuse has no bearing upon the evaluation of the correctness of these or any other principles. Just as one should keep distinct considerations of the truth of some proposition from considerations of its misuse by some, e.g., by drawing conclusions that are unwarranted—the recent controversy over the genetic basis of differences in intelligence bear this out—so, with respect to moral principles, their possible misuse should be kept distinct from the question of whether the principle is correct or not. To use an analogy from the law, the fact that if we allow sleepwalking as a defense to criminal charges, that will allow a certain number of guilty people to evade punishment either through fraud or the reasonable errors of juries, should not affect the claim that it is unjust to any individual who did sleepwalk not to be able to claim it as a defense. If the possibility of misuse is great enough and the results of misuse grave enough, we may have to abandon a principle of justice; but this does not show in the least that the principle was in any way illegitimate. Similarly, it may be that a principle shows when we are right and when we have a right to break the law, but the possible consequences of the misuse of the principle are great enough and grave enough to lead us not to exercise our rights. Now it is also true that sometimes perfectly correct uses of a principle may lead to very bad consequences and, in such circumstances, we may be well advised to abandon the principle though not denying its correctness. One needn't accept the notion that justice should triumph even if the world perishes—although one needn't agree this shows the proposed course of action was unjust. But I am arguing that the two cases are not on a par. The possible evil consequences of the correct use of a principle must, at least in some and probably most cases, be taken into account and given some weight in the assessment of the correctness of the principle. Whereas the misuses of a principle and their evil consequences only come into consideration after the moral calculation has occurred.

But surely, it will be objected, in deciding the legitimacy of policies such as governmental censorship, questions of the misuse of such powers by officials must be directly relevant in making the whole calculation. Of course, in discussing the question of what powers to give to officials, the possible bad uses they can make of those powers is very important. Nevertheless, it remains true that if the only objection made to the censoring of opinions is that it is likely to be misused by authorities, then one is not objecting to censorship but to governmental misuse. This makes a difference, surely, for what one should try to do in this case is institute censorship and look for ways of lessening the quantity and quality of the abuses. It is not an argument against censorship but rather mistaken censors, just as Socrates' argument against giving back a knife to a madman is not an argument against knives but madmen.

II

I now want to consider a quite different argument, one which is based on considerations that are independent of an assessment of the consequences (in particular the misuse) of non-neutral principles. It is a position developed by Rawls in *A Theory of Justice* and is basically a corollary of the contractualist view developed there. An issue very much related to the more general problem we have been discussing arises for Rawls in his discussion of toleration and the equal liberty of conscience. It is interesting to consider his argument both because the specific issue of toleration is an important one and because the discussion will reveal a more general difficulty that faces contractualist schemes.

For a contractualist, the argument against non-neutral principles is not an argument against their misuse but rather against the possibility of their being used at all. By this is meant the following. The principles are formulated in such a way that application to particular cases will be controversial.

Given that views as to what is true/false, moral/immoral, just/unjust are so varied and given that there is no common decision procedure for determining who is correct, no person would be willing to let others limit his freedom in accordance with such principles. Such principles cannot command general agreement and, therefore, must be replaced by principles whose application to particular cases commands widespread agreement or for which neutral procedures for determining the correct application can be devised.

To develop this argument let me use a distinction made by Rawls between pure procedures on the one hand and perfect or imperfect ones on the other. In the case of pure procedures we cannot specify what outcome is desirable independently of the procedure which generates the outcome. Thus the distribution of cash after a fair series of bets is by definition fair no matter what the actual distribution. In the case of perfect/imperfect procedures we can specify the desirable outcome independently of the procedure used to reach it, e.g., a fair distribution of a cake is defined as an equal distribution and the problem is to devise methods for reaching that outcome. In the case of dividing a cake we can devise a method which is guaranteed (given plausible empirical assumptions) to produce the desired outcome—namely, having the person who cuts the cake get the last piece—and we have perfect procedures. If we have an independent notion of the desired outcome but cannot specify procedures guaranteed to achieve it, we have imperfect procedures.

The concepts that occur in non-neutral principles (justice, truth, etc.) are defined independently of the processes by which they are achieved. This is what makes the conception of unjust but constitutionally valid laws or untrue but reasonably arrived at opinions possible. But to produce these outcomes we must use imperfect procedures since we know of no way of guaranteeing that just laws or true opinions will emerge. According to the view we are now considering, men choosing principles to regulate their disputes would not choose non-neutral principles. Thus they wouldn't adopt the principle that a man should be punished if he is guilty of an offense. Or, more precisely, even if they adopted that principle, they would do so only if its application to a particular man was determined by a procedure which they all can agree to. So, in practice, the principle would be replaced by one which states that a man is to be punished if he is found guilty by a jury of his peers in accordance with rules accepted by all the parties in advance. Unless this were so, it is argued, it would be irrational for men to agree to have their freedom restricted by a legal system.

This position is argued for by Rawls with respect to equal liberty of conscience. Departures from liberty of conscience are justified only when there is a 'reasonable expectation that not doing so will affect the security and order which the government should maintain' (p. 213). Reasonable expectation is defined as one 'based on evidence and ways of reasoning acceptable to all. It must be supported by ordinary observation and modes of thought (including the methods of rational scientific inquiry where these are not controversial) which are generally recognized as correct' (p. 213). Thus, for example, Aquinas' argument for the death penalty for heretics, which relies on the premise that corrupting the soul is far graver than counterfeiting money and that crime is punished by death, is unacceptable since it 'cannot be established by modes of argument commonly recognized'. On this view one is saved the labor of investigating specific details of Aquinas' position such as whether a heretic does corrupt the soul or whether, if he does so, this is bad or whether, if it is bad, it deserves death as a penalty. Now Rawls does not think that those who advocate non-neutral principles are making a formal error in the sense of violating some constraint on what could count as moral principle or that they are inconsistent in any way. His position is put forward as a substantive moral position.

This is the strongest substantive argument that has been developed against non-neutral principles. I want to suggest, however, that it ultimately depends on an unargued epistemological assumption.

Rawls denies that any skeptical theses are involved in his argument.

> ... the argument does not rely on any special metaphysical or philosophical doctrine. It does not presuppose that all truths can be established by ways of thought recognized by common sense.... The appeal is indeed to common sense, to generally shared ways of reasoning and plain facts accessible to all, but it is framed in such a way to avoid these larger presumptions. Nor, on the other hand, does the case for liberty imply skepticism in philosophy or indifference to religion. Perhaps arguments for liberty or conscience can be given that have one or more of these doctrines as premises. There is no reason to be surprised at this, since different arguments can have the same conclusion. But we need not pursue this question. The case for liberty is at least as strong as its strongest argument; the weak and fallacious ones are best forgotten [p. 214].

I want to suggest that the specific application of the notion of the 'veil of ignorance' to this argument does involve an epistemological assumption. For the stipulation as to what the choosing agents know about their political or religious views—in particular, they 'do not know, of course, what their moral or religious convictions are, or what is the particular content of their moral or religious obligations as they interpret them' (p. 206)—does lead to neutrality because it deprives the agents of relevant information which might lead to other decisions. If all they know is that people have different views, and not, say, that some are more adequately grounded or more likely to be true or more reasonable, then *laissez-faire* seems the obvious choice. But suppose the parties to the agreement knew (or believed) the following: With respect to some area of life (morals, science, aesthetics) there are people who are better placed than others to know the truths on these matters. Further, there are signs by which we may know them (they have red hair or they have an IQ of 200 or ...). Would it not then be rational to choose these people as authorities with the right to suppress false views? If there were a truth and it could be ascertained, would those in the original position who contemplated the possibility that they would be holders of false views regard their integrity as harmed by choosing that it should be suppressed?

It might be argued that the same result would be reached if we lifted the 'veil of ignorance' and allowed the parties to know the content of their religious and moral beliefs. For since it is contingently true that the parties will not agree as to which views are true and who are genuine authorities and since unanimity is required, no policy of suppression will be adopted and toleration will emerge as the solution of the decision-making problem. But this is not at all obvious, for those who are convinced that erroneous opinions will 'pollute' the environment and corrupt their children will rationally oppose a policy of toleration. It looks as if there will be one policy which is agreeable to all.

Now if we had independent reasons for suppressing the relevant beliefs, the matter would be different. But we don't find the same rationale for the exclusion of knowledge which is present at the earlier stages when the principles of justice are chosen. The justification for what information should be available to the choosers is given by Rawls as follows:

> The flow of information is determined at each stage by what is required in order to apply these principles intelligently to the kind of question of justice at hand, while at the same time any knowledge that is likely to give rise to bias and distortion and to set men against one another is ruled out [p. 200].

With respect to the first two grounds—intelligent application and bias—it would seem that one can only argue against the choosers having more specific information on skeptical grounds. For while it may be biased to put one's own interests and desires in a favored position, it is not bias to give preference to the truth or the highly probable. It is only where the views put forward are all on the same epistemological plane that giving preference to some may be arbitrary. As to the last ground, it reflects an ideological preference for harmony over conflict that is quite different from the conditions on knowledge which are intimately linked with the notion of a rational and impartial application of principle.

This specific line of criticism can now be added to those arguments presented by a number of recent critics of Rawls that show how many of the substantive claims that emerge from hypothetical contractualism follow only because implicit and controversial assumptions are built in a non-obvious fashion into the structure of the theory.[5] All these arguments tend to show that unless further assumptions of a controversial nature are made, there is no argument from the original position which will justify the particular substantive conclusions for which Rawls argues. Usually the implicit assumptions are of a moral nature, e.g., that a non-controversial theory of the good can generate as desirable just those goods which Rawls designates as 'primary goods'. In my argument, the relevant assumption is an epistemological one and it is a great merit of Rawls' theory that one can investigate more clearly than has yet been possible until now the connection—which has seemed so obvious to some and so obscure to many—between one's epistemological views and one's political views. The trite reply that the connection is 'psychological' and not 'logical' can now be seen to be both correct and misleading. For by showing the connections between rational decision making and moral theory, Rawls makes it clear that contrast serves no useful purpose. An agreement schema is both about how men, under certain conditions, would choose and what it is 'logical' for them to choose.

Although the particular argument Rawls uses works only if one assumes in effect that one cannot arrive at justified belief in religious matters, the basic idea behind the argument does seem to capture the essence of the persistent worry that non-neutral principles are somehow biased. There is an analogy to be drawn with egocentric principles. Just as one cannot envisage giving special consideration and advantage to Smith just because he's Smith, so one cannot envisage giving special preference to the true and the just where these are given particular content by individuals whom one has no reason to suppose better placed to decide these matters than oneself. Non-neutral principles are common to all only in the dimension of truth as abstract propositions, but given that we stuff them with different fillings and given that they are not common to us in the dimension of application, they do seem incompatible with an ideal of social cooperation.

[5] Cf. Adina Schwartz, 'Moral Neutrality and Primary Good', *Ethics* (July 1973), pp. 294–307; and Tom Nagel, 'Rawls on Justice', *Philosophical Review*, LXXXII, No. 2 (April 1973), pp. 220–34.

The analogy with egocentric principles is just an analogy, however, and while egocentric principles violate the requirements of justice, the most we can say about non-neutral principles is that it would be preferable if we could avoid them. But sometimes we cannot and we do not act unjustly if we adopt them. Rawls himself admits that civil disobedience and conscientious refusal on the grounds of the injustice of laws is sometimes legitimate. He takes a much stronger position, as we have seen, with respect to liberty of conscience. I happen to agree with his position, but that is because I don't believe there is any True Religion, belief in which is necessary to eternal salvation. So I don't think any important harm results from allowing complete freedom of conscience. But in other areas we do not act in this way and do not think it unjust. Christian scientists are forced to have blood transfusions both for their children and themselves. To do so is to override their conception of harm on the basis of modes of reasoning which are not acceptable to them. If one is to defend such laws, and I think they can be defended, at least with respect to third parties, then one will have to appeal at some point to the falsity of various beliefs about 'impure substances', 'salvation', etc.[6]

One concluding point. It is sometimes objected that the proponents of non-neutral principles cannot expect their viewpoint to be victorious through rational persuasion and must rely on political pressure and other kinds of force to gain their ends. But this may simply be a reflection of the facts of life and not a demerit in their morality. In the history of ethics we have come a long way recently in an attempt to rescue reason from its emotivist enemies. Perhaps at this point we must pay more attention to some of the limitations of reason.[7]

[6] Cf. Gerald Dworkin, 'Paternalism', *Monist* (January 1972).

[7] I would like to express indebtedness to Baruch Brody for many helpful conversations on this topic, a topic which grew out of a course on contemporary moral issues which we designed and taught together. As usual, anything I write is improved by the members of SELF. Earlier (and later) versions of this essay were read at the following universities: University of California (San Diego), University of California (Santa Barbara), University of Washington, University of Pennsylvania, Johns Hopkins, University of North Carolina (Greensboro), University of Maryland, Simon Fraser University.

7 Nature and Soundness of the Contract and Coherence Arguments[1]

DAVID LYONS*

John Rawls' long-awaited *Theory of Justice* is complex and rich. It will warrant and inspire not only commentaries and criticisms but also searching interpretations. As these begin to appear, however, some elementary and important aspects of his position seem to be misunderstood. My purpose in this paper is to discuss what I take to be Rawls' main, most explicit arguments for his principles;[2] for until we are clear about matters like these we cannot profitably go much further. And I wish especially to consider them in light of some plain but frequently neglected complexities in his principles.

Section I concerns the principles themselves. Section II discusses Rawls' claim that his principles of social justice express our most basic, shared values. We turn to the famous 'contract argument' in Section III, where this is given a fairly straightforward interpretation.[3] Section IV deals with difficulties for Rawls' use of that argument. I shall maintain that Rawls' two arguments are, as they stand, unsubstantiated at best, and therefore unsound. I shall also express some reservations about them as justifications for moral principles.

I

Rawls emphasizes and is noted for advancing two principles that

[1] This paper is based largely on my 'Rawls versus Utilitarianism', *Journal of Philosophy* 69 (1972), pp. 535–45, and 'The Nature of the Contract Argument', *Cornell Law Review* 59 (1974), pp. 1064–76.

* Professor of Philosophy, Cornell.

[2] Rawls has other arguments (or additional considerations supporting one or the other of the two arguments that I examine here) which I do not discuss. Some limits must be drawn for this paper, and I think the arguments (or parts thereof) that I examine are clearly the main props supporting Rawls' substantive position.

[3] My reading may be contrasted with the very interesting suggestions of Dworkin and Scanlon. See Ronald Dworkin, 'The Original Position', *University of Chicago Law Review* 40 (1973), pp. 500–33, this volume, p. 16; and Thomas M. Scanlon, Jr., 'Rawls' Theory of Justice', *University of Pennsylvania Law Review* 121 (1973), pp. 1020–69; this volume, p. 169.

diverge significantly from utilitarianism, which is seen as the most important alternative moral view. The *Greatest Equal Liberty Principle* says (roughly) that 'each person is to have an equal right to the most extensive basic liberty[4] compatible with a similar liberty for others' (p. 60; see also p. 302). One can assume that some restrictions are inevitable: as a matter of contingent fact, societies need some social controls, which limit liberty, but which are also needed to secure it. On Rawls' principle, liberty may not be restricted save to secure the maximum equal liberty that is possible. Any infringement of this principle is supposed to be an unacceptable injustice, not justifiable on any grounds: liberty is thus said to be inviolable.

This certainly departs from utilitarianism. For the latter is committed, basically, to promoting the general welfare (which I shall assume hereafter means average *per capita* welfare) and is not obliged to serve either equal or maximum liberty. It is conceivable that inequalities in liberty or general limitations on it that are not needed to secure liberty would sometimes serve the general welfare. Utilitarianism would then require such arrangements, but the Greatest Equal Liberty Principle would not permit them.

The second principle in Rawls' system concerns goods other than liberty, specifically the so-called 'primary goods', such as income, wealth, power, and authority, so far as their distribution is affected by the basic social institutions. Concerning such goods, the second principle says that 'social and economic inequalities are to be arranged so that they are both (a) reasonably expected to be to everyone's advantage, and (b) attached to positions and offices open to all' (p. 60; see also pp. 83, 302). Although (b) is relevant to the arguments of this paper, we can hereafter ignore it: part (a), also called the *Difference Principle,* is the more famous and, perhaps, the more important part, and the relevant effects of (a) on our arguments are duplicated either by (b) or by the Greatest Equal Liberty Principle.

The Difference Principle also represents a departure from utilitarianism. Inequalities can be justified by it, but never just because they serve the general welfare; for the average welfare might be served by benefiting some at others' expense. Rawls' model for justified inequalities involves incentives, when the

[4] Rawls includes under 'liberty' not merely protected areas of unrestricted action but also rights that presuppose institutional enablements, such as 'the right to hold (personal) property' (pp. 61, 201–4; see also pp. 221–7).

extra benefits offered some increases their productivity in such a way that everyone benefits—though some may wind up much better off than others (pp. 78, 151). But utilitarianism would allow inequalities that do not benefit all, so, once again, utilitarianism permits, even requires, what Rawls' principles do not allow.

These contrasts between Rawls' principles and utilitarianism are widely appreciated, but they are also exaggerated. One finds it said, for example, that utilitarianism 'could easily justify' suppressing the 'political rights that are basic to our notions of constitutional government', such as speech, assembly, and the franchise, which Rawls' principles hold inviolable—that utilitarianism could even 'justify slavery', which is almost certainly excluded by the Greatest Equal Liberty Principle.[5] Utilitarianism is thus supposed to 'violate our sense of justice', while Rawls' principles are supposed to express, or at least come much, much closer to it.[6]

These contrasts are misleading and inaccurate in important ways. The true contrast is not nearly so sharp; and this tends, I think, to undermine, or at least throw into question, Rawls' arguments for his principles, as I shall go on to show. Rawls quite rightly emphasizes the two principles we have been discussing, but they are part of only one of Rawls' two distinct 'conceptions' of justice, and the 'special' one at that. Rawls is explicit about this. The 'special conception' applies only in certain circumstances, for it is a special case or application of Rawls' 'general conception', which applies universally. Within the special conception, liberty is held inviolable; but it does not always apply. Under the general conception, liberty is not inviolable; it is, so to speak, mixed in with the other socially

[5] As well as by part (b), the *Fair Equality of Opportunity Principle.*

[6] Marshall Cohen, 'Review of *A Theory of Justice*', *New York Times Book Review* (16 July 1972) 1, 16. Compare the review of Rawls' book by Hugo Adam Bedau, 'Founding Righteousness on Reason', *The Nation* (11 Sept. 1972) 180–1. 'The trouble with utilitarian liberalism', says Bedau, 'has always been the facility with which it rationalizes the sacrifices of freedom.' He illicitly invokes against utilitarianism the implications of all sorts of consequentialistic attitudes, regardless of their values, and only qualifies the unwarranted criticism by noting 'that a scrupulous and accurate utilitarianism' might avoid commitment to the policies criticized. Rawls' significant work has, unfortunately, seemed to encourage such excessive attacks on utilitarianism. I fear the other side of the coin is an insufficiently critical acceptance of Rawls' claims. Thus my attempts in this paper on behalf of a more balanced view.

distributed primary goods. The general conception is like an expanded Difference Principle, requiring that everyone benefit from inequalities but allowing liberty to be restricted in exchange for other benefits, such as a (universally) higher standard of living (pp. 62–3, 83, 150–1, 303, 541–3).

Rawls may have inadvertently contributed to some confusion concerning his principles. In his famous earlier paper, 'Justice as Fairness', he had but one conception of justice, the two leading principles of which resemble the two we have discussed. In that paper he also claimed that his principles '*always*' condemn such institutions as serfdom and slavery—which are assumed to be unjust—while utilitarianism does not, because, on a utilitarian approach, it is always logically an open question whether they would serve the general (average) welfare.[7] This is the contrast that others seem to see between Rawls and utilitarianism. It looks as if the contrast suggested by Rawls turns entirely on the differences between his principles and the principle of utility; Rawls seemed to be saying that clearly unjust institutions are *logically* compatible with utilitarianism but not with his principles. At any rate, such a claim would be false. For, in 'Justice as Fairness', Rawls' principles did not absolutely condemn either serfdom or slavery, or any other such social arrangement. For example, his principles never made liberty inviolable. They gave only a 'presumption' favoring equal maximum liberty, which could be rebutted if everyone could be expected to benefit in other ways from restrictions on liberty.[8] It is logically an open question what specific institutions are justifiable on those principles, just as it is for utilitarianism—no more, no less.

Even though Rawls' earlier principles never contrasted so sharply with utilitarianism as was suggested by Rawls and has also been assumed, the divergence suggested may nevertheless seem to express the true relation between utilitarianism and Rawls' new (or revised) principles, as developed in *A Theory of Justice*. For the Greatest Equal Liberty Principle now states more than a presumption favoring equal maximum liberty; serfdom and slavery would presumably be excluded. But, as I have said, this appearance is misleading. For the Greatest Equal Liberty Principle does not apply universally. As part of the special conception of justice, it applies only when liberty can be 'effectively' established and exercised (pp. 15, 244 ff.); other-

[7] *Philosophical Review* 67 (1958) p. 188.
[8] *Ibid.*, pp. 166–7.

wise, the general conception applies. It is unclear what these conditions amount to (though our society is assumed to satisfy them). But that is irrelevant here. The point for us is simply that liberty is *not* generally held inviolable to Rawls: it thus remains logically an open question, in Rawls' system, whether limitations of it can be justified.

This is not to deny the contrasts that are real. Rawls requires, for example, that social arrangements benefit everyone, while utilitarianism requires that the general interest be served; and the latter might conceivably involve some persons' benefiting at others' expense. But let us see what difference all this makes to Rawls' arguments.

II

It is often supposed that, if moral principles are subject to any sort of justification, it is by showing that they match our 'intuitive' lower-level judgments about specific cases. On this sort of view, justified principles serve as the ground of such judgments and extend them to other cases in an intuitively acceptable way. The result is a 'coherence argument' (as I shall call it) for moral principles. Two further conditions should be noted. First, not all the judgments that we make are to be used—only those that reasonably seem to express our basic moral convictions. They should, for example, be based on true beliefs, be unaffected by interfering factors such as special interests, be made without hesitation, and be reaffirmable on reflection. Second, alternative accounts of such judgments should be entered and the one chosen that is, on reflection, most faithful to the given moral data.

Rawls suggests a coherence argument along these lines for his own principles.[9] That is to say, he suggests that a coherence argument serves to *justify* principles when the moral data—the considered moral judgments against which proposed principles are to be checked—are held by a number of individuals in

[9] See pp. 19–21, 46–53, 578–82: Rawls' current use of the coherence argument seems actually to be more complex than this brief sketch suggests, but I shall note some of the more important complications later. The pure coherence argument idea is presented more fully in Rawls' earlier paper, 'An Outline of a Decision Procedure for Ethics', *Philosophical Review* 60 (1951), pp. 177–97. For another account, see Richard B. Brandt, *Ethical Theory* (Englewood Cliffs, N.J.: Prentice-Hall, 1959), chap. 10.

common (p. 580). Even if this is not precisely Rawls' view, many philosophers have reasoned in some such way,[10] and it will be useful to examine such an argument made on behalf of Rawls' principles. I begin by expressing some general reservations about this kind of justification, and then turn to the substance of claims made for Rawls' principles.

As Rawls fully recognizes, no coherence argument can be conclusive. For none of our moral judgments—not even those 'fixed points' which we are prepared to embrace—can be considered incorrigible. And it is, moreover, unlikely that any set of considered judgments will mesh neatly with an initially proposed, intuitively attractive and illuminating set of principles that might seem to ground them and to extend them in an acceptable way. To achieve a satisfactory fit between judgments and principles, some judgments must be discounted, initial principles must be modified, or both. And further changes may result from entertaining alternative conceptions. As a consequence, and because our shared, considered judgments cover but a limited range of cases, alternative explications are always possible, and no clear rules can tell us which is best. Thus Rawls does not and cannot claim that his principles match our considered judgments perfectly or that utilitarianism is ruled out conclusively.

Much more important, the justificatory force of coherence arguments is unclear. Suppose one assumes that there are such things as valid principles of Justice which can be justified in some way; suppose one believes, moreover, that a coherence argument explicates our shared sense of justice, giving precise expression to our basic moral convictions: one may still doubt whether a coherence argument says anything about the validity of such principles. For pure coherence arguments seem to move us in a circle, between our current attitudes and the principles they supposedly manifest. We seem to be 'testing' principles by comparing them with given 'data'. Because the latter (our shared, considered moral judgments) are impartial, confidently made, and so on, we can, indeed, regard them as reliably reflecting our basic moral convictions. But we can still wonder whether they express any more than arbitrary commitments or sentiments that we happen now to share. To regard such an argument as

10 Arguments of this type have been used from Plato onwards, though sometimes only to discredit moral views that conflict with lower-level judgments which are assumed. I have argued this way myself, against utilitarianism and other views, and so my comments here are intended to raise doubts about my own practice as much as anyone else's.

justifying moral principles thus seems to assume either a complacent moral conventionalism or else a mysterious 'intuitionism' about basic moral 'data'.

It may be objected that my qualms are ill-founded because I am demanding more than it is reasonable to want. It may be said that coherence arguments really do justify moral principles because they provide the strongest arguments that are possible for them—and this constitutes justification, properly understood. But such a reply seems to assume a fallacious principle: that something is justified when all possible arguments for it have been given.[11] This implies the possibility of justifying unjustifiable assertions.

My point is that we need to know whether a given form of argument constitutes anything like justification for an item. To say that the best we can do in ethics is, say, to show which principles are most congruent with our considered moral judgments is not to imply that principles are then justified, in the sense that they are shown not to be fundamentally arbitrary or accidental. Thus I am inclined to view a pure coherence argument as a kind of justification whose legitimacy has never clearly been established.

These comments may suggest the relative importance which I think should be attached to Rawls' contract argument. Before we turn to it, however, let us see how substantively weak a coherence argument for Rawls' principles over utilitarianism would be anyway.

If a coherence argument is to favor Rawls' principles over utilitarianism, then we must be prepared to make confident, impartial, intuitive, informed judgments to the effect, say, that certain institutions approved by utilitarianism but not by Rawls' principles are unjust, or otherwise morally intolerable. Slavery and serfdom have been mentioned already, and they are important examples because they are among the few social institutions concerning which we can expect the relevant sort of moral consensus; but I use them here merely as illustrations.

As we have seen, it may be thought that a relevant contrast exists between Rawls' principles and utilitarianism—a contrast that is philosophically certain because it does not rely upon

[11] Compare R. M. Hare, *The Language of Morals* (New York: Oxford University Press, 1964), p. 69. Rawls tends rather to emphasize the importance of *agreement* about premisses, though he also insists that they concern 'reasonable conditions' (pp. 580–7).

contingent, empirical considerations, but turns entirely on logical possibilities. Thus, some have rejected utilitarianism because it does not flatly exclude slavery, assuming, of course, that such an institution is unequivocally unjust, morally unacceptable in all circumstances. I sympathize with this assumption. It seems to me, at any rate, that serfdom and slavery could hardly be justified in (what I shall vaguely call) ordinary circumstances and that they have hardly been justified in modern history. But I am also uncertain how to judge that they are unacceptable under *all possible* conditions—in all 'possible worlds'. It is not clear what features are (logically or contingently) essential to such institutions, and what the alternatives might be. But suppose we assume that serfdom and slavery *are* always unjustifiable: there would be no contrast between utilitarianism and Rawls' principles here, for, as we have seen, both moral views are logically compatible with such institutions. It should be noted, then, that anyone rejecting utilitarianism because it is compatible with slavery is bound, on pain of inconsistency, to reject Rawls' principles too. And if he modulates his criticism of slavery, his attitude toward utilitarianism should change accordingly.

But someone might go further, not relying on mere logical possibilities, and claim that utilitarianism is defective because it actually condones slavery in real, historical circumstances. I must confess I find this claim implausible. Perhaps, however, this is because I think of slavery as it has existed in the modern world, for it is difficult to believe that such arrangements ever would have served the general welfare *better than any social alternative*. It may be said that these recent examples do not exhaust the possibilities; perhaps there have been, or could be, freer, more humane, and less exploitative forms of serfdom and slavery. If so, perhaps utilitarianism would sometimes condone them. But the less objectionable we imagine that such institutions might be, the less plausible it is to suppose that they must always be unacceptable. Nevertheless, I doubt that we are likely to form considered judgments favoring such institutions. It seems more likely that we would fail to have considered judgments in such cases, for we probably could not confidently judge which of the unhappy alternatives would be preferable.

These difficulties aside, we still find no relevant contrast here. Rawls acknowledges that his principles are not only compatible with such institutions but might approve them in some actual circumstances (p. 248). If that is right, then it cannot be argued, as a matter of contingent fact, that utilitarianism sometimes actually does, while Rawls' principles never actually do, allow such institutions. But I think in fact that we lack the necessary knowledge of human history, including the social alternatives at each historical juncture, to say with confidence that either moral view would ever actually have allowed them.

We have restricted our attention so far to Rawls' general conception of justice. What about his special conception, which, we have assumed, would rule out serfdom and slavery? One may believe he sees a relevant contrast with utilitarianism here. But we really cannot tell. It is unclear what precise conditions make the special conception operative, so it is impossible to determine what utilitarianism would require or allow in all the relevant circumstances. We know that these conditions favor the effective establishment and exercise of liberty, and Rawls thinks that (in the circumstances of the social contract) individuals would prefer to keep certain effective freedoms intact and not allow them to be traded for other social goods. None of this implies, or even suggests, that utilitarianism would then approve of serfdom or slavery. When Rawls unfolds the implications of his special conception for concrete social institutions, he does not contrast it with those of utilitarianism. Here, as elsewhere, the coherence argument remains unsubstantiated, at best.

If someone thinks that a coherence argument clearly favors Rawls, he is probably subject to one of the following errors. He may exaggerate the contrast between the two competing views, most likely by ignoring restrictions on the special conception. This is to argue unsoundly. Or, he may compare the two conceptions themselves, without regard to their practical implications. Thus, one may prefer the Rawlsian requirement (most generally, that everyone benefit) to the utilitarian (that the average welfare be maximally served) without considering what they have to say about specific social arrangements. This is not a coherence argument, for it does not involve comparing the implications of competing views with our considered judgments about concrete cases. It is a direct appeal to moral intuition, and not an argument at all.

III

Rawls' main argument for his principles goes well beyond the sort of pure coherence argument just discussed and involves the notion of a 'social contract'. He asks us to imagine that a number of individuals who realize that they can benefit from cooperation seek agreement on the distributive ground rules for their social arrangements. If they can all agree on one set of principles, then, Rawls claims, these are certified as *the* principles of justice. And he argues, of course, that his principles would be selected. I shall deal with the nature of this argument first, and later turn to the soundness of Rawls' specific claims on behalf of his own principles.

One qualification should be kept in mind. I shall begin by regarding Rawls' contract argument as if it were self-sufficient and independent. But this is not precisely the way that Rawls presents it. As I shall later indicate, Rawls' version of the contract argument seems closely connected with (perhaps even a special branch of) the coherence argument. I leave these points for later, not because they would make the present argument unmanageable, but rather because of my qualms about coherence arguments in general. The more the contract argument is subordinated to coherence, the less justificatory force it seems to have. So, I shall suggest an interpretation of the contract argument that first regards it as independent as possible and later I will add the further Rawlsian modifications. We are, after all, interested not just in understanding Rawls but most generally in possible arguments for moral principles.

What bearing could a Rawlsian social contract, an imaginary agreement, have on us? Why should we think ourselves in any way bound by it, obliged to judge our institutions by the principles agreed to, and to act accordingly? I shall try to suggest some answers to these questions by putting together the Rawlsian idea of a contract argument in stages.

An obvious difficulty for a hypothetical contract argument is that any group of individuals is likely to be misinformed about, or at least ignorant of, some relevant facts. How absurd to suppose ourselves bound by principles that may be grounded on ignorance or bad reasoning! Rawls avoids these objections by assuming that the deliberators have full knowledge of all the relevant general facts and scientific laws (pp. 137–8) and that they also are rational, at least in the sense that they can decide on the basis of their long-term self-interest (pp. 142–3).

Even so, an arbitrarily selected group of individuals are unlikely to agree, or may well agree on distributive ground rules that specially favor some rather than others. Each person will seek principles to serve him best, given his own special talents, interests, needs, and condition in society. To avoid this source of contention, Rawls places the deliberators behind a 'veil of ignorance' that deprives them temporarily of information about themselves, their specific conditions, and their social circumstances (pp. 136–7). Then they cannot serve their separate, divergent interests, so they must select principles on the basis of their general knowledge of human beings and social institutions. Accordingly, they consider only the distribution of the primary goods.

These features of Rawls' hypothesis simplify the argument enormously, for they mean that the deliberators reason alike from the exact same premises. An incidental effect is that this is a 'contract argument' in the most attenuated sense, since no room is left for disagreement, bargaining, or even relevant differences among the parties. At any rate, an important consequence is that unanimity is guaranteed if any of the hypothetical deliberators can rank alternative principles on the information that is made available to him (pp. 139–40).

The problem of choice would still be extremely complex, and Rawls simplifies it further, increasing the likelihood that it has a rational solution. For example, instead of considering all questions of social justice, the deliberators limit their attention to the basic institutions of society (p. 7); instead of choosing principles to suit all possible circumstances, they initially assume that their society actually conforms to whatever principles they select and that everyone there tries his best to serve justice (pp. 245–6).

When all such qualifications[12] are imposed, deliberations can proceed. Rawls has the deliberators compare alternative principles. He argues, first, that, given the special conditions that have been imposed, rational individuals who are choosing basic distributive ground rules would adopt a 'maximin' strategy, which aims at guaranteeing that the worst condition one might find oneself in is the least undesirable of the alternatives (pp. 152–7). Given this connecting link, Rawls reasons that his principles would be preferred to others, since they favor the least advantaged members of society.[13]

[12] For example, the principles actually concern the condition of 'representative persons in the various social positions' (p. 64). I am ignoring here many features of Rawls' theory that seem irrelevant to the present argument.

[13] I am glossing over complex problems here. For example, the maximin rule favors the worst-off, but the general conception of justice does not,

Let us now step back and ask what all this accomplishes. Although Rawls provides only an elaborate sketch, he suggests that a relatively rigorous argument is possible. If that is right, then, from a logical standpoint, the contract argument is more powerful than a standard coherence argument. More important for us here, it might also be thought to possess greater justificatory force, because it avoids the suspicious circularity of the coherence argument by grounding principles, not on moral convictions that we happen to have, but on the independent theory of (self-interested) rational choice and facts about the human condition. Rawls, at any rate, claims that this argument justifies his principles (pp. 17, 21, 577–87).

However, it is not transparently obvious that the principles of justice are to be viewed as the solution to a problem of (self-interested) rational choice—or that, when they are so viewed, they have been certified as *moral* principles. Rawls does not seem to explain this aspect of his argument adequately. And there are at least some possible grounds for thinking that he may have missed the mark and brought forward something other than a conception of justice.

The reason is that one can construe Rawls' principles as a rational (self-interested) departure from an egalitarian norm, where equality (and not Rawls' explicit principles of distribution) serves as the conception of justice *per se*. This interpretation is encouraged by the fact that one can find in Rawls the suggestion of an argument for egalitarianism plus another argument, on rational (self-interested) grounds, to depart from that norm. In the first place, Rawls maintains that distributions flowing from or based on natural or social contingencies alone are 'arbitrary from a moral point of view' (pp. 15, 72, 312). Although he seems to believe there is a valid distinction between just and unjust distributions, he seems at first to deny that there is any valid *moral* basis for *discriminating* among persons when conferring benefits and imposing burdens. This points (at least on the sur-

requiring instead that all benefit from inequalities; and the Difference Principle (in the special conception) sometimes seems to favor the least advantaged, sometimes not. The gaps between these alternative conceptions, or formulations, are undoubtedly mediated by Rawls' 'natural assumptions', 'chain-connection' and 'close-knittedness' (pp. 80–3), which seem more controversial than he suggests. The upshot seems to be that favoring-the-worst-off is theoretically basic in Rawls' system, benefiting-all turning upon further contingent assumptions. Confusion arises in part because Rawls never formulates the general conception in the basic way, favoring the worst-off, though he takes pains to do this for the Difference Principle.

face) to a strict egalitarianism. But, in the second place, Rawls also believes that devices such as incentives can benefit all, though they do so unequally (p. 151). When they do, it seems rational to accept them and irrational (from a self-interested point of view) to refuse them. Thus, it seems rational, in general, from a self-interested point of view, to accept such departures from strict egalitarianism. One cannot lose: one stands only to gain. In this way, Rawls' position can seem like the amalgam of a moral egalitarianism and a non-moral acceptance of beneficial inequalities.[14]

The temptation so to view Rawls' principles is reinforced by (what I imagine would be) our shared, considered moral judgment of a test case: Suppose that a society has been organized on egalitarian lines by unanimous agreement, freely entered into. Suppose, further, that its members realize they could improve their material conditions by accepting Rawlsian inequalities, which benefit everyone. But, despite this, they freely and unanimously reaffirm their commitment to egalitarian institutions, thus refusing possible benefits. Now, from a self-interested standpoint, they might well be regarded as irrational. But there seems little reason to call them, or their institutions, unjust, or in any way defective from the standpoint of justice. Since they would be defective, according to Rawls' principles,[15] those principles seem miscast as principles *of justice*, even if they can be supported by the argument from rational self-interest in the original position.

There is, of course, another notion of social justice that is closer to the surface, and more faithful to the official spirit, of Rawls' principles. This is what he calls 'reciprocity' (pp. 102–3), which corresponds closely to the notion of a 'fair exchange'. It seems a fair exchange, indeed, for the less advantaged to allow others extra benefits when everyone will profit as a consequence; and for the more advantaged to restrict their extra benefits to whatever will be useful to others. But Rawls' principles go well

[14] The general conception is at one point said to be arrived at by a similar argument that begins differently (pp. 150–1). But I am uncertain how this squares with my reconstruction of Rawls' argument for the general conception, discussed below, and accepted by Rawls in our 1972 Eastern Division APA symposium.

[15] Rawls' principles require any departure from equality that will benefit everyone (see note 16 below). But his special terminology seems to acknowledge our contrary intuitions. Thus, egalitarian arrangements when inequalities would benefit all are to be called 'just throughout, but not the best just arrangement' (p. 79).

beyond this intuitive notion of fairness too. The intuitive notion merely *allows* such fair exchanges and does not require them; Rawls' principles require them, in the name of justice.[16]

These remarks are intended here, not as objections to Rawls' substantive principles, but as suggestions that his contract argument lacks moral force. It must be granted, however, that I have only given reasons for doubting that Rawls' principles fall neatly into the traditional category of *justice*; even if this were true, it would not follow that they do not describe the most important virtue of social institutions, as Rawls claims they do. Furthermore, I believe that much can be said in defense of Rawls' idea of a contract argument.

Let me deal with a couple of unsatisfactory defenses of it first. Someone might concede that the contract argument does not generate moral principles and hold that it is not supposed to. The coherence argument identifies certain principles as expressing our shared sense of justice, and the contract argument is supposed only to confer on those principles independent rational force, not moral certification. But, if nothing more were said, even that rational force would be problematic, since none of us seems likely to be found in the original position.

Alternatively, it might be held that the contract argument has moral force because of certain constraints imposed on it that I have not yet mentioned. For example, Rawls restricts the alternatives that are to be considered by the hypothetical deliberators to what he calls 'recognizably ethical' conceptions (pp. 125, 130–6). But this would not help the argument at all. It would not explain how such principles are binding on us, or at least rational for us, here and now. Indeed, from a self-interested point of view, such restrictions would only serve to weaken the contract argument, for they would limit the choices of deliberators and might exclude principles that would otherwise be favored. Finally, these restrictions might account for and thus reinforce the impression that I have already described, that Rawls' principles are an amalgam of morality and self-interest.

To develop a more satisfactory account of the contract argument, one must combine some suggestions made by Rawls that are either never explicitly put together, sufficiently emphasized, or

[16] Unless his principles required such 'reciprocity', thus guaranteeing that prospects under them would be better than under egalitarianism, they might also compare unfavorably with utilitarianism and thus fail to be selected in the original position.

adequately developed. Rawls maintains that the principles of justice can be regarded as emerging from a fair procedure: the original position is supposed to guarantee just that (pp. 12, 120). But, to understand the force of this claim, one must exploit what Rawls has to say about 'pure procedural justice' (p. 86). Finally, one must take seriously Rawls' assertion that we can enter the original position at any time (p. 19).

'The idea of the original position', Rawls says, 'is to set up a fair procedure so that any principles agreed to will be just. The aim is to use the notion of pure procedural justice as a basis of theory' (p. 136). In the original position, no one enjoys an unfair advantage over another; no one is able, for example, to exploit his knowledge of the facts in order to serve his own special interests at others' cost. The veil of ignorance prevents that. When full knowledge of general facts and sound reasoning ability are also conferred on the deliberators, they all stand as equals. If they freely concur, that agreement will be fair. But what they agree to is how goods should be distributed in their society. In this way, a fair procedure is used to determine just distributions. That, I take it, is the root idea of 'justice as fairness'.

Let us suppose, for the sake of argument, that a fair agreement is reached in the original position. Why should it follow that distributions flowing from it are unjust? Or, rather, why should we say that distributions conflicting with the principles accepted there are unjust? According to Rawls, 'pure procedural justice obtains when there is no independent criterion for the right result: instead there is a correct or fair procedure such that the outcome is likewise correct or fair, whatever it is, provided that the procedure has been properly followed. This situation is illustrated by gambling' (p. 86). We might try another example: Suppose that during an epidemic medical supplies are scarce relative to need; they cannot usefully be divided among all the persons who need them. We can think of everyone in need having an equal claim; or, if this violates the conditions assumed for pure procedural justice, we can suppose that no one has any claim to the supplies. In such circumstances, a fair lottery might legitimately be used to decide who shall obtain the medicine. Whatever the procedure chosen, if it is fair and properly followed, then the outcome can be regarded as 'fair, or not unfair' (p. 86). Rawls' idea must then be that his hypothetical deliberators cannot invoke any independent criterion of just distribution. If they are to have one, they must forge their own.

They themselves must choose among the possible bases for social organization.

But this seems to imply that *there simply are no* independent criteria of social justice. It is not that the veil of ignorance deprives one of, or prevents one from discovering, moral knowledge, but rather that it must be *created.* To see the weight of these remarks, we must apply them to our own case.

My reconstruction of Rawls' idea of a contract argument does not yet explain how it could have a bearing on us. We are not in the original position; we are imperfect reasoners; we lack full general knowledge; and we know at least some of our own special circumstances. Why should we suppose that the principles some imaginary deliberators would accept under conditions very different from ours are the principles of justice that *we* should judge with, and act by, here and now? Why should we suppose they have any rational force for us? Rawls' reply seems to be, that we can enter the original position at any time, and that the contract argument represents needed constraints upon, as well as ingredients for, our own satisfactory deliberations concerning the principles of justice (pp. 18–19, 138–9, 516–18).

Can we truly enter the original position, in the sense that we can assume the corresponding conditions for our moral arguments? Rawls seems to suggest that we can do so if we have the will, and, to some extent, this may be possible; for we can, perhaps, constrain ourselves to reason and deliberate impartially. In our own deliberations, for example, we must not allow ourselves to be swayed by considerations of special interest. (This mirrors the veil of ignorance and is, presumably, part of its justification.) But it is not true that we can reproduce at will other central features of the original position. Even with the best will in the world we cannot simply confer on ourselves either full rationality or full knowledge of all the general facts and scientific laws. I do not believe, however, that these limitations indicate defects in the contract argument idea. They acknowledge, in effect, that even our best efforts at deliberation are subject to correction in the light of scientific discovery and better reasoning. One would expect some such room for error in an account of anything aspiring to be moral knowledge. We should therefore think of the hypothetical contract argument (as distinct from Rawls' claims about the principles it generates) as an ideal that we can approximate in practice.

So far, I think, so good. Let us suppose that we can, so to speak, enter the original position. What significance does that have for us? I am not asking what (nonmoral) reason we might have for *participating* in fair deliberations designed to generate distributive principles. I am asking what reason we have for *believing* that the outcome of such deliberations are *the* principles of justice. Rawls answers this question by invoking the notion of pure procedural justice. But what can that mean for us? I see two ways of understanding it: (a) Outside the contract argument (or our approximations of it), there is no objective basis for social justice. (b) There is no alternative mode of argument for principles of social justice.

(a) We have noted that Rawls regards distributions flowing from or based on natural or social contingencies alone as morally arbitrary, not eligible to serve as valid bases for just distributions. And I suggested that this view seems at first to imply a strict egalitarianism. My point now is that Rawls avoids such a conclusion by invoking the notion of pure procedural justice. We will not—cannot—find the principles of justice in the natural world; we cannot discover them, for there is nothing to be discovered. There is no objective basis for just distributions—outside the contract argument. For the contract argument incorporates a fair procedure, and (the notion of pure procedural justice tells us) when there is no independent criterion of justice, we must turn to fair procedures if we are to have just, or fair, distributions at all. In other words, we must make our own moral principles. Moral skeptics conclude that our principles must be arbitrary. But, on Rawls' view, this does not follow. We can not only make our principles, we can also certify them ourselves, at least in the sense that we can rationally choose among the alternatives. Indeed, if Rawls is right about what is 'morally arbitrary', then this is the *only* way we can validly get moral principles—and that alone entitles us to use it, for that allows us to invoke the notion of pure procedural justice.

It is unfortunate, therefore, that Rawls merely claims, without supporting argument, that distributions flowing from natural or social contingencies alone are arbitrary from a moral point of view. An adequate defense of this claim is required for the very idea of a contract argument, that is, to warrant the notion that pure procedural justice is relevant to our case. Rawls evidently believes that the claim would be accepted as one of our more general, abstract considered moral judgments. But there is room for disagreement here. As I understand Rawls' claim, so that it entitles him to invoke the notion of pure procedural justice, considerations of, say, merit or desert based on such qualities as talent or intelligence are not morally basic. More important, Rawls' dismissal of the notion of desert seems sometimes confused. He says, for example, that we do not deserve the advantages or disadvantages we receive in the 'natural lottery' or as a consequence of social accident. It does not follow, however, as Rawls seems to suggest (pp. 103–4), that some features of the natural or social 'lotteries' could not serve as the *just basis* of deliberate distributions: one need not deserve what is itself a ground of desert. But I wish only to point out that difficult problems remain here, and I shall not pursue the matter further; it requires and will undoubtedly receive careful attention.

(b) Rawls' use of the notion of pure procedural justice also seems to assume that no alternative mode of argument for moral principles is possible. Suppose another line of reasoning were possible: then the principles emerging from the contract argument might clash with those certified in other ways. The problem here is not merely that of conflict between principles (though Rawls, unlike some other philosophers, is unprepared to accept it). It is, roughly, this: a sound argument for moral principles employing premisses other than those found in the contract argument amounts, in effect, to an *independent* criterion of social justice. But the very possibility of such an independent criterion is contrary to the conditions of pure procedural justice. If there are grounds external to the contract argument for judging the justice of social arrangements, then Rawls' 'justice as fairness' notion would seem to be discredited.

As we have seen, however, Rawls seems to embrace the coherence argument too. But if the notion of pure procedural justice is to do its job of validating the contract argument, then Rawls cannot regard the coherence argument as anything like a justification or defense of moral principles. It may be seen as an explication of our shared sense of justice, but its conclusions cannot be regarded as rationally binding. In fact, if the contract argument serves to justify moral principles, while the coherence argument is used to define our actual convictions, the results of the former cannot be applied in criticism of the latter.

This does not seem to be Rawls' view of the relations between these arguments, however (see note 9). His method is, instead, to design the contract argument so that its results agree with the outcome of the coherence argument, or at least to provide the best fit possible between our considered moral judgments and plausible premisses for the argument, mediated by the principles of justice. On my attempt to suggest how the contract argument could have justificatory force, however, Rawls' approach looks as if it puts the cart before the horse.

One can go even further. Rawls may not merely try to fit the contract and coherence arguments together into a larger, consistent view. It is possible to conceive of Rawls' contract argument as a special branch of the coherence argument, that is to say, as a means of working out the implications of certain very basic values that we happen to share. I have refrained from suggesting this before because of my qualms about coherence arguments and my fear that such a way of viewing the contract argument would compromise it, undermining its independence and integrity, stripping it entirely of justificatory force. But, as I am not at all confident that Rawls would accept my view of these matters, I must suggest this way of looking at them now.

Suppose it can be said that justice rests on fairness, as Rawls believes. The contract argument *presupposes* the value of fairness (which, in turn, commits us, on Rawls' view, to other values, such as impartiality). The significance of this presupposition can be understood as follows. It sometimes looks as if the contract argument shows that, given general contingent facts about the human condition, certain principles of justice are, so to speak, rationally inescapable; at least that reason prefers them to the alternatives. But, on the alternative account that I am now suggesting, the contract argument shows, instead, that one who accepts the *prior* claims of fairness (impartiality, etc.) is rationally committed (given the relevant facts) to certain principles of justice.

In the contract argument, of course, fairness is initially construed, not as a way of dispensing benefits and burdens, but as a set of constraints upon arguments, deliberation, reasoning, and procedures generally; and this may seem to mitigate the argument's evaluative presuppositions. But this is misleading, for, under the notion of pure procedural justice, procedural fairness is valued precisely because it leads to social outcomes.

At any rate, the contract argument rests upon an unargued commitment to fairness and impartiality. On the assumption, then, that these are among the most 'fixed' or least 'provisional' of the 'fixed points' in our system of shared values—perhaps fundamental to our (or a) moral outlook—the contract argument can retain some force. But it would nevertheless appear to be fundamentally weakened by its subordination to the coherence argument. It remains to be seen whether such evaluative assumptions are unavoidable; and, if so, whether it is reasonable to regard the resulting principles as justified.

IV

For the sake of further argument right now, I shall assume that Rawls' notion of a contract argument (or something like it) has moral force—can be used as a basic defense of moral principles, or at least to show that some embody the 'moral point of view' (p. 136) more fully than others. I turn now from the very idea of a contract argument to Rawls' claim that his principles would be chosen in the original position and thus that they are certified as *the* principles of justice. I wish especially to indicate how the full character of Rawls' principles and argument tends to undermine, or at least throw into question, this claim too.

The parties to Rawls' hypothetical contract are supposed to reason from a self-interested standpoint, considering how they are likely to fare if society were regulated by this or that conception of justice. They are blessed with extraordinary knowledge of human psychology, economics, and social relations, but they are handicapped by the veil of ignorance: no one of them can take into account his own special conditions. Initially, at least—that is, within the arguments of *A Theory of Justice*—they also assume that the ground rules they unanimously select will be in force when they return to ordinary life. Their decision therefore has a direct and profound effect on their most fundamental interests, since it concerns the central institutions of society, their basic rights and benefits, burdens and obligations. Rawls believes that, in such circumstances, rational individuals would be averse to taking risks—at least, when they understand what the alternative conceptions have to offer. They would choose principles according to a 'maximin' strategy, looking, that is, for the least undesirable outcomes possible. There is, obviously, a close affinity between this strategy and the general conception of justice (or, under the special conception, between it and the Difference Principle). So, it looks as if no argument is needed once that strategy is adopted (but see note 13).

I wish to question both these steps: first, the maximin approach; second, even granting it, the selection of Rawls' principles.

Rawls is not unaware that his attribution of the maximin approach to the hypothetical deliberators can seem arbitrary or unreasonable. He admits that 'the maximin rule is not, in general, a suitable guide for choices under uncertainty' and that a more 'natural' rule would have us maximize expectations (pp. 153–4). The difference is this: On the maximin approach, we worry only about avoiding the worst outcomes; we do not consider how likely they might be (and thus the actual risks involved) or the probability of doing better. On the expectation-maximizing rule, we take all outcomes into account, weighted in accordance with their likelihood, and choose the course that carries the highest probable-utility. But reasoning of the latter kind leads directly to utilitarianism; so Rawls must defend his alternative strategy at all costs. I shall limit my remarks here to comments on his actual argument for the maximin approach.

Rawls says that 'there appear to be three chief features of situations that give plausibility to this unusual rule' (p. 154):

(1) In the original position, 'knowledge of likelihoods is impossible, or at best extremely insecure' (p. 154). Therefore, one could not even try to maximize expectations at all, or could not do so in a reasonable way. For the deliberators are ignorant of their individual circumstances, and, while they know what the possible outcomes are, they do not know how likely they may be.

I am troubled by this argument. The veil of ignorance of course prevents one from basing his calculations on his own actual talents, interests, needs, and social condition, since one does not know what they are. But it does not follow that one is unable to calculate what these are likely to be. If one can, then one can try to maximize utilities on the basis of one's likely condition. And it would seem that such initial calculations would be possible. For the hypothetical deliberators have full knowledge of the relevant general facts and scientific laws, and these could presumably support actuarial calculations concerning, for example, the likelihood of having certain natural endowments. What about the likelihood of being in this or that social condition? It is not clear that this could be calculated, unless (a) their general knowledge includes knowledge of the full course of history (though not, of course, their own place in it); or (b) the laws concerning human nature and society take an historical, developmental character, and they know how long human history will last (though, once again, they do not know their actual place within it). Either sort of knowledge would enable them to make calculations about the sort of social condition they are likely to be in.

Rawls seems explicitly to exclude this, however. He says: 'While they know the first principles of social theory, the course of history is closed to them; they have no information about how often society has taken this or that form, or which kinds of societies presently exist' (p. 200).[17] I assume the rationale for excluding such information is not to protect the maximin approach, for that would beg the question. Rawls must believe that some general information, as well as information about one's actual condition, would somehow undermine the argument. Well, we can be sure that such knowledge would not prevent agreement among the parties, since all would still reason in the same way from shared premisses; it would simply change some premisses. Perhaps the worry is that such information would undermine fairness: it would bias their deliberations by favoring the statistically average person rather than, say, the worst-off person. Now, there may well be a moral argument preferring Rawls' approach, favoring the worst-off person rather than the statistical norm; but employing it here would beg the question, since it rests on the principles that Rawls is supposed to be justified.[18]

(2) Rawls claims, secondly, that 'the person choosing has a conception of the good such that he cares very little, if anything, for what he might gain above the maximum stipend that he can, in fact, be sure of by following the maximum rule' (p. 154). The most obvious difficulty here is that this claim is unsubstantiated: Rawls does not show that utilitarianism (for example) actually yields worse outcomes than his own principles in the exact same social conditions. We shall return to this point in a moment. More important here, this argument assumes that relevant probabilities are knowable to the parties. Primary goods, the basis of their ('thin') conception of the good, are the things it is reasonable to believe that anyone would want, whatever his actual interests (p. 92). So primary goods *assume* the very actuarial calculations that I suggested may be possible in the original position. Thus, Rawls' points (1) and (2) seem incompatible. If they are both necessary (see p. 154), then his defence of the maximin rule collapses.

[17] I am grateful to Gerald Dworkin for pointing out this passage to me.

[18] If we deprive the hypothetical deliberators of all general information that could support actuarial calculations needed to maximize expectations, the result may be that we are giving them only a distorted or impoverished

In reply to this objection,[19] Rawls has suggested that the hypothetical deliberators do not themselves identify the primary goods, and so do not require the relevant general information and need not engage in the relevant actuarial calculations. The list of primary goods is, rather, imposed by stipulation. But that is arbitrary, and weakens the argument. In any event, it is based on Rawls' educated guess about the results of such actuarial calculations. But, if we can identify the primary goods, then surely so can the hypothetical deliberators, with their immeasurably greater knowledge and understanding of the human condition. The only reason for preventing their doing so would seem to be a desire to preserve the maximin rule, and thus avoid utilitarianism.

The conflict between Rawls' first and second points can also be seen as follows. Rawls fully recognizes that the list of primary goods is based on a kind of statistical norm, and that there are individuals who do not value income, wealth, power, authority, liberty, and self-esteem so much as to regard them all, and only them, as *primary* goods. The list of primary goods therefore favors those closest to the norm. Now, in discussing (1), we imagined that Rawls' exclusion of certain historical information was motivated by a desire to prevent such bias. So, either Rawls should not exclude such information under (1), or else he should exclude it also under (2) and construct the argument differently. Consistency would seem to demand this.

(3) Rawls claims, finally, that 'the rejected alternatives have outcomes that one can hardly accept'. I see no basis whatever for this claim, at least so far as utilitarianism is concerned, and it will pay to make this point emphatically. We can do it, however, by moving on to Rawls' use of the maximin approach to get his own principles of justice.

For the sake of further argument, then, we shall assume that the maximin strategy is adopted. Does it follow that Rawls' principles would be chosen over, say, utilitarianism? One might argue as follows. No alternative could possibly be chosen over Rawls' principles, since they require that everyone benefit but also represent a self-interested improvement over mere egalitarianism. But, if the alternatives are truly different from Rawls' principles,

understanding of the human condition, with corresponding effects upon their deliberations.

[19] 'Reply to Lyons and Teitelman', unpublished, in response to Lyons, 'Rawls *versus* Utilitarianism'.

then they must allow worse outcomes. So, the hypothetical deliberators would necessarily select Rawls' principles.

This is a bad argument. Let us note in passing that it must be restricted to Rawls' general conception of justice and could not hope to account for the special treatment accorded liberty under ideal conditions. Even so, it is invalid, for alternatives to Rawls' principles could conceivably differ from the latter without having worse outcomes in contingently possible circumstances.

A strengthened form of the argument is this. Under the maximin rule, it would be irrational for the hypothetical deliberators to risk faring worse than they would under the general conception. But they would do this if they ranked it no higher than, say, utilitarianism, even if they did so on the basis of their knowledge that utilitarianism would not *in fact* have worse outcomes than the general conception. Such knowledge would be contingent, and the deliberators would reason that too much is at stake to rest on something about which they might be mistaken. (Rawls confers knowledge on them, but not absolute rational confidence in that knowledge.) They might conceivably be mistaken about the equal riskiness of the alternatives, so they should prefer the alternative about which they could not be mistaken, namely, the general conception.[20]

This argument assumes that the choice between principles can be made on *a priori* grounds alone. But Rawls seems to deny this: 'Contract theory agrees, then, with utilitarianism in holding that the fundamental principles of justice quite properly depend upon the natural facts about men in society. This dependence is made explicit by the description of the original position: the decision of the parties is taken in the light of general knowledge' (p. 159). Substantive knowledge is supposed to affect the choice of principles. The argument suggested, however, made such matters irrelevant, so I assume that Rawls could not intend it.

Let us note what sort of argument Rawls requires. Given the general knowledge that one has in the original position, one would presumably restrict one's attention to contingent social possibilities, within the limits of, say, the laws of human psychology, economics, and social relations. Theoretically, one could determine how the worst-off members of societies regulated by Rawlsian principles would fare under the contingently possible conditions, and could compare that with the fate of the worst-off

[20] I owe this suggestion to Richard Miller. (Once again, see note 13.)

in utilitarian societies. (Neither probabilities nor mere logical possibilities are to be considered.) Under the maximin rule, utilitarianism is riskier, then, if, and only if, there is *some contingently possible* condition of the worst-off members of a utilitarian society that is worse than *any contingently possible* condition of the worst-off members of Rawlsian societies. Rawls' principles are preferred to utilitarianism if, and only if, the latter are riskier in this maximin sense.

Now, it may look quite certain that Rawls' principles would be chosen in the original position, partly because two stages in the deliberations are conflated. To construct a manageable argument, Rawls draws up a short list of alternative principles to be considered by the hypothetical deliberators (pp. 122–4) and then stipulates that each is to be regarded as holding 'unconditionally, that is, whatever the circumstances of society' (p. 125). The list includes utilitarianism (which of course *is* supposed to cover all cases) and also Rawls' special conception of justice (that is, the set of principles including, among others, the Greatest Equal Liberty Principle and the Difference Principle). As we have noted, social arrangements like serfdom and slavery are logically compatible with utilitarianism, while they are, presumably, excluded by the special conception (e.g. by the Greatest Equal Liberty Principle). It therefore might look as if the worst-off members of society could fare much worse under utilitarianism than under Rawlsian restrictions, and thus that the former are riskier in the relevant sense.

But this reconstruction of the argument is unacceptable. Once again, the deliberators' general knowledge of the human condition plays no role in it; it turns entirely on logical possibilities. Also, two stages of the argument have been conflated. To see the latter is to underscore the importance of the former; so the complexity of the argument is worth going into a bit further.

As we have noted, Rawls generates not one but two distinct conceptions of justice, one (the special conception) being a special case or application of the other (the general conception) in certain conditions. Accordingly, Rawls' contract argument must be developed in two parts. In the first stage, the deliberators select principles that are to govern a 'well-ordered' society under ideal conditions, and the special conception of justice is supposed to result. In the second stage, they lay down guidelines for less happy circumstances, and the general conception emerges. At first they assume (i) that the basic structure of the society to which they will return will satisfy the principles they choose; (ii) that all members of it have a firm commitment to the principles of justice; and (iii) that conditions actually favor the effective establishment and exercise of liberty. In the second part of the argument, the last assumption, (iii), is dropped.

It can be seen (a) that the contingently possible conditions relevant to the first part of the argument are significantly different from those relevant to the second part, and (b) that even those relevant to the second part do not exhaust the historical possibilities. Thus, some *actual* implications of utilitarianism will not be relevant to the argument.[21] Most important, (c) the deliberators never compare utilitarianism *simpliciter* with Rawls' *special* conception of justice. When the latter is contrasted with utilitarianism, in the first stage, ideal conditions are assumed; so the relevant implications of utilitarianism cover only those conditions. In the second stage, Rawlsian and utilitarian societies are compared more generally, though still within a limited range of contingently possible conditions. In each stage, the competing principles must be compared under identical ranges of conditions.

The first question, then, is whether utilitarianism is riskier (in the relevant sense) than Rawls' special conception under ideal conditions. The remarkable thing is that no evidence is forthcoming to suggest this. (For example, we have no reason to believe that utilitarianism *under ideal conditions* is compatible with serfdom or slavery.) This is the sort of question that the hypothetical deliberators must answer, and they presumably have the understanding needed to answer it. But we are given no clue as to why they should reject utilitarianism.

Since Rawls does not actually appeal to substantive factors, in the way his argument requires, I am tempted to speculate that the two stages of deliberation really have been conflated—that the contrast assumed by Rawls, and by anyone who is persuaded by the argument as it has been presented, is the illicit one between utilitarianism (without restrictions) and the special conception (under ideal conditions).

Similar gaps appear in the second part of the contract argument. We no longer assume ideal conditions, so utilitarianism is

[21] I am referring here, of course, to implications concerning institutions, which, for utilitarianism, are problematical, since it is often understood to cover acts, not institutions. The present point is that certain sorts of extreme cases (e.g. involving extreme scarcity) are irrelevant to the contract argument, since the deliberators are not choosing principles for such circumstances.

compared with the general conception, but only within the relevant circumstances. The illusion of a striking contrast should not arise in this case because, for example, both alternatives seem compatible with serfdom and slavery (though whether they are in fact compatible with such institutions *in the relevant conditions* is open to question). The hypothetical deliberators can, presumably, figure out how they would fare under one type of regime as compared with the other. But we do not know. And Rawls gives us no reason to believe that utilitarianism *actually* works out worse for the worst-off members of society.

It may be said that many agree, or at least that some have asserted, that serfdom or slavery sometimes promotes the 'common good', which may be understood here in utilitarian terms; while it is never claimed that such institutions benefit the worst-off members of society.[22] But this surely counts for little. Such claims are not evidence for themselves, especially when they are likely to be made by apologists for serfdom and slavery or by unreflective critics of utilitarianism. We need to know whether such institutions actually served, or would have served, the general welfare *better than any social alternative*—and whether, indeed, the general (average) welfare was in fact invoked and not (as is often the case) some class-biased conception of, say, the national interest. Most important, we need real, hard evidence for each stage of the argument.

I conclude, therefore, that the contract argument for Rawls' principles is unsubstantiated. *Even if* we think the contract argument has moral force and that the maximum rule applies in the original position, *we still have no reason* to believe that Rawls' principles would be chosen over utilitarianism.

Obituaries for utilitarianism would be premature.

[22] This was suggested by Rawls in his 'Reply to Lyons and Teitelman'.

THE PRINCIPLES OF JUSTICE

8 Rawls' Theory of Justice[1]

T. M. SCANLON*

A leading characteristic of the ideal of social life underlying Rawls' theory is the primary role assigned to a shared conception of justice as the basis of social cooperation. Rawls says,

> Now let us say that a society is well-ordered when it is not only designed to advance the good of its members but when it is also effectively regulated by a public conception of justice. That is, it is a society in which (1) everyone accepts and knows that the others accept the same principles of justice, and (2) the basic social institutions generally satisfy and are generally known to satisfy these principles. In this case while men may put forth excessive demands on one another, they nevertheless acknowledge a common point of view from which their claims may be adjudicated. If men's inclination to self-interest makes their vigilance against one another necessary, their public sense of justice makes their secure association together possible. Among individuals with disparate aims and purposes a shared conception of justice establishes the bonds of civic friendship; the general desire for justice limits the pursuit of other ends. One may think of a public conception of justice as constituting the fundamental charter of a well-ordered human association [pp. 4–5].

The task of the parties to Rawls' hypothetical Original Position

[1] A slightly revised version of Parts II and IV of my 'Rawls' Theory of Justice', *University of Pennsylvania Law Review*, Vol. 121 (1973), pp. 1020–69.

* Associate Professor of Philosophy, Princeton University.

is to choose principles of justice which well play this central role in their own society. As is well-known, Rawls puts forward the following Two Principles as those the parties would prefer.

> [*First Principle*]
> Each person is to have an equal right to the most extensive total system of basic liberties compatible with a similar system of liberty for all [p. 250].
>
> [*Second Principle*]
> Social and economic inequalities are to be arranged so that they are both (a) to the greatest benefit of the least advantaged and (b) attached to offices and positions open to all under conditions of fair equality of opportunity [p. 83].[2]

In what follows I will discuss two leading features of Rawls' theory: the special place assigned to the basic liberties protected by the First Principle and the particular formula of distributive justice presented in the Second Principle. In each instance I will try to show how the case for the conclusions Rawls wishes to defend is related not only to the details of the Original Position construction but also to features of the ideal of social cooperation that underlies his theory.

I. *Liberty*

The ideal of social life expressed in Rawls' notion of a well-ordered society is strongly pluralistic: mutually accepted principles of justice are to provide a common bond for cooperation between persons with disparate aims and purposes, and these principles are to be accepted as limiting conditions on the pursuit of these other ends. It is not obvious that the parties to a social contract would assign this kind of priority to any principles of justice. Rather, it might be expected that they would insist on principles grounded in their own most firmly held values, e.g. in their religious ideals or in their conception of human excellence.

[2] This principle is advanced as the favored interpretation of the more ambiguous principle that 'social and economic inequalities are to be arranged so that they are both (a) reasonably expected to be to everyone's advantage, and (b) attached to positions and offices open to all' (p. 60). On the relation of these two formulations of clause (a), see note 15 *infra*.

Rawls seeks to prevent the parties to his Original Position from making such a choice by depriving them of the knowledge of their own religious beliefs and conceptions of the good. Forced to choose behind such a veil of ignorance, it is argued, the best way for them to secure a place for their own ideals and values will be to choose principles of toleration.

But the idea of principles chosen behind a veil of ignorance itself requires justification. If the best that can be said for Rawls' principles of justice and for the conception of social cooperation on which they are based is that they would be preferred by parties forced to choose without the knowledge of factors which many would consider to be of crucial importance, then the case for these principles has scarcely been made. To complete the argument some motivation for the construction of the Original Position is required, ideally a motivation which connects the device of choice behind a veil of ignorance with intuitively plausible arguments for a pluralist form of social cooperation. I believe that this further motivation can be found in Rawls' theory in the notion that a person's good is to be identified not with the particular goals and commitments which he may at any given time have adopted but rather with his continuing status as a rational agent able to adopt and modify these goals. In Section A below I will discuss this notion as it figures in Rawls' defense of his theory against that class of alternative theories that he calls perfectionist. Then, in Section B, I will examine the connection between this rather abstract notion and Rawls' argument for the priority of specific constitutional liberties.

A. THE ARGUMENT AGAINST PERFECTIONISM

Those theories which Rawls calls perfectionist direct us 'to arrange institutions and to define the duties and obligations of individuals so as to maximize the achievement of human excellence in art, science, and culture' (p. 325). In my previous remarks I have grouped such theories together with theories which take as the ruling aim of social institutions the promotion of a particular religious ideal. This grouping may seem somewhat unfair since there is in theories of the first sort a strong tendency toward elitism—i.e., toward placing much greater emphasis on the needs and interests of some members of society than on those of others—and while some religious-based theories may exhibit a tendency of this kind in singling out a small group (e.g., 'the elect' or the clergy) for special privileges, this need not be regarded as a characteristic feature of the type.

What all of these theories, religious and secular, share is first of all a teleological structure:[3] once the value of a certain end is established, social institutions are to be appraised strictly on the basis of their tendency to promote this end. In addition, quite apart from tendencies to elitism, all of these theories raise serious problems concerning individual liberty: institutions which preserve the opportunity for each person to adopt and pursue his own interests and ideals and to try to persuade others to follow him will be justified on perfectionist grounds only to the extent that they are the most effective means to the promotion of the given end.

Now it would be possible to reject theories of this kind simply on the basis of their tendency to support institutions which conflict with our considered judgments of justice, and then to design the Original Position in such a way that the offending theories are ruled out. Adopted alone, however, this strategy is not wholly satisfying. If we can give no independent rationale for the design of the Original Position then this manoeuver appears somewhat *ad hoc*. To provide such a rationale, based on a non-perfectionist ideal of social cooperation, would not constitute a refutation of perfectionism; but without such a rationale we are left with no response to the basic theoretical challenge which these theories raise: If there is an objective difference in the intrinsic value of different talents, goals and pursuits why should not information about these differences be used by the parties in the Original Position as the basis for their choice of the principles by which social institutions will be judged? How, in short, can we defend an egalitarian or libertarian position without embracing some form of skepticism about values?

Rawls' response to this challenge (and his rationale for the design of the Original Position) is grounded in the notion that social institutions are just only if they can be defended to each of their members on the basis of the contribution they make to his good as assessed from his point of view. We must be able to say to each member that the arrangements he is asked to accept provide as well for him as they possibly can, consistent with satisfying the parallel demands of others. In order to spell out this idea more fully it is necessary first to consider Rawls' analysis of the notion of an individual person's good (see pp. 60–5).

[3] The notion of a teleological theory is discussed more fully in Section II *infra*.

Those experiences, ends and activities are components in the good for a particular individual, Rawls argues, which have an important place in a plan of life which it would be rational for him to choose. Now it may seem that a person could be said rationally to choose a plan of life (if at all) only after he has developed a conception of his own good, on the basis of which he can judge and rank alternative plans of life. But Rawls argues, persuasively I think, that this is not the case. In real life our deliberations about those actual choices which, taken together, determine our plan of life proceed on the basis of knowledge of our present tastes and capacities, knowledge of what things we have in the past found satisfying, and knowledge of general principles governing the ways in which our tastes and capacities are subject to growth and change over time. This information allows us to decide on course of action not only with the aim of satisfying our current desires but also with the knowledge and intent that our choices will be instrumental in determining what interests, talents and desires we will come to have in the future. Long-range choices such as the choice of a career or a place to live give perhaps the best example of choices which, because they may be foreseen to have far-reaching effects on our interests and objectives, must be made on some basis which goes beyond the satisfaction of our current desires and specific interests.

Rawls puts forward a negative and a positive thesis about this process of deliberation. The negative thesis consists of an attack on the idea that there must be some single overriding general goal (e.g., the maximization of satisfaction or happiness) which underlies all of our deliberations and explains how we can compare and choose between disparate alternatives (see sect. 83–4).[4] The positive thesis consists of a sketch of standards of rationality with reference to which our choices, particularly those most general and far-reaching choices described as choices between alternative life plans, can be criticized. This sketch consists of two parts. First, there are general principles of rational choice according to which it is irrational, e.g., for anyone to prefer plan of life *A* to plan of life *B* if *B* involves the development of exactly the same interests and desires as *A* and provides for their satisfaction at a markedly higher level. Not all of these principles, which

[4] This negative thesis has an important role in Rawls' argument against utilitarianism. See Sect. 83–5, esp. 562–3. This argument is discussed in Scanlon, 'Rawls' Theory', pp. 1046–56.

Rawls calls 'counting principles', are as uncontroversial as this example, but are all fairly weak, and taken together they by no means can be expected to determine a unique plan as the only rational choice for a person to make. A choice from among the plans not ruled out by these principles (the set of maximal plans) will involve such things as comparing the relative intensity of different desires and the relative value for us of different kinds of accomplishments. For this choice there are on Rawls' view no principles of rationality which directly require a choice of some plans over others. The only relevant standards concern the manner in which the choice is made—whether the relevant evidence has been duly weighed, the possible sources of uncertainty and error properly discounted for, etc. These criteria are grouped together by Rawls under the heading 'deliberative rationality'.

Thus, to say that a certain thing is, objectively, a good for a certain person is, on Rawls' analysis (p. 417), to say that it would be a prominent feature in a plan of life which that person would hypothetically choose, with deliberative rationality, from among the class of maximal plans. Under any actual conditions, of course, not only the means for attaining those things which are good for us, but also ideal conditions for determining what things are such goods, will in some measure be lacking. What the parties in Rawls' Original Position look for in a society is not only the means for securing those things, whatever they may be, which are objectively components of their good, but also the conditions necessary for determining what these goods are.

From the fact that the parties in Rawls' Original Position suppose that as members of a society they will choose their own plan of life, and hence also determine their own conception of the good, it should not be thought that they suppose themselves to be independent of social forces which will in large part shape and influence the choices they make. It would be idle to deny that such influences exist, and irrational to object to all such influences as interfering with one's liberty. But it is still reasonable to prefer some institutions to others on grounds of the conditions they provide for rationally forming a conception of one's good. Obviously one may reasonably object, simply on grounds of efficiency, to institutions which place arbitrary obstacles and difficulties in the way of individuals' attempts to get a clear view of the alternatives open to them, of their own potentialities, and of what they and others can expect from various courses of action.

A more difficult case is presented by the fact that some features of institutions will not merely be random inferences but can be seen clearly to favor certain choices and to discourage others, and to do this not by just enlarging people's views or by approaching 'ideal conditions' thereby favoring 'the correct answer', but rather by skewing the evidence available or by restricting the alternatives likely to be considered, or by affecting people's deliberations in other more subtle and indirect ways. Systematic inference of this kind might be the result of relatively fixed impersonal features of institutional arrangements. Alternatively, certain individuals may be charged with overseeing and maintaining these influences through censorship or other devices.

It is one of the features of perfectionist views which strike us intuitively as objectionable that such views may authorize the use of means of this sort in order to produce individuals conforming to a particular ideal. Now we cannot simply reject as involving unacceptable 'conditioning' all social institutions which mold a person's choices and beliefs without his consent with the aim of bringing him closer to some ideal. Certainly Rawls cannot do this. For as he himself says, his own view involves a certain ideal of the person, and he is at some pains to show (sect. 51–9) that there are psychological laws which give us reason to believe that persons growing up in a well-ordered society governed by his Two Principles of justice will naturally acquire what he calls a sense of justice—the tendency to understand and be motivated by considerations of justice as specified by those principles. The action of these psychological laws is in part dependent upon the intellectual activity of the person on whom they are acting, but is also in large part something which happens to a person without his knowledge or rational scrutiny.

How then is one to distinguish among the various ways in which social institutions may be arranged to influence the choices and beliefs of their members without each member's consent? Can one distinguish acceptable from unacceptable influences of this kind on any basis other than an appraisal of the relative value of the particular types of persons these influences produce? The appropriate standards for making this distinction on Rawls' theory seem to me to be suggested by the criteria he offers for distinguishing justifiable from unjustifiable paternalism (pp. 249–250). The relevant principles here require first that paternalistic interventions, i.e., interventions in a person's life 'for his own sake' which are pursued contrary to his wishes or without his knowledge, have to be rationally justifiable *to him* after the fact. Second, such interventions must be justified on the grounds that the subject's evident failure or absence of reason and will at the time rules out a direct presentation of the issues to him for his own rational consideration and decision. A third requirement is that the intervention 'must be guided by the principles of justice and what is known about the subject's more permanent aims and preferences' or, failing such knowledge, by some neutral standard such as that provided by the primary goods.

While Rawls formulates these requirements specifically for the case of paternalistic action by one person toward another, they seem to be applicable as well to the broader class of interventions we are considering. This is indicated, for example, in the fact that Rawls' defense of the process by which a sense of justice is inculcated in persons who grow up in a well-ordered society governed by his Two Principles of justice advances considerations essentially parallel to these requirements (pp. 514–15). One can maintain here, first, that the principles which form the content of this sense of justice are ones the person can later come to see as justified. (This fact alone, of course, would not be an adequate defense since any successful piece of indoctrination, or at least any successful indoctrination of justifiable beliefs, could make this claim.) Further, the practices of moral education in a well-ordered society proceed as far as possible by appeal to the subject's reason, and rely upon other factors only insofar as the natural limitations of childhood make necessary. Finally, the acquisition of a sense of justice is, it is argued, not inconsistent with a person's good. Since the conception of justice (i.e. Rawls') which is the content of the sense of justice in question provides a secure protection for each person's interests and for his desire to determine his own conception of the good, the acquisition of such a sense of justice is not something which leaves a person open to exploitation or manipulation by others. In addition, having a sense of justice is a necessary condition for sharing fully in the life of a well-ordered society (p. 571) and a necessary condition as well for susceptibility to the natural attitudes of friendship, love and trust (p. 570). These are things, Rawls argues, which almost[5] anyone has reason to want.

Without going fully into the arguments for these claims, we may compare them to the case which might be made on per-

[5] Rawls does allow for the possibility that there may be 'some persons for whom the affirmation of their sense of justice is not a good' (p. 575).

fectionist grounds for features of social institutions designed to mold or restrict the choices of their members so as to promote a particular secular or religious ideal. There is a clear sense in which such features will have a rational justification: they will be justifiable on the basis of the objective value of the particular ideal in question. A perfectionist might thus maintain that the interferences with a person's liberty which these features represent are ones which he should, rationally, come to accept. But the justification which is offered by the perfectionist will not necessarily be one which claims that these features promote the good of the person whose liberty is restricted or which claims that they are consistent with his desire to determine his own conception of the good; it is apt to appeal instead to some impersonal scheme of values. Moreover, this justification need not be based on considerations which would be agreed upon by almost anyone regardless of his conception of the good. Rather, it is likely to be based on one specific conception of the good which, even if it is objectively correct, may nonetheless be something which is a matter of some disagreement among rational adults in the society in question. Indeed, it is just the fact that this conception of the good, though correct, does not compel general agreement, which may be taken on perfectionist grounds to make necessary the intervention in question. On Rawls' theory, however, such interventions are permitted only when there is 'evident failure or absence of reason and will', a phrase intended to cover cases such as infancy, insanity or coma which involve major diminution of rational capacities relative to the standard of 'a normal adult in full possession of his faculties'.

Thus, while Rawls' theory bases principles of justice on a hypothetical choice made by persons who may appear to be standing temporarily outside any particular society, the point of view which the theory takes as fundamental is actually that of a person *in* society. The parties in the Original Position do not act from special wisdom or knowledge which enables them to make choices which they later, as persons under the limiting and distorting conditions of real life in an actual society, will have to take on faith. Rather, the parties' aim is to make choices which they, as real citizens, will have reason to accept. Each party therefore regards his own judgment as a real citizen as sovereign—not as infallible or immune from limitations, but as the basis from which his life will be lived, his choices made and his work as ideal contractor appraised.

Rawls remarks that 'embedded in the principles of justice there is an ideal of the person that provides an Archimedean Point for judging the basic structure of society, (pp. 261–5, 584). Although I have not described this ideal in full, the preceding argument seems to me to illustrate part of the force of this remark. The ideal of each person as a rational chooser of his own ends and plans provides an Archimedean Point partly in virtue of the fact that this conception of a person is taken to be prior to any particular independently-determined conception of his good. One need not be a skeptic about values or truth to hold that each of us does in fact look at himself in this way. If this is so, then the assumption that the parties in the Original Position adopt this view of themselves should seem a natural one, and the fact that certain principles of social cooperation involve the recognition of each member of society as in this sense a sovereign equal, while others involve the denial of this status to at least some members, should seem a fact of some importance.

The conception of the person described by Rawls is of course not an Archimedean Point in the sense of being itself a notion formed outside of or independent of particular social and historical circumstances. It may well be that this conception of the person and the ideal of social cooperation founded on it are typical of particular historical eras and civilizations. But this is not in itself an objection to Rawls' theory, particularly if, as it seems to me, the conception of the person in question is one that has a particularly deep hold on us and is not a matter of great controversy or of significant variation across the range of societies to which the theory should be expected to apply. The question is not whether this conception of the person is in some sense absolute, but whether the particular features of this conception that are appealed to in Rawls' argument are more controversial than the conclusions they are used to support.

Certainly this conception of the person involves a number of important parameters which must be fixed before the notion can be appealed to in support of conclusions about the justice or injustice of particular institutions. The most obvious of these is the standard of rationality: what is to count, for example, as 'evident failure or absence or reason or will'? Other parameters are represented by the general facts of social science which the parties in the Original Position use in reaching their conclusions, by the notion of the primary social goods, and by other appeals to the idea that certain goods or circumstances are to be desired 'no matter what one's conception of the good may be'. The latter appeals depend upon some idea of the normal range of variation in conceptions of the good and upon some idea of the means and conditions required for the pursuit of these goods. All of these may be subject to some variation over time and social circumstances. But here again the theory need make no claim to *absoluteness* in these matters. It is sufficient to ask whether the appeals the theory makes to facts about persons and the circumstances of human life are controversial *for us*; in particular, whether the facts appealed to are more controversial than the conclusions at issue; and finally, whether the ways in which conclusions about the justice of institutions are made to depend on such facts strike us as plausible.

Much of the preceding discussion has been internal to Rawls' particular conception of social cooperation and is thus not in any proper sense a *refutation* of perfectionism. It is, rather, a description of an alternative ideal of social life, one which might be called 'cooperation on a footing of justice'. The development of this ideal enables Rawls to move beyond the observation that perfectionism seems to support arrangements which are at variance with our intuitive judgments of justice to a theory which explains why this should be so and provides a point of view from which we can see how the perfectionist challenge can be answered.

B. THE PRIORITY OF LIBERTY

I turn now from these theoretical issues to consideration of Rawls' more specific conclusions concerning the place of liberty in just institutions. Rawls' substantive account of justice is put forward in two forms which he calls respectively the General Conception and the Special Conception of Justice as Fairness. The General Conception of Justice as Fairness provides that '[a]ll social primary goods—liberty and opportunity, income and wealth, and the bases of self-respect—are to be distributed equally unless an unequal distribution of any or all of these goods is to the advantage of the least favored' (p. 303). The Special Conception is expressed in the two principles of justice stated earlier, with the proviso that the First Principle is to be held prior to the Second in a sense to be discussed more fully below. The Second Principle allows for inequalities in the distribution of goods other than basic liberties on terms similar to those specified by the General Conception, but the First Principle lays down a more stringent requirement of equality in basic liberties, a requirement which is not to be set aside for the sake of greater economic or social benefits. This principle and the rule specifying its priority receive their final statement in the following form.

> *First Principle*
> Each person is to have equal right to the most extensive total system of equal basic liberties compatible with a similar system of liberty for all.
>
> *Priority Rule*
> The principles of justice are to be ranked in lexical order[6] and therefore liberty can be restricted only for the sake of liberty. There are two cases: (a) a less extensive liberty must strengthen the total system of liberty shared by all, and (b) a less than equal liberty must be acceptable to those citizens with the lesser liberty [p. 250].

The 'basic liberties' with which the First Principle is concerned are specified by Rawls as follows.

> The basic liberties are, roughly speaking, political liberty (the right to vote and to be eligible for public office) together with the freedom of speech and assembly; liberty of conscience and freedom of thought; freedom of the person along with the right to hold (personal) property; and freedom from arbitrary arrest and seizure as defined by the concept of the rule of law. These liberties are all required to be equal by the first principle, since citizens of a just society are to have the same basic rights [p. 61].

A liberty in the sense in which Rawls uses the term is defined by a complex of rights along with correlative duties of others to aid or not to interfere. Thus, by a restriction on liberty or an unequal liberty he means a restriction or inequality in what people are legally entitled to do (or, perhaps, entitled to do by the nonlegal rules defining the basic institutions of their society). In-

[6] 'This is an order which requires us to satisfy the first principle in the ordering before we can move on to the second . . . A principle does not come into play until those previous to it are either fully met or do not apply' (p. 43). For Rawls' initial statement of the Second Principle, see text accompanying note 2 *supra*.

equalities in people's ability to take advantage of their rights due, e.g., to unequal economic means do not count as inequalities in liberty for Rawls but rather as inequalities in what he calls the 'worth' or 'value' of liberty. While the basic liberties must be held equally, the worth of these liberties may vary since any significant inequality in wealth, income or authority (allowed under the Second Principle) will represent an inequality in the ability of citizens to make use of their liberty in order to advance their ends (p. 204). Rawls stresses at a number of points (e.g., pp. 224–6, 277–8) the importance of preserving 'the fair value' of the basic liberties, particularly political liberties, but strict equality in the worth of these liberties is not required by the First Principle itself.

Two examples, frequently cited by Rawls, of restrictions on basic liberties that are justified on the ground that they strengthen the total system of basic liberty are the restrictions on the scope of majority rule imposed by a bill of rights and the restrictions on the freedom to speak imposed by a system of rules of order. In the first case a restriction of the legal powers of citizens is justified by the fact that more extensive powers could legally erase other basic liberties. In the second case what are sometimes called restrictions as to time, place and manner are imposed on the exercise of a basic liberty in order, Rawls says, to preserve the worth of that liberty to all (p. 204). Thus it appears that while equal worth of the basic liberties is not required by the First Principle, securing the worth of these liberties is one of the goals which can justify restrictions on basic liberties under the Priority Rule.[7]

1. The Preference for Basic Liberties over Other Primary Goods

Given the degree to which the content of the Priority Rule, and hence the claim of Rawls' theory to provide a secure basis for liberty, depends upon the distinction between the basic liberties and other goods and opportunities, it may seem surprising that no theoretical account of this distinction is offered. The list of familiar constitutional categories given above is offered by Rawls not as a precise enumeration of the class of basic liberties but

[7] Both of the cases considered here are examples of a lesser but still equal liberty as provided for by clause (a) of the Priority Rule. Primary examples of unequal liberty allowed by clause (b) seem to be cases of 'justifiable paternalism' in which a less than equal liberty is 'acceptable to those whose liberty is restricted' in the sense spelled out in the requirements discussed above.

only as indicating 'roughly speaking' what this class is to include. I suspect that here, as with the class of primary goods itself, no precise theoretical demarcation can be given. What is claimed for these liberties is just that, due both to the importance for anyone of the interests they safeguard and to their great instrumental value for the enjoyment of other goods, they are not only things it is rational for anyone to want but also things it is rational for anyone to value particularly highly relative to other primary social goods.

It is not claimed that these liberties are always to be valued more highly than any other goods. Rawls allows that under particularly dire conditions, when bare survival or the pursuit of the means for a minimally comfortable life is the dominant concern, and when the necessary prerequisites for the effective exercise of the basic liberties are lacking, it may be rational to sacrifice basic liberties for the sake of other goods such as increased security or economic development. It is under such conditions that the General Conception of Justice as Fairness applies. Rawls argues (sect. 82), however, that as conditions improve and the possibility for the effective exercise of the basic liberties becomes real, people will set an increasingly high marginal value on basic liberties relative to other goods. After the most urgent wants are satisfied, people come to set greater importance on the liberty to determine and pursue their own plans of life. They will therefore insist on the right to pursue their own spiritual and cultural interests, seek to 'secure the free internal life of the various communities of interests in which persons and groups seek to achieve . . . the ends and excellences to which they are drawn' and, in addition, 'come to aspire to some control over the laws and rules that regulate their association, either by directly taking part themselves in its affairs or indirectly through representatives with whom they are affiliated by ties of culture and social situation' (p. 543). Recognizing these tendencies, the parties in the Original Position will see that '[b]eyond some point it becomes and then remains irrational [for them] . . . to acknowledge a lesser liberty for the sake of greater material means and amenities. . . .' (p. 542). Thus the position of liberty under the Special Conception makes explicit the priority that emerges under the General Conception as the natural preference for basic liberties over increases in other primary social goods asserts itself.

There are a number of questions one might raise concerning this argument. First, since the appeal to an increasing preference for basic liberties over other primary social goods represents Rawls' most detailed claim about the way in which the parties in the Original Position would order bundles of primary social goods, it naturally gives rise to questions of the sort considered above under the heading of 'parameters'. Rather than to consider the general question of whether this preference is in some suitable sense 'universal', however, it seems to me more profitable to ask whether an appeal to such a preference provides adequate and interesting answers to those questions about liberty (and about the particular basic liberties listed by Rawls) that one would want a philosophical theory of liberty to answer.

Foremost among these is the question to what extent the basic liberties have some kind of absolute status and to what extent, and within what limits, they are to be understood and interpreted in terms of a balancing of competing interests. Rawls appears to have two answers to this question. The first, given by the Priority Rule, makes the limitation on acceptable balancing depend upon the distinction between basic liberties and other primary social goods: basic liberties are to be limited only for the sake of the total system of basic liberty itself. The second answer, and the one most often used by Rawls to indicate when a lesser but still equal liberty is just, is given by what he calls the Principle of the Common Interest:

> According to this principle institutions are ranked by how effectively they guarantee the conditions necessary for all equally to further their aims, or by how efficiently they advance shared ends that will similarly benefit everyone. Thus reasonable regulations to maintain public order and security, or efficient measures for public health and safety, promote the common interest in this sense. So do collective efforts for national defence in a just war [p. 97].

Rawls does not formulate this principle explicitly, but his discussion (p. 213–14) suggests the following formulation: basic liberties may be restricted only when methods of reasoning acceptable to all make it clear that unrestricted liberties will lead to consequences generally agreed to be harmful for all.

Rawls seems to hold (pp. 246–7, 212–13) that these two doctrines are consistent, i.e., that cases in which a restriction of basic liberties is justified by the Principle of the Common Interest are also cases in which basic liberty is being limited for the sake of the total system of basic liberty itself. This appears to be true in the most apocalyptic cases, e.g., cases in which a restriction of basic liberties is necessary as part of the common defense against an invasion. It may be true as well in some more mundane cases, such as Rawls' example of the restrictions imposed upon the right to speak by fair rules of order (taking into account, as was noted above, that what is protected in this case is not, strictly speaking, liberty but rather the worth of liberty). But if the restrictions on utterances imposed by such a set of rules count as restrictions on a basic liberty, then so also must similar restrictions on the time, place and manner of political demonstrations, religious festivals, parades, the placing of posters and the use of loudspeakers and sound trucks. Regulation of these activities is normally thought to be acceptable, and appears to be justified by something like the Principle of the Common Interest, but it seems to me difficult to maintain (without considerable stretching of the notion of a basic liberty) that in these cases basic liberties are being restricted only for the sake of the same or other basic liberties. It seems to me much more plausible and straightforward to say that in order to arrive at a policy in these cases we must balance the value of certain modes of exercise of a basic liberty not only against the exercise of other basic liberties but also against the enjoyment of other goods (uninterrupted sleep, undefaced public buildings, etc.). Something like this is surely true in the case of the restriction of expression by laws against defamation: different standards of defamation for, on the one hand, private, artistic or cultural expression and, on the other, political debate, seem to me obviously appropriate, and I take this to be the reflection of the differing values we place on the unfettered exercise of these forms of expression relative to, among other things, the value placed on safeguarding the primary good of self-respect.

One could of course maintain that what is balanced against liberty in these cases is not liberty itself but the *worth* of liberty. Since almost anything, including any significant increase or decrease in material well-being, can affect the worth of liberty, the general principle that basic liberties may be restricted only when this brings an increase (or is necessary to avoid a decrease) in the worth of the total system of basic liberties appears to be a weaker principle than Rawls wishes to defend. I suggest that Rawls' response here would be that while a great number of things can contribute to the worth of liberty, not every restriction of basic liberty yields gains in other goods or will yield sufficient gains to constitute a net increase in the worth of the total system of such liberties. This is what a restriction must do in order to be acceptable.

Conceivably, this principle can be made to fit the most obvious cases in which a restriction of basic liberty is justified. Given the rather diffuse character of the notion of 'the worth of the total system of basic liberties', however, it is not a principle that is easy to apply. Under any account the decision as to when a restriction on basic liberty is justified will involve some difficult balancing, but I do not think that a clear guideline between acceptable and unacceptable balancing is obtained by describing everything in terms of 'the worth of liberty'. Such an approach might seem inviting if one thought that the notion of an increasing preference for basic liberties over other goods represented the most important theoretical element in the case for liberty. But I do not think that this is so. On the contrary it seems to me that the idea of an increasing preference for basic liberties leaves out or obscures the most important factors in the case for certain of the basic liberties, factors which Rawls' own discussion of these particular basic liberties brings out quite clearly.

2. Freedom of the Person

The argument from the increasing marginal preference for liberties over the other primary goods is most appropriate as an account of the basis of freedom of the person. It is not completely clear from Rawls' discussion what this category of basic liberties is to encompass other than the protections against arrest and seizure embodied under 'the rule of law', but I take it to include at least freedom of movement within the country and across its borders, freedom of choice in aspects of one's personal life, and perhaps also freedom from surveillance. The increasing preference for these liberties claimed by Rawls can be seen as deriving in part from the fact that they represent important conditions for the use and enjoyment of other goods. Beyond this, however, there is the fact that the interventions these liberties are intended to preclude constitute particularly deep intrusions into a person's life which anyone has strong reasons to want to avoid, both because of the real disruption they cause and because of their great symbolic impact.

We can of course imagine people who felt quite differently about these matters. To the extent that such differences are not merely the object of speculative imagination but the subject of real disagreement and controversy, the force of Rawls' argument for the priority of freedom of the person will be seriously weakened. But in such an event it seems clear that the case for these liberties will be genuinely in doubt. Rawls' analysis of the case for the freedoms of the person as a matter of relative preference thus seems quite appropriate; there is no obvious theoretical element in the case for these liberties that his analysis leaves out.

3. Liberties of Expression, Thought and Conscience

Freedom of speech and assembly, liberty of conscience and freedom of thought present a slightly different case. The argument for the priority of these liberties rests upon the recognition by the parties in the Original Position that as material conditions improve there will be a 'growing insistence upon the right to pursue our spiritual and cultural interests' (p. 543). As Rawls says in arguing for freedom of conscience, the parties 'must assume that they may have moral, religious, or philosophical interests which they cannot put in jeopardy unless there is no alternative' (p. 206).

Now this argument contains two distinguishable elements. The first is the recognition by the parties in the Original Position that, for the reasons discussed in connection with the argument against perfectionism, they cannot concede to the government any authority in matters of religious, moral or philosophic doctrine. As Rawls says,

> The government has no authority to render religious associations either legitimate or illegitimate any more than it has this authority in regard to art and science. These matters are simply not within its competence as defined by a just constitution. Rather, given the principles of justice, the state must be understood as the association consisting of equal citizens. It does not concern itself with philosophical and religious doctrine but regulates individuals' pursuit of their moral and spiritual interests in accordance with principles to which they themselves would agree in an initial situation of equality [p. 212].

The second element is the recognition by the parties that they will come to set a particularly high value on the pursuit of their 'spiritual and cultural interest'.

These two elements are clearly independent. To take the case of religion, the value that a group of people place on keeping their religious commitments will be reflected in such things as the amount of economic loss and disruption of the pattern of life they are willing to undergo to allow everyone to observe the holidays of his religion, attend services, etc. and in the lengths to which they are prepared to go to recognize and respect the religious scruples of individual members against taking part in certain necessary tasks and activities. It is certainly possible that the cost a society is willing to bear in order to allow full freedom of religious observance might vary widely while the principle of the lack of governmental authority to decide between particular religious doctrines remained quite fixed.

This kind of variation in the value attached to religious observance, while possible, may in fact be unlikely if, as Rawls says, '[a]n individual recognizing religious and moral obligations regards them as binding absolutely in the sense that he cannot qualify his fulfillment of them for the sake of greater means for promoting his other interests' (p. 207). This extraordinary importance attached to religious matters tends to overshadow the distinction I have tried to draw and makes it inviting to rest the case for toleration entirely on the claim that the parties in the Original Position can forsee that they will come to set an incomparably higher value on religious liberty (i.e., on the freedom to meet their religious commitments) than on other primary social goods. But this approach becomes less attractive if we think not only of religious liberty but of freedom of thought and expression more broadly construed. A society is apt to set rather different values on the fulfillment of religious commitments, the pursuit of scientific knowledge and the pursuit and enjoyment of excellence in the arts, and these differences will be reflected in the price the society is willing to bear in order to allow these activities to go forward. But in a society which recognizes freedom of thought and expression the regulation of these pursuits will be guided by a common principle that governments lack the authority to decide matters of moral, religious or philosophic doctrine (or of scientific truth) and hence also lack the authority to restrict certain activities on the grounds that they promulgate false or corrupting doctrines. Let me call this principle, which I have formulated only very crudely, the Principle of Limited Authority.

Taken alone such a principle does not constitute a complete doctrine of liberty or even of freedom of thought and expression.

But it seems to me that this principle is the most important element in such a doctrine that can be established from the point of view of the Original Position. It is not possible to determine from that standpoint exactly what relative values are to be assigned to these pursuits and to other interests with which they may conflict. Nor is it possible to forsee from that standpoint what will be the best way of regulating these pursuits so that they do not conflict. These are problems that can be dealt with only at a later stage when the full facts about a society and the preferences of its members are known. (I suspect that this process of balancing and coordination is what Rawls has in mind when he speaks of restricting particular basic liberties in order to strengthen the total system of basic liberties.) While it may be possible for the parties in the Original Position to forsee that in general they will attach a high value to their spiritual and cultural interests, such a general preference, or a resultant general principle that in the balancing process these liberties are to take precedence over other goods, seems to me to be less useful as the basis for a doctrine of freedom of thought and expression than the idea that the process of balancing must take place within the constraints imposed by something like the Principle of Limited Authority.

A doctrine of freedom of expression founded on this idea is suggested by Rawls on a number of occasions, in particular in his principle of the common interest, with its emphasis on the distinction between what might be called 'neutral' and 'non-neutral' grounds for restricting liberty. I think that some account of freedom of expression of this general type must be correct, although there are a number of difficulties in formulating such a view.[8] While I have some misgivings about Rawls' particular formulation (misgivings, e.g., as to whether too much may be conceded to the doctrine of clear and present danger by his blanket allowance that liberty of conscience may be limited 'when there is a reasonable expectation that not doing so well will damage the public order which the government should maintain' (p. 213), it seems to me one of the strong points of Rawls' theory (as described in the first part of this section) that it provides a philosophical basis for an account of liberty of this type. It therefore seems to me important to ask whether this strength is adequately represented in his doctrine of the priority of liberty.

[8] I have myself put forward a view of this kind in 'A Theory of Freedom of Expression', *Philosophy and Public Affairs*, Vol. 1 (1972).

While it is not explicitly stated in the Priority Rule, the Principle of Limited Authority will be implied by clause (b) of that rule if (as seems plausible on the basis of the argument of the preceding section) we take governmental authority over matters of religion, etc., to represent an unequal liberty which would not be acceptable to those whose liberty is restricted. It is unclear, however, how this principle is related to the argument for the priority rule based on the increasing marginal value of liberty. There seem to be two possible interpretations of this argument.

While the parties in the Original Position might readily agree that there are conditions under which the pursuit of spiritual and cultural interests may be severely curtailed for the sake of other more pressing needs, it may seem unlikely, given the close relation between the Principle of Limited Authority and the conception of individual autonomy underlying the argument against perfectionism, that the parties would ever concede to a government the right to decide matters of moral, religious or philosophic doctrine. This suggests an interpretation of Rawls' argument according to which the Principle of Limited Authority applies under the General Conception of Justice as Fairness as well as under the Special Conception. What distinguishes the Special Conception, on this view, is just the increased importance that is attached to spiritual and cultural interests as the opportunity to pursue these interests presents itself and the demands of mere survival become less pressing. This interpretation is faithful to Rawls' description of the transition from the General Conception to the Special Conception as consisting of a shift in the ordering of primary social goods. But the Principal of Limited Authority is not a factor in this shift; it stands instead as a constant element of the theory. Given the importance of his principle from Rawls' point of view, it seems somewhat surprising on this interpretation that nothing resembling this principle is either stated or implied in Rawls' account of the General Conception.

An alternative, somewhat more extreme interpretation, and one which seems to me more likely to represent Rawls' view, would identify the Principle of Limited Authority as one of the distinguishing elements of the Special Conception. This means that there must be circumstances to which the General Conception of Justice as Fairness applies but in which the parties in the Original Position would not only allow the severe curtailment of expression on the grounds allowed under the Principle of the Common

Interest but would also suspend the Principle of Limited Authority itself. I am not quite certain what such situations would be like. Presumably they would be situations in which cooperation on certain common tasks is not merely mutually advantageous but essential for survival or for the amelioration of intolerable conditions. If deep disagreements were to exist which made the basis of this cooperation fragile, and if close and uninterrupted cooperation were required to avoid consequences that would be disastrous for all, then perhaps it would be rational not only to accept rigid regulation on the time, place and manner of expression to prevent interference with essential work, but also to grant to the government the power to ban the expression of views likely to give rise to dangerous controversy or to dissension and doubt.

It seems to me most accurate to describe such situations as ones in which the circumstances of justice would be present only to a limited degree. Cooperation in certain tasks may be feasible and profitable and in these areas of common purpose considerations of justice may apply, dictating, e.g., that the benefits and burdens of this cooperation (including liberties and constraints) should be shared in accordance with Rawls' Second Principle of Justice. But if the basis of this cooperation is quite shaky, and if the ends at which it aims are truly vital, then it might be rational for the parties involved to regard each other primarily as means to these ends. This attitude would be reflected, for example, in the parties' placing the smooth functioning of their institutions ahead of the right of individual members to raise and discuss with each other questions about the wisdom, viability or propriety of these institutions. I have some inclination to say that such a case would not represent cooperation on a footing of justice at all; collective actions to quell controversy in such circumstances are best seen not as the exercise of the distinctive authority of a just government in the sense defined by the Original Position, but rather as acts which must be justified on a case-by-case basis by appeal to the residual rights of the individuals involved to undertake those measures necessary to their self-defence and survival.[9]

However this may be, it is at least clear that the justification I have offered for limited tolerance in what might be called situations of partial justice depends upon the presence of conditions under which anything which undermines effective cooperation represents an immediate threat to all. When these conditions are

[9] Such a view is suggested in Scanlon, 'A Theory of Freedom of Expression', pp. 224–6.

lacking, such justification is also lacking and in addition it becomes rational for people to seek to establish cooperation on a footing that gives full recognition to the status of the participants as autonomous equals, i.e., to something like Rawls' Special Conception.

One thing making this transition rational is the fact that under improving conditions individuals will develop religious, moral and philosophical interests and will want their institutions to safeguard their pursuit of these interests. But on the interpretation I have been discussing the Special Conception of Justice as Fairness can no longer be seen simply as what emerges under the General Conception once these interests begin to develop. For the transition to the Special Conception involves a fundamental change in the basis of cooperation, namely a move to what I called in the first part of this section cooperation on a footing of justice. Cooperation on this basis would be less apt to be rational for people if they did not place a high value on certain kinds of opportunity, but the defining elements of this form of cooperation go beyond this configuration of preferences, just as the defining elements of cooperation in the economic sphere go beyond the structure of needs and interests that make such cooperation inviting.

II. *Distributive Justice and the Difference Principle*

Rawls is concerned with justice in only one of the many senses of the term. For him, questions of justice are questions of how the benefits and burdens of social cooperation are to be shared, and the principles of justice he develops are to apply in the first instance not to arbitrary distributions of goods but to the basic institutions of society which determine 'the assignment of rights and duties and . . . regulate the distribution of social and economic advantages' (p. 61). Rawls' principles apply to particular distributions only indirectly: a distribution may be called just if it is the result of just institutions working properly, but the principles provide no standard for appraising the justice of distributions independent of the institutions effecting them (p. 88). Conceived of in this way, principles of justice are analogous to a specification of what constitutes a fair gamble. If a gamble is fair then its outcome, whatever it may be, is fair and cannot be complained of. But the notion of a fair gamble provides no standard for judging particular distributions (Smith and Harris win five dollars, Jones loses ten dollars) as fair or unfair when these are considered in isolation from particular gambles which bring them about.

The principle which Rawls offers for appraising the distributive aspects of the basic structure of a society is his Second Principle of Justice which, considerations of the priority of liberty aside, is equivalent to what he calls the General Conception of Justice as Fairness. This principle is stated as follows: 'Social and economic inequalities are to be arranged so that they are both (a) to the greatest benefit of the least advantaged and (b) attached to offices and positions open to all under conditions of fair equality of opportunity' (p. 83).[10]

According to clause (a) of this principle, which Rawls refers to as the Difference Principle, a system of social and economic inequalities is just only if there is no feasible alternative institution under which the expectations of the worst-off group would be greater. The phrase 'fair equality of opportunity' in clause (b) requires not only that no one be formally excluded from positions to which special benefits attach, but also that persons with similar talents and inclinations should have similar prospects of attaining these benefits 'regardless of their initial place in the social system, that is, irrespective of the income class into which they are born' (p. 73). The rationale behind this principle, particularly the motivation for clause (a), will be discussed at length below. First, however, I will consider briefly how the principle is to be applied.

A. DIFFERENCE PRINCIPLE AND ITS APPLICATION

The most natural examples of inequalities to which Rawls' principle might be applied involve the creation of new jobs or offices to which special economic rewards are attached or an increase in the income associated with an existing job. But the intended application of the principle is much broader than this. It is to apply not only to inequalities in wealth and income but to all inequalities in primary social goods, e.g., to the creation of positions of special political authority. Further, its application is not limited to 'jobs' or 'offices' in the narrow sense but includes all the most general features of the basic structure of a society that give rise to unequal shares of primary social goods. In the case of economic goods these will include the system of money and

[10] Rawls' final formulation of this principle (p. 312) incorporates considerations of justice between generations which the present discussion leaves aside.

credit, the laws of contract, the system of property rights and the laws governing the exchange and inheritance of property, the system of taxation, the institutions for the provision of public goods, etc.

It is fairly clear how Rawls' principle is to apply to the creation of one new office to which special rewards are attached (or to the assignment of new rewards to one existing position) in an otherwise egalitarian society: such an equality is just only of those who do not directly benefit from this inequality by occupying the office benefit indirectly with the result that they too are better off than they were before (and than they would be if the benefits in question were distributed in any alternative way). It is less obvious how the principle is to apply in the more general case of complex institutions with many separable inequality-generating features. Rawls deals with this problem by specifying that institutions are to be appraised as a whole from the perspective of representative members of each relevant social position. The Difference Principle requires that the total system of inequalities be so arranged as to maximize the expectations of a representative member of the class which the system leaves worst off.

The notions of relevant social position and the expectations of a representative person in such a position require explanation. Relevant social positions in Rawls' sense are those places in the basic structure of society which correspond to the main divisions in the distribution of primary social goods. (He mentions the role of 'unskilled worker' as constituting such a position (p. 98).) Rawls believes that the distribution of other primary social goods will be closely enough correlated with income and wealth that the latter can be taken as an index for identifying the least advantaged group. Accordingly, he suggests that the class of least advantaged persons may be taken to include everyone whose income is no greater than the average income of persons in the lowest relevant social position (or alternatively everyone with less than half the median income and wealth in the society (p. 98)). To compute the expectations of a representative member of a given social position one takes the average of the shares of primary social goods enjoyed by persons in that position. Thus, while the parties in the Original Position do not estimate the value to them of becoming a member of a given society by taking the likelihood of their being a member of a particular social position to be represented by the proportion of the total population that is in that position, they do estimate the expected value (in primary social goods) of being a member of a particular social position by taking the likelihood that they will have any particular feature affecting the distribution of primary social goods within that position to be represented by the fraction of persons in the position who have that feature. Rawls does not explicitly discuss his reasons for allowing averaging within a social position when he has rejected it in the more general case. A more extreme position eschewing averaging would require maximizing the expectations of the worst-off individual in society. The Difference Principle occupies a position somewhere between this extreme and the principle of maximizing the average share of primary social goods across the society as a whole, its exact position within this range depending on how broadly or narrowly the relevant social positions are defined. The resort to averaging seems to some extent to be dictated by practical considerations: a coherent and manageable theory cannot take into account literally every position in a society (p. 98). In addition, the theoretical case against the use of averaging (as opposed to some more conservative method of choice) is weaker when we are concerned with differences in expectation within a single social position rather than differences between such positions. For here we are not concerned with a single 'gamble' with incomparably high stakes: intraposition differences are, by definition, limited, and each person's allotment is determined by a large number of independent factors, many of which are of approximately equal magnitude (cf. 169–71).

There is a further problem about the notion of expectations which requires consideration. Rawls refers to the relevant social positions as 'starting places', i.e., as the places in society people are born into (p. 96). Now the expectations of a person born into a family in a certain social position can be thought of as consisting of two components. First, there is the level of well-being he can expect to enjoy as a child. Presumably we may identify this with his parents' allotment of primary social goods. Second, there are his long term prospects as a member of society in his own right. If perfect fair equality of opportunity were attained then this latter component would not be substantially affected by the social and economic position of one's parents. As Rawls notes, however, such perfect equality of opportunity is unlikely, at least as long as the family is maintained (pp. 74, 301), so we may suppose that in general the second component will be heavily influenced by the first. One might conclude that the second component can be neglected entirely, reasoning that the distribution of social and economic advantages will influence the long term life prospects of a representative person born into the worst-off class mainly through its effect on the conditions in which such a person grows up. Taking this course would have the same consequences as deciding that what should be considered in applying the Difference Principle are not the expectations of a representative person born into the worst-off social position but the expectations of a representative person who winds up in that position after the social mechanism for assigning people to social roles has run its course.

But the principle which results from ignoring long term expectations seems to me unsatisfactory. Suppose we have a society in which there are 100 people in the lowest social position and twenty-five people in each of the two higher positions, and suppose it becomes known that the basic institutions of the society could be altered so that in later generations there would be fifty people in each of the three social positions, with the levels of wealth, income, authority, etc., associated with these positions remaining the same as they are now. Now it seems to me that a person in the lowest social position in this society is apt to be strongly in favor of this change. And such a person could plausibly support this preference by saying that the expectations of a representative person born into this social position (in particular, the expectations of his children) would be better if this change were made than if it were not. This increase in expectations will not be captured by the interpretation of the Difference Principle just suggested or by any principle which focuses only on the levels of income, wealth, etc., associated with various positions in society while ignoring the way in which the population is distributed among these positions. Examples of this kind convince me that considerations of population distribution have to be incorporated in some way into Rawls' theory, and the most natural way to do this seems to me to be to bring them in through the notion of long term expectations.

But how is this to be done? The rule mentioned above that the expectations of a representative person in a given social position are to be determined by averaging the benefits enjoyed by persons in that position suggests that in a society with three relevant social positions whose average levels of income, wealth, authority, etc., can be indexed by p_1, p_2 and p_3, the long term prospect of a person born into the worst-off position should be represented by

$a_1p_1 + a_2p_2 + a_3p_3$, where a_1, a_2 and a_3 are the fractions of people born into the worst-off position who wind up in each of the three places.

But the adoption of averaging as the method for computing long term expectations has unpleasant consequences for Rawls' theory. To the extent that the inequalities in childhood expectations resulting from the unequal economic and social positions of different families are eliminated (perhaps by eliminating the institution of the family itself), the first component in the expectations of a representative person will become the same for everyone regardless of the social position into which he is born, and Rawls' requirement that the expectations of a representative person in the lowest social position be maximized becomes the requirement that we maximize the second component of these expectations, i.e., the long term expectation $a_1p_1 + a_2p_2 + a_3p_3$. Moreover, to the extent that fair equality of opportunity is achieved (and barring the formation of a genetic elite) the coefficients a_1, a_2 and a_3 in this polynomial will become the same for every representative person regardless of social class, and the polynomial will thus come to express the average share of primary goods enjoyed by members of the society in question. It follows that on the interpretation just suggested Rawls' Difference Principle will be distinct from the principle requiring us to maximize the average share of primary social goods only so long as the inequalities resulting from the institution of the family persist or the fair equality of opportunity required by clause (b) of the Second Principle is otherwise not achieved. Even if fair equality of opportunity is an unattainable ideal this conclusion seems to me unacceptable for Rawls' theory. The principle of maximum average primary social goods is an extremely implausible one, much less plausible than the principle of maximum average utility.[11] I see no reason to think that this principle would be acceptable even if perfect equality of opportunity were to obtain.

The problem here is how to give some weight to the way in which the population is distributed across social positions without introducing aggregative considerations in such a way that they take over the theory altogether (or would do so but for the 'friction' introduced by imperfect equality of opportunity). One way of dealing with this problem which seems to me in the spirit of Rawls' theory would be to modify the Difference Principle to require the following:

[11] See Scanlon, 'Rawls' Theory', p. 1050.

> First maximize the income, wealth, etc. of the worst-off representative person, then seek to minimize the number of people in his position (by moving them upwards); then proceed to do the same for the next worst-off social position, then the next and so on, finally seeking to maximize the benefits of those in the best-off position (as long as this does not affect the others).[12]

This seems to me a natural elaboration of what Rawls calls the Lexical Difference Principle (p. 83).[13] It also has the advantage of dealing with the problem of population distribution without introducing the summing or averaging of benefits across relevant social positions. There are obviously many variations on this theme as well as many altogether different approaches.[14]

B. THE ARGUMENT FOR THE DIFFERENCE PRINCIPLE

I return now to the central question of the rationale behind the Difference Principle. The intuitive idea here is that a system of inequalities is just only if we can say to each person in the society, 'Eliminating the advantages of those who have more than you would not enable us to improve the lot of any or all of the people in your position (or beneath it). Thus it is unavoidable that a certain number of people will have expectations no greater than yours, and no unfairness is involved in your being one of these people.' The requirement that we be able to say this to *every* member of society, and not just to those in the worst-off group, corresponds to what Rawls calls the Lexical Difference Principle:

> [I]n a basic structure within *n* relevant representatives, first maximize the welfare of the worst-off representative man;

[12] This solution was suggested to me by Bruce Ackerman.

[13] See text accompanying note 15, *infra*.

[14] One would be to take the position a person is 'born into' to be defined not only by the social and economic status of his family but also by his inborn talents and liabilities, i.e., those features which will enable him to prosper in the society or prevent him from doing so. Given this definition of the 'starting places', one could employ averaging as a method for representing the long term expectations of a representative person born into the worst-off such place without fear that the theory would collapse into the doctrine of maximum average primary social goods if the institution of the family were eliminated. Modifying the Difference Principle in this way would bring Rawls closer (perhaps too close) to what he calls 'the principle of redress', the principle that the distribution of social advantages must be arranged to compensate for undeserved inequalities such as the inequalities of birth and natural endowment. See pp. 100–2.

> second, for equal welfare of the worst-off representative maximize the welfare of the second worst-off representative man, and so on until the last case which is, for equal welfare of all the preceding n-1 representatives, maximize the welfare of the best-off representative man [p. 83].[15]

This form of the principle is called 'lexical' since 'lexical priority' is given to the expectations of the worse-off: the fate of the second worst-off group is considered only to decide between arrangements which do equally well for the worst-off, and so on for the higher groups, working always from the bottom up. This asymmetry of concern in favor of the worse-off is a central feature of the theory. Rawls remarks a number of times in contrasting his theory with utilitarianism that under the Difference Principle no one is 'expected . . . to accept lower prospects of life for the sake of others' (pp. 178, 180). But what this means, as Rawls himself notes (p. 103), is that no one is expected to take *less than others receive* in order that the others may have a greater share. It seems likely, however, that those who are endowed with talents which are much in demand will receive less in a society governed by Rawls' Difference Principle than they would if allowed to press for all they could get on a free market. Thus, in a Rawlsian society these people will be asked to accept less than they might otherwise have had, and there is a clear sense in which they will be asked to accept these smaller shares 'for the sake of others'. What, then, can be said to these people?

Rawls' stated answer to this question consists in pointing out that the well-being of the better endowed, no less than that of the other members of society, depends on the existence of social cooperation, and that they can 'ask for the willing cooperation of everyone only if the terms of the scheme are reasonable'. The Difference Principle, Rawls holds, represents the most favorable basis of cooperation the well-endowed could expect others to

[15] I will regard this as the canonical formulation of Rawls' principle. When this version of the principle is fulfilled there is a clear sense in which prevailing inequalities are 'to everyone's advantage' since there is no one who would benefit from their removal. Fulfillment of the simple Difference Principle (that inequalities must benefit the worst-off) insures fulfillment of the lexical principle only if expectations of members of the society are 'close knit'—it is impossible to alter the expectations of one representative person without affecting the expectations of every other representative person—and 'chain connected'—if an inequality favoring group *A* raises the expectations of the worst-off representative person *B* then it also raises the expectations of every representative person between *B* and *A* (pp. 80–2).

accept. Taken by itself this does not seem an adequate response to the complaint of the better endowed, for the question at issue is just what terms of cooperation are 'reasonable'.

The particular notion of 'reasonable terms' that Rawls is appealing to here is one that is founded in the conception of social cooperation which he is propounding. The basis of this conception lies not in a particular bias in favor of the less advantaged but in the idea that economic institutions are reciprocal arrangements for mutual advantage in which the parties cooperate on a footing of equality. Their cooperative enterprise may be more or less efficient depending on the talents of the members and how fully these are developed, but since the value of these talents is something that is realized only in cooperation the benefits derived from these talents are seen as a common product on which all have an equal claim. Thus Rawls says of his Two Principles that they 'are equivalent . . . to an undertaking to regard the distribution of natural abilities as a collective asset so that the more fortunate are to benefit only in ways that help those who have lost out' (p. 179).

This same notion of the equality of the parties in a cooperative scheme is invoked in the following argument for the Difference Principle.

> Now looking at the situation from the standpoint of one person selected arbitrarily, there is no way for him to win special advantages for himself. Nor, on the other hand, are there grounds for his acquiescing in special disadvantages. Since it is not reasonable for him to expect more than an equal share in the division of social goods, and since it is not rational for him to agree to less, the sensible thing for him to do is to acknowledge as the first principle of justice one requiring an equal distribution. Indeed, this principle is so obvious that we would expect it to occur to anyone immediately.
>
> Thus, the parties start with a principle establishing equal liberty for all, including equality of opportunity, as well as an equal distribution of income and wealth. But there is no reason why this acknowledgment should be final. If there are inequalities in the basic structure that work to make everyone better off in comparison with the benchmark of initial equality, why not permit them [p. 150–1]?

If one accepts equality as the natural first solution to the problem of justice then this argument strongly supports the conclusion that the Difference Principle marks the limit of acceptable inequality. More surprisingly, it also appears to show (whether or not one accepts equality as a first solution) that the Difference Principle is the most egalitarian principle it would be rational to adopt from the perspective of parties in the Original Position. It is of course a difficult empirical question how much inequality in income and wealth the Difference Principle will in fact allow, i.e., how many economic inequalities will be efficient enough to 'pay their own way' as the principle requires. The only theoretical limitation on such inequalities provided by Rawls' theory appears to be the possibility that glaring inequalities in material circumstances may give rise to (justified) feelings of loss of self-respect[16] on the part of those less advantaged, offsetting the material gains these inequalities bring them. One can thus make the Difference Principle more (or less) egalitarian by introducing a psychological premiss positing greater (or lesser) sensitivity to perceived inequality. But as far as I am able to determine there is no plausible candidate for adoption in the Original Position which is distinct from the Difference Principle and intermediate between it and strict equality. Since the inequalities allowed by the Difference Principle, while not great, may nonetheless be significant, this strikes me as a surprising fact. What it shows, perhaps, is that if one wishes to defend a position more egalitarian than Rawls' then one must abandon distributive justice as the cardinal virtue of social institutions, i.e., one must abandon the perspective which takes as the dominant moral problem of social cooperation that of justifying distributive institutions to mutually disinterested persons each of whom has a fundamental interest in receiving the greatest possible share of the distributed goods.[17]

[16] Inequalities give rise to loss of self-respect in Rawls' sense to the extent that they give a person reason for lack of confidence in his own worth and in his abilities to carry out his life plans (p. 535). Whether given inequalities have this effect will depend not only on their magnitude but also on the public reasons offered to justify them. Rawls believes that effects of this kind will not be a factor in a society governed by the Difference Principle since the inequalities in wealth and income in such a society will not be extreme and will 'probably [be] less than those that have often prevailed' (p. 536). In addition, the justification offered for those inequalities that do prevail will be one which supports the self-esteem of the less advantaged since this justification must appeal to the tendency of these inequalities to advance their good.

[17] A position of this kind was put forward, for example, by Kropotkin. See

The ideal of social cooperation which Rawls presents is naturally contrasted with two alternative conceptions of justice. The first of these is what Rawls calls the system of natural liberty (p. 72). This conception presupposes background institutions which guarantee equal liberties of citizenship in the sense of the First Principle and preserve formal equality of opportunity, i.e., 'that all have at least the same legal rights of access to all advantaged social positions' (p. 72). But no effort is made to compensate for the advantages of birth, i.e., of inherited wealth. Against the background provided by these institutions individuals compete in a free market and are free to press upon one another whatever competitive advantages derive from their different abilities and circumstances.

The second alternative is that of utilitarianism, understood broadly to include the two modified views presented at the end of the last section. The last of those two views differed from the versions of utilitarianism criticized by Rawls in that it incorporated Rawls' principle that no one may be asked to accept a less than equal share in order that some others may enjoy correspondingly greater benefits. But even though it is not simply a maximizing conception, this view is like other forms of utilitarianism in holding it to be the duty of each person to make the greatest possible contribution to the welfare of mankind. Any asset one may have control over, whether a personal talent or a transferable good, one is bound to disburse in such a way as to make the greatest contribution to human well-being.[18] Utilitarianism is in this sense an asocial view; the relation taken as fundamental by the theory is that which holds between any two people when one has the capacity to aid the other. Relations between persons deriving from their position in common institutions, e.g., institutions of production and exchange, are in themselves irrelevant. It would be possible to maintain a view of this kind which focused only on the well-being of members of a particular society, but such a restriction would appear arbitrary.

P. Kropotkin, *The Conquest of Bread*, ch. 13, p. 62, *et passim* (Penguin ed. 1972). Kropotkin holds that if one accepts, as Rawls appears to, the view that the productive capacities of a society must be seen as the common property of its members, then one must reject the idea of wages (or any other way of tying distribution to social roles). Rather, the social product is to be held in common and used to provide facilities which meet the basic needs of all.

[18] This aspect of utilitarianism is most clearly emphasized by William Godwin. See W. Godwin, *Enquiry Concerning Political Justice*, Bk. VIII (3d ed., 1797) (facsim. ed., F. Priestley, 1946).

The natural tendency of utilitarian theories is to be global in their application.

Rawls' Difference Principle can be seen as occupying a position intermediate between these two extremes. Like the system of natural liberty and unlike utilitarianism, Rawls' conception of justice applies only to persons who are related to one another under common institutions. The problem of justice arises, according to Rawls, for people who are engaged in a cooperative enterprise for mutual benefit, and it is the problem of how *the benefits of their cooperation* are to be shared. What the parties in a cooperative scheme owe one another as a matter of justice is an equitable share of this social product, and neither the maximum attainable level of satisfaction nor the goods and services necessary, given their needs and disabilities, to bring them up to a certain level of well-being.

The qualification 'as a matter of justice' is essential here since justice, central though it is, is not the only moral notion for Rawls, and other moral notions take account of need and satisfaction in a way that justice does not. Rawls speaks, for example, of the duty of mutual aid, 'the duty of helping one another when he is in need or jeopardy, provided that one can do so without excessive risk or loss to oneself' (p. 114). Now it seems likely that those to whom we are bound by ties of justice will fare better at our hands (or at least have a stronger claim on us) than those to whom we owe only duties of mutual aid; for justice, which requires that our institutions be arranged so as to maximize the expectations of the worst-off group in our society, says nothing about others elsewhere with whom we stand in no institutional relation but who may be worse off than anyone in our society. If this is so, then it may make a great deal of difference on Rawls' theory where the boundary of society is drawn. Are our relations with the people of South Asia, for example (or the people in isolated rural areas of our own country), governed by considerations of justice or only by the duties which hold between any one human being and another? The only satisfactory solution to this problem seems to me to be to hold that considerations of justice apply at least wherever there is systematic economic interaction; for whenever there is regularized commerce there is an institution in Rawls' sense, i.e., a public system of rules defining rights and duties etc. (p. 55). Thus the Difference Principle would apply to the world economic system taken as a whole as well as as to particular societies within it.

In distinguishing justice from altruism and benevolence and taking it to apply only to arrangements for reciprocal advantage Rawls' theory is like the system of natural liberty. But a proponent of natural liberty takes 'arrangements for reciprocal advantage' in the relevant sense to be arrangements arising out of explicit agreements. Such arrangements are just if they were in fact freely agreed to by the parties involved, and the background institutions of the system of natural liberty are designed to ensure justice in this sense. Since Rawls' Difference Principle constrains people to cooperate on terms other than those they would arrive at through a process of free bargaining on the basis of their natural assets, it is to be rejected. As Rawls says, the terms of this principle are equivalent to an undertaking to regard natural abilities as a common asset, and a proponent of natural liberty would say, I believe, that the terms of the principle apply only where such an undertaking has in fact been made.

Rawls holds, on the other hand, that one is born into a set of institutions whose basic structure largely determines one's prospects and opportunities. Background institutions of the kind described in the system of natural liberty are one example of such institutions; the various institutions satisfying the Difference Principle are another. Within the framework of such institutions one may enter into specific contractual arrangements with others, but these institutions themselves are not established by explicit agreement; they are present from birth and their legitimacy must have some other foundation. The test of legitimacy which Rawls proposes is, of course, the idea of hypothetical contract, as it is embodied in his Original Position construction.

The argument sketched here is obviously parallel to a familiar controversy about the bases of political obligation. The doctrine of natural liberty corresponds to the doctrine which seeks to found all political ties on explicit consent, and seems to me to inherit many of the problems of that view. For Rawls, on the other hand, the legitimacy of both political and economic institutions is to be analyzed in terms of merely hypothetical agreement. (Indeed, Rawls does not separate the two cases.) The parallel between the problems of political institutions and those of economic institutions is often obscured because the political problem is thought of in terms of *obligation* while economic justice is thought of in terms of *distribution*.[19] But

[19] For a discussion of political obligation relevant to economic contribution as well, see M. Walzer, *Obligations* (1970).

economic institutions, no less than political ones, must be capable of generating obligations, viz. obligations to cooperate on the terms these institutions provide in order to produce shares to which others are entitled (p. 313).[20]

The idea of such economic obligations raises a number of interesting issues which I can only mention here. Such an obligation to contribute would be violated, e.g., by a person who, while wishing to receive benefits derived from the participation of others in a scheme of cooperation satisfying the Difference Principle, refused to contribute his own skills on the same terms, holding out for a higher level of compensation than the scheme provided. Presumably obligations of this kind do not in general prevent a person from opting out of a scheme of economic cooperation, any more than political obligation constitutes a general bar to emigration; but this does not mean (in either case) that people are always free to simply pick up and go. Further, there obviously are limits to what a just scheme can demand of those born into it and limits to how far their freedom to choose among different forms of contribution can be restricted. It seems likely that these limits would be defended, on Rawls' view, by appeal to an increasing marginal preference for various species of liberty in the economic sphere relative to other goods. Thus one might claim, following Rawls' argument in Sect. 82, that as soon as a certain level of basic well-being is attained it becomes and then remains irrational for persons to accept lesser control over the terms and conditions of their working lives 'for the sake of greater material means and amenities'. Indeed, such an appeal to increasing preference seems to me more satisfactory as an argument for industrial democracy than as an account of the priority of traditional constitutional liberties.

As I have argued above, the central thesis underlying the Difference Principle is the idea that the basic institutions of society are a cooperative enterprise in which the citizens stand as equal partners. This notion of equality is reflected in Rawls' particular Original Position construction in the fact that these parties are prevented by the veil of ignorance and the requirement that the principles they choose be general (i.e., contain no proper names or token reflexives) from framing prin-

[20] The contribution side of the problem of economic justice is forcefully emphasized in R. Nozick, *Anarchy, State and Utopia* (New York and Oxford, 1974). Nozick criticizes Rawls from the prospective of a purely contractarian view much more sophisticated and subtle than the system of natural liberty I have crudely described here.

ciples which ensure them special advantages.[21] But the fact that it would be chosen under these conditions is not a conclusive argument for the Difference Principle since a person who favored the system of natural liberty would undoubtedly reject the notion that principles of justice must be chosen under these particular constraints. The situation here is similar to that of the argument against perfectionism: Rawls' defense of the Difference Principle must proceed in the main by setting out the ideal of social cooperation of which this principle is the natural expression. The advantages of this ideal—e.g., the fact that institutions founded on this ideal support the self-esteem of their members and provide a public expression of their respect for one another—can be set out, and its ability to account for our considered judgments of justice can be demonstrated, but in the end the adoption of an alternative view is not wholly precluded. A person who, finding that he has less valuable talents, wishes to opt for the system of natural liberty is analogous to the person who, knowing his own conception of the good, prefers a perfectionist system organized around this conception to what I have called 'cooperation on a footing of justice'. In both cases one can offer reasons why cooperation with others on a basis all could agree to in a situation of initial equality is an important good, but one cannot expect to offer arguments which meet the objections of such a person and defeat them on their own grounds.

I do not regard this residual indeterminacy as a failing of Rawls' book or as a source for skepticism. The conception of justice which Rawls describes has an important place in our thought, and to have presented this conception as fully and displayed its deepest features as clearly as Rawls has done is a rare and valuable accomplishment. Almost no one will read the book without finding himself strongly drawn to Rawls' view at many points, and even for those who do not share Rawls' conclusions will come to a deeper understanding of their own views as a result of his work.

[21] These considerations alone, of course, do not ensure that the parties in the Original Position will arrive at a principle of equal distribution even as a first solution. Given that they have no way to ensure a larger share for themselves the question remains whether they should settle for the maximum solution represented by the Difference Principle or gamble on receiving a larger share under some other rule.

9 Rawls and Marxism

RICHARD W. MILLER*

In *A Theory of Justice*, John Rawls claims that all of the most fundamental questions about justice, including questions about an individual's duty to help achieve justice, can be settled from the standpoint of the original position (pp. 11, 17, 115, 333–4). He proposes, as a criterion of social justice, the so-called 'difference principle', the principle that basic institutions ought to maximize the life prospects of the worst-off (pp. 60–1; 75 ff). And he presents this standard as a morally realistic one, in that people in societies that do not yet fulfill it ought to accept advances toward it (at least if these advances do not conflict with the maximization of equal basic liberties or with fair equality of opportunity), and ought, to some extent, to help achieve such advances (pp. 115, 246, 288–9, 334).[1]

I shall try to show that these claims, taken together, presuppose a relatively low estimate of the extent and consequences of social conflict. In particular, if a Marxist analysis of social conflict is right in certain respects, a commitment to accept advances toward the difference principle in societies that do not embody it would not emerge from the original position. Thus, if these Marxist ideas are correct, either the difference principle is unrealistic, in that people do not have a duty to accept and to further its realization, or there are fundamental issues concerning justice (for example, this issue of moral realism) that cannot be resolved from the standpoint of the original position.

In Rawls' 'ideal contractualism', principles concerning justice are seen as agreements that would be made by rational deliberants seeking to pursue their interests behind a veil of ignorance which excludes knowledge of what one's place in society and one's special interests are. This is, of course, a very rough statement, but most refinements on it go beyond the needs of this essay. There are, however, a few details of Rawls' theory that

* Assistant Professor of Philosophy, Cornell University.

[1] On the priority of the principles of equal liberties and of equal opportunity, see note 7, below.

need further elaboration, for the sake of my subsequent arguments.

In the original position, one does not know what one's social position or one's special needs and interests are. And in deciding the most general questions about justice, one does not know what particular form of society (e.g., slave-holding, feudal, capitalist) one lives in (p. 137). But one does know 'the general facts about human society' (p. 137).[2] Thus, if Marxist social theory is correct, the general facts contained in this theory would be known in the original position, and could affect its outcome.

In defining the agreements Rawls believes would emerge from the original position, it will be helpful to distinguish the outcomes Rawls thinks he *has* established in his book, from the outcomes he thinks *can* be established by ideal contractualism. In particular, Rawls claims to have actually derived a commitment to support the difference principle in circumstances of 'strict compliance', i.e., in any society the basic institutions of which conform to the difference principle and the members of which regard the latter as a principle of justice and willingly do what it requires.[3] But he surely regards it as *possible* to derive a commitment to accept and, to some extent, to further advances toward the difference principle in any society within the 'circumstances of justice', i.e., in any society with respect to which questions of justice can appropriately be raised. For he says the difference principle applies throughout the circumstances of justice.[4] As noted before, he regards it as a duty to help realize principles of justice. And he sees all fundamental questions concerning justice as being decidable by determining what commitments would emerge from the original position. Thus some commitment to uphold the difference principle, even when strict compliance has not been achieved, ought to emerge from the original position.

[2] Rawls goes on to say, 'They understand political affairs and the principles of economic theory. . . . Indeed the parties are presumed to know whatever general facts affect the choice of the principles of justice' (p. 137).

[3] The derivation is contained in sections 26, 29, and 51. For the restriction to circumstances of strict compliance, see pp. 8, 288–9, 334. This restriction was also imposed in unpublished comments of Rawls in response to David Lyons' paper at the 1972 A.P.A. symposium on *A Theory of Justice*. I should add that a reader who favors a less restrictive interpretation of Rawls' claims will find the conflict I sketch between Rawls and Marx all the more acute.

[4] Strictly speaking, the difference principle is always to be fulfilled in so far as this does not conflict with the principle of greatest equal liberty or

Of course, that the text of Rawls' book implies that there is an ideal contractualist account of the duty to uphold Rawlsian justice does not show that this claim is central to Rawls' theory. But in fact this contention of Rawls' cannot be rejected without casting doubt on the significance and the validity of the whole ideal contractualist approach.

The view that people in less than just societies have no duty to help achieve justice, or even to accept advances in this direction, would be not just false but monstrous and more than a bit absurd. To claim, for example, that slavery is unjust while denying that a slave-owner should let his slaves go free at the cost of moderate financial losses is a moral absurdity, and almost a logical one. If someone were to make such conflicting claims, we would hold the belief in the injustice of slavery either to play no real role or to play only a trivial role in his sense of justice. By the same token, a theory of justice that assigns ratings to institutions but provides no principles governing the behavior of individuals with respect to social change, would be a radically limited theory of justice, if it counts as a theory of justice at all.

Not only would Rawls' ideal contractualism be severely limited in its scope if it did not provide appropriate principles of individual duty, its reasoning in support of principles of justice for institutions would also be cast into doubt. For if Rawls' form of moral reasoning had to be confined to ratings of institutions, it would be inferior in important ways to what Rawls regards as its chief rival, utilitarian reasoning based on the standpoint of the impartial, sympathetic observer. Utilitarianism does provide a rule to guide an individual's response to defects in society (a rule which, in Rawls' view, conflicts with his own principles of justice). And the utilitarian treatments of social institutions and of the individual's responses to societal defects suggest and reinforce each other. In both of these respects ideal contractualism would be greatly inferior to utilitarianism, if no general duty to uphold appropriate principles of justice emerged from the former doctrine. Indeed, when utilitarians assert the unique realism and

the principle of fair equality of opportunity (pp. 61, 302–3). Also, an expanded version of the difference principle, in which liberty, opportunity, and self-respect are counted among the relevant goods, holds universally (pp. 62–3). In this essay, all discussions of the difference principle are to be understood as bearing on these two restricted claims. There is little danger of resulting oversimplification, for we shall not be concerned with conflicts between the (narrower) difference principle and principles that override it.

rationality of their view they often seem to be suggesting, in just this way, that rival moralities are only attractive at the level of abstract institutional ratings.

In sum, if ideal contractualism is supposed to tell us how basic institutions should operate, but must be abandoned when we ask how people should behave in less than ideal circumstances, it cannot be the satisfactory end result of moral reflection that Rawls wants it to be. Thus, the importance of an ideal contractualist account of the duty to help achieve justice is much greater than might be thought from the few paragraphs Rawls devotes directly to this aspect of his theory.

I shall be arguing that certain aspects of Marxism (and not very hard-line ones) would preclude the requisite agreement to uphold the difference principle throughout the circumstances of justice. In particular, I shall argue that this commitment would not be made if *some* societies in the circumstances of justice display the following three features: no social arrangement that is acceptable to the best-off class is acceptable to the worst-off class; the best-off class is a ruling class, i.e., one whose interests are served by the major political and ideological institutions; the need for wealth and power typical of the best-off class is much more acute than that typical of the rest of society. This piece of Marxism is, of course, less controversial than Marx's whole theory, in which these features are said to hold true of all nonprimitive societies. But certain of the claims I have sketched can be so understood that it is implausible that they should hold true of any society. And, of course, the question of whether there is an obligation to uphold the difference principle in a society like our own will be an implicit concern in my arguments, even though I do not explicitly address myself to it. For these reasons, I would like to spell out in more detail the ways in which the above three properties might obtain in various societies.

To begin with, Marx claims that in any society, from the dissolution of primitive communism to the overthrow of capitalism, there is no social contract that the best-off class and the worst-off one will acquiesce in, except as a result of defeat in class struggle or a tactical retreat to preserve long-term advantages. For example, Marx would say that no aristocracy has reduced feudal obligations, no bourgeoisie has reduced the length or pace of the working day, except in response to the actual or potential militancy of peasants or of workers, usually in alliance with other classes. And no peasantry or proletariat has accepted an economic arrangement for long without fighting against it, whether in peasant uprisings, militant strikes, or revolutions. For Marx, this determination of social affairs by what Rawls would call 'threat advantage' reflects people's rational pursuit of their self-interest. Moreover, improvements in the relative position of the worst-off class cannot, in Marx's view, be brought about by appeals to any universal sense of justice. Even when such a sense exists, no appropriate consensus can be achieved as to whether the demands of justice have in fact been fulfilled. For instance, capitalists, as a class, have always insisted that a proposed reduction of the working day, e.g., from twelve or more hours to ten, would do immeasurable harm to workers by destroying the capitalist economy on whose existence workers' welfare depends.[5]

The second Marxist idea I shall emphasize is the notion that the best-off class is a ruling class, one whose interests are served by all major institutions. Marx and Engels emphasize two aspects of this rule, the repressive and the ideological. In their view, the official instruments of coercion are employed, in almost all crucial instances of class conflict, in favor of the best-off class. Thus, to take some dramatic examples, the police, the army, and the courts were used in the United States and Great Britain to break up meetings in support of the ten-hour day, not meetings against it. In the fight for industrial unionism in this country in the twenties and thirties, they were used to protect strikebreakers and threaten sit-down strikers, not the other way around. In addition to institutions of repression, ideological institutions, in the Marxist view, help to maintain the special status of the best-off class. For example, in the Middle Ages, the Church tended to teach that submission to the dominant social order was an expression of piety. In nineteenth-century England, according to Marx, as class struggle became more intense, academic economists mostly became 'hired prizefighters' of the bourgeoisie,[6] arguing, e.g., that abolishing tariffs on grain would immensely enrich workers and that a ten-hour workday would reduce profits to zero.[7] To take a present-day American example, many Marxists would now argue that the media and the schools

[5] See Karl Marx, *Capital* (Moscow, n.d.), I, chap. 9, sec. 3, 'Senior's "Last Hour"'.

[6] *Ibid.*, I, 'Preface to the Second German Edition', p. 25.

[7] *Ibid.*, I, p. 25 and chap. 9, sec. 3.

foster anti-Black racism, because it serves to divide those having common interests against American big business.

The third Marxist idea to which I shall refer is, strictly speaking, an extrapolation from Marx's writings, and not taken directly from them. Marx seems to regard a typical member of the best-off class in an exploitative society as having an especially acute need for wealth and power. He would, I think, have accepted the following estimate of how acute these needs usually are: the need for wealth and power of a typical member of the best-off class is sufficiently great that such a person would be miserable if his society were transformed to accord with the egalitarian demands of the difference principle. Indeed, this misery would be so great that the possibility of such unhappiness would dissuade someone in the original position from committing himself to help realize the difference principle, when the veil of ignorance is lifted. (I shall spell out this claim in more detail later on.)

While this specific estimate of ruling-class needs cannot, of course, be found in Marx, it is, I think, a reasonable extrapolation from Marx's writings. Some such radical estimate of how much greater a lord's needs are than a serf's, or how much greater a capitalist's are than a worker's, is suggested by Marx's comments on the immense differences between what workers in different societies regard as 'necessities of life'.[8] It also seems implicit in his historical writings. For example, in Marx's account, the factions of the French bourgeoisie who overthrew Louis Philippe risked large-scale social disorder to escape a subordination to other factions that allowed the worst-off factions wealth and power beyond the dreams of most of the French working class, or, indeed, most workers in the Paris Commune, twenty years later. The general idea that classes differ in their needs,

[8] See, for example, *ibid.*, I, chap. 6, 'The Buying and Selling of Labor-Power', p. 168: 'On the other hand, the number and extent of his [the wage-laborer's] so-called necessary wants, as also the modes of satisfying them, are themselves the product of historical development, and depend therefore to a great extent on the degree of civilization of a country, more particularly on the conditions under which, and consequently on the habits and degree of comfort in which, the class of free laborers has been formed.'

There is a discussion to the same effect in *Wages, Price and Profit*, in which Marx speaks of 'a *traditional standard of life*' as 'the satisfaction of certain wants springing from the social conditions in which people are placed and reared up', and contrasts 'the English standard of life' with 'the Irish standard'; 'the standard of life of a German peasant' with that 'of a Livonian peasant' (*Selected Works in One Volume* [New York, 1970], p. 225). See also *Wage-Labor and Capital*, *ibid.*, pp. 84–5.

and not just in the degree to which their needs are satisfied, is almost explicitly stated in one of Marx's last writings, his notes on Wagner's *Lehrbuch der politischen Oeknomie*, when he criticizes some remarks of Wagner's on the 'natural' needs of Man: '[If Man] means Man, as a category, he has ... no ... needs at all. If it means man confronting nature by himself, one has in mind a non-social animal. If it means a man who is already to be found in some form of society ... one must begin by presenting the particular character of this social man, i.e. the particular character of the community in which he lives, since here production, and thus *the means by which he maintains his existence* already have a social character'.[9] This idea is, in turn, a special case of Marx's general thesis that 'the social being [of men] determines their consciousness'.[10]

If some societies in the circumstances of justice have the three features I have sketched, then, I shall try to show, no commitment to uphold the difference principle throughout the circumstances of justice would emerge from the original position.[11] To organize my argument, I shall take advantage of a certain feature of Rawls' exposition. His arguments for the difference principle often look like arguments for a commitment to participate in the immediate realization of the difference principle, once the veil of ignorance is lifted.[12] As I have mentioned, I do not think Rawls actually means to put forward such an argument within the confines of his book. He means only to argue for a commitment to uphold the difference principle in a society in which the principle has already been stably realized and everyone has a psychology supporting compliance with it. Nevertheless, the considerations

[9] Marx's emphasis. See Marx and Engels, *Werke* (Berlin, 1958), 19, p. 362. There are somewhat similar passages in the *Grundrisse* (New York, 1973) concerning the production of needs (pp. 92–3) and the social determination of the content of 'private interests' (p. 156). The idea that needs differ significantly among different classes or socio-economic groups is not, of course, by any means confined to Marxist or left-wing writers. See, for example, Emile Durkheim, 'Anomic Suicide', in *Suicide* (New York, 1951), pp. 249 ff.

[10] Preface to the *Critique of Political Economy*, in Marx and Engels, *Selected Works*, p. 182.

[11] One clear and immediate corollary will be that if present-day society has these features, no commitment to uphold the difference principle in the present day would emerge.

[12] These arguments, in sections 26 and 29, are actually presented as arguments for Rawls' two principles of justice, in serial order. But they are fairly characterized as 'Rawls' arguments for the difference principle', since no other derivation of that principle is given in Rawls' book.

Rawls brings forward in these arguments, considerations concerning the strains of commitment, rationality, self-respect, and stability, seem the ones that would be appealed to in support of a commitment to uphold the difference principle in less than ideal circumstances. In fact, it will be helpful to examine Rawls' arguments in turn, first taking each as if it *were* an argument for a commitment to realize the difference principle as soon as possible, then seeing if such an argument could be modified so as to generate a commitment to gradual realization. It might seem that working in this way I will tie my arguments too closely to special features of Rawls' text. But I think the reader will find that if the putative arguments I construct conflict with Marxist social theory, it seems quite unlikely that any argument for a commitment to uphold the difference principle throughout the circumstances of justice could accord with relevant aspects of Marxism.

One final comment may be necessary, before setting out my main arguments, in order for their import not to be misunderstood. I shall be maintaining that Rawls is tacitly committed to a social hypothesis that can reasonably be argued either way, i.e., that the Marxist analysis I have just sketched is wrong throughout the circumstances of justice. I should note that I do not at all intend to criticize Rawls for developing a moral theory based on controversial factual assumptions. Indeed, it seems unlikely he could have written a book of such merit and importance if he had not been willing to commit himself to empirical claims. My intention in this essay is, rather, to show what some of Rawls' implicit assumptions are. By that token, I shall be arguing that some of his claims must be rejected if these assumptions are held to be false.

I shall now consider Rawls' main attempts to derive the difference principle from the original position. In the chapter entitled 'Some Main Grounds for the Two Principles of Justice' Rawls begins with the argument that his standard of social justice uniquely satisfies the following constraint: '. . . they [the bargainers in the original position] consider the strains of commitment. They cannot enter into agreements that may have consequences they cannot accept. They will avoid those that they can adhere to only with great difficulty' (p. 176). Rawls argues along these lines against the principle of utility, noting that it will probably be a standard one cannot adhere to if it turns out to demand that one make great sacrifices simply to create more happiness for mankind as a whole. In contrast, by accepting Rawlsian principles of justice, participants in the original position are said to insure that they will be able to keep their agreement 'in actual circumstances' (p. 176).

Such reasoning from the strains of commitment cannot produce a rational agreement to realize the difference principle as soon as possible, if the difference principle is subject to the same defect of potential intolerability that Rawls ascribes to utilitarianism. In particular, this reasoning cannot succeed if there are societies in the circumstances of justice in which the best-off people generally find it intolerable to give up great advantages in order to maximize the well-being of the worst-off.[18] Marxists, of course, would claim that such situations do obtain. They would claim that the best-off people in any exploitive society cannot be made to give up their privileges except by force. In support of this claim, they would argue, for example, that no dominant exploitive class has voluntarily given up its rule, no matter how unjust. If Marxist social theory is right, at least when applied to some societies, someone in the original position would foresee that the difference principle may be intolerable for him, if he turns out to be a typical member of a dominant exploitive class. Thus, he could not accept, as grounds for a commitment to help realize the difference principle, an argument that this commitment, unlike its rivals, will be one he can live up to, no matter what social position he turns out to occupy.

In his book Rawls advances certain considerations that might be taken to show that in the original position one can foresee oneself as fulfilling a commitment to the difference principle, even if one turns out to be among the best-off. Most notably, in the chapter 'On the Tendency to Equality', Rawls argues, in the following terms, that the difference principle is a 'principle of mutual benefit':

> ...to begin with, it is clear that the well-being of each [i.e. both the best-off and the worst-off] depends on a scheme of social cooperation without which no one could have a satisfactory life. Secondly, we can ask for the willing cooperation of everyone only if the terms of the scheme are reasonable.

[18] In the original position one is to choose between the difference principle and alternative standards under the assumption that one will turn out to have the life prospects of a typical member of some significant income-group. See p. 96.

> The difference principle, then, seems to be a fair basis on which those better endowed, or more fortunate in their social circumstances, could expect others to collaborate with them when some workable arrangement is a necessary condition of the good of all [p. 103].

Such a line of reasoning is inadequate to establish the tolerability of commitment to the difference principle, if the best-off are a *ruling class* in an exploitive society. In a system which is in fact thoroughly exploitive, a ruling class can, for centuries, maintain as much willing cooperation as it needs, because ideological institutions serve its interests, while restraining most who do not cooperate and dissuading most who are tempted not to, by employing the coercive apparatus of the state. In such a situation, the rewards of exploitation for the ruling class far outweigh the costs to it of maintaining cooperation in an exploitive society. (Note that most of these costs are not supplied by the ruling class at all, but by workers who supply taxes and, in times of war, their lives.) Thus, if the best-off are sometimes a ruling class, of the sort just described, someone in the original position would foresee that if he turns out to be one of the best-off, his interests may not lie in the realization of Rawls' standard of social justice.

Of course, there is nothing in Marxist social theory to indicate one is *likely* to find it impossible to live up to a commitment to help realize the difference principle. To the contrary, the sort of exploitive ruling class I have described is supposed to be a small minority in any society. But an argument that one is quite unlikely to find commitment to the difference principle intolerable could not be successfully advanced behind the veil of ignorance. For, as Rawls indicates on several occasions, if reasoning from the respective likelihoods of one's occupying various social positions were admitted in the original position, the social ideal chosen would be some version of the principle of average utility (pp. 154, 164–5, 168–9). Thus, if probabilistic reasoning were admitted (and Rawls thinks it should not be), the resultant commitment would be to help maximize the welfare of the average person, or something of the sort.

Let us suppose that if the sort of exploitive ruling class Marx describes has, at times, existed, the reasoning about the strains of commitment that Rawls uses against utilitarianism would also count against a commitment to help realize the difference principle immediately, or, in any case, as soon as possible after the veil of ignorance is lifted. It might still seem possible that the argument from the strains of commitment might persuade one to help maintain some *gradual* course of development toward full realization. Ideal contractualism might thus lead to a view according to which one sometimes has a duty to accept (and, perhaps, to promote) a certain incremental advance toward the realization of the difference principle without having a duty to accept any advance beyond this increment. An example of such a gradualist moral claim, though admittedly an extreme one, is the view that slave holders are obliged to accept emancipation with compensation, but not to accept uncompensated emancipation involving considerable financial sacrifice.

One thing to be said about the gradualist position is that it is repugnant to many people's considered moral judgments. Many would say that if the difference principle defines the requirements of perfect social justice, everyone should willingly accept its immediate realization. Indeed, many would say that everyone who can be helpful in realizing perfect social justice should take an active part in this process, if no great effort is involved.

In any case, an argument from the strains of commitment would not in fact support a gradualist commitment to the difference principle, if exploitive classes such as Marx describes have existed. For on the latter assumption every course of development leading to the difference principle will be more than some social class can willingly accept. Roughly speaking, there will be at least some phase in any course of social development in which any rate of change is either too fast for the best-off or too slow for the worst-off. The gradual realization of Rawls' principle will require a narrowing of the gap between best-off and worst-off, until Rawls' standard is satisfied. In the process, the wealth, power, and status of the best-off will be progressively reduced. There are, presumably, definite limits to the size of the reduction an upper class can be expected to accept willingly in a given generation. They may accept a greater advance toward equality as a concession to force or the prospect of it. (Thus, the prospect of further social turmoil is sometimes said to be the cause of ultimate corporate acceptance of New Deal legislation.) But in a society for which a Marxist analysis holds true, the best-off will not accept an advance toward equality beyond a certain limit as a voluntary expression of their sense of justice. On the other hand, there are definite lower limits to the rate of advance typical members of the worst-off class find acceptable. They may settle for less out of fear, or a misguided perception of social reality, but not out of a feeling that in the actual social setting justice requires no quicker rate of progress. According to the view of class conflict characteristic of, though by no means special to, Marxist social theory, these two ranges of tolerability do not, as a rule, intersect. An advance that is not too much for the best-off class is, characteristically, too little for the worst-off class. Thus Marx (e.g. in the beginning of Part I of the *Manifesto*) characterizes the whole course of social development as a more or less veiled civil war. And a great many people who do not discern such conflict in the most advanced societies would still claim that in some less advanced societies (where the circumstances of justice obtain) there is no course of action that is relevantly tolerable to all classes. Given such assumptions as these, the reasoning from the strains of commitment that Rawls directs against utilitarianism could also be directed against a commitment to accept any course of development leading to the difference principle.

I have tried to show that if Marxist social theory precludes a certain argument for a commitment to the immediate realization of the difference principle, it precludes the use of this argument in support of a gradualist commitment, as well. While this discussion of gradualism as a punitive 'way out' has been tied to reasoning from the strains of commitment, analogous considerations can readily be developed in connection with arguments we shall subsequently examine, which concern rationality, self-respect, and stability. In considering these arguments, rather than repeating my discussion of gradualism, I shall, on the whole, assume that gradualism does not remove conflicts between Rawls and Marxist social theory.

One further objection to my arguments about tolerability concerns supposed strains of pursuing something less than justice. I have assumed that members of an exploitive ruling class would oppose the changes that the difference principle requires, even if they had a sense of justice defined by Rawls' principles. It might be felt that this would produce a considerable burden of either self-hatred or self-deception, a burden so great that it would not, in fact, be in one's interests to resist the consequences of the difference principle, if on leaving the original position one turns out to be a member of an exploitive ruling class. The answer to this, I think, is that while a ruling class does, in a sense, practice considerable self-deception, this self-deception is of a peculiarly unburdensome kind.

In the following sense, the capacity of self-deception of an exploitive ruling class would be said by Marx to be practically infinite: The long-term nonmoral interests of a typical member of such a class often sharply conflict with the moral principles which he puts forward without conscious hypocrisy; and when this conflict obtains, no reasoning from those moral principles can, in a typical case, dissuade an exploiter from doing what his non-moral interests demand. Thus, suppose a decision to speed up the pace of work without installing safety equipment will keep profits up at the cost of hundreds of workers' lives. According to Marx, a typical capitalist will make the decision that serves the needs of profit, in spite of an appeal to common principles of justice, even if that appeal is backed by the best arguments in the world. It should also be noted that the emotional strain associated with the instances of self-deception most of us have encountered will typically be lacking in such cases. In Marx's view, the unresponsiveness of members of an exploitive ruling class to arguments conflicting with their class interests is supported by falsehoods, e.g., 'What's good for Business (or—what preserves feudal bonds) is good for everyone', which the immense variety of ideological institutions operating in the interests of the ruling class promote, and which everyone in the social circle of a typical member of that class acknowledges. Indeed, even if a revolution were to destroy the basic institutions that contribute to these falsehoods, the survival of bourgeois ideologues, and of bourgeois ideas among many people, including working people, together with the survival of exclusive networks of social acquaintance would continue to insulate a member of the bourgeoisie from the truths on which the correct application of principles of justice depends. Thus, if one wants to speak of self-deception in situations like the one described (where the self gets so much help in its deception), it is not the sort of self-deception that imposes such strains as to make it preferable for a member of an exploitive ruling class to accept the objective fulfillment of his sense of justice.

I have sketched some reasons for rejecting the claim that the realization of Rawls' standard of social justice would be tolerable to a typical member of every income group in every society in the circumstances of justice. In particular I tried to show how a Rawlsian argument from a common interest in social cooperation conflicts with the conception of the best-off group in society as a ruling class. There are, however, a number of derivations of the difference principle that Rawls presents which do not directly rely on the idea that this standard is tolerable from the standpoint of every social position. I shall now consider these arguments in turn, and try to establish that they would not support a commitment to uphold the difference principle in less than just a circumstances, if Marxist social theory is true.

In what is probably the most celebrated derivation of the difference principle, Rawls argues for it on the basis of an assessment of the standard of rational choice appropriate to the original position. According to this argument, an agreement to maximize the life prospects of the worst-off is the outcome of a conservative attitude toward risk-taking that the circumstances of the original position demand. It represents a rational effort to hedge one's bet in a situation which displays the following three features:

(1) The factual basis for an estimate of one's probable prospects under a given social contract is extremely slender.
(2) '. . . the person choosing has a conception of the good such that he cares very little, if anything, for what he might gain above the minimum stipend that he can, in fact, be sure of by following the maximin rule. It is not worthwhile for him to take a chance for the sake of a further advantage, especially when it may turn out that he loses much that is important to him' (p. 154).
(3) The most unlucky choices of a social contract would have outcomes 'one can hardly accept' (p. 154).

Rawls' discussion of rational choice in the original position is often criticized on the grounds that it arbitrarily takes a single, conservative attitude toward risk to be rational, when, in fact, more daring attitudes would be equally rational. I shall suppose, to the contrary, that it is not at all arbitrary of Rawls to favor a conservative attitude toward risk, given the three motivational assumptions he makes. I shall indicate why the second assumption, if applied to typical occupants of social positions in societies violating the difference principle, might be rejected as underestimating the importance, for some of these representative individuals, of wealth and power such as the difference principle excludes. At the same time, I shall try to show that the failure of Rawls' second assumption could preclude accepting a commitment to help achieve the difference principle as a uniquely rational commitment, in the original position.

Rawls' second assumption amounts to the claim that in the original position one does not care enough about the difference between standing at the top of the heap in a relatively non-egalitarian society, and enjoying the minimum standard of living in a society satisfying the difference principle, to choose a less egalitarian principle, with the added risks of suffering that choice entails. On the face of it, some such assumption concerning what people regard as important seems crucial to a commitment to accept the less risky, more egalitarian social rule. After all, because of the importance success would have for him, most of us feel Orville Wright was rational in making the risky choice and taking off at Kitty Hawk, even though the basis for estimating the prospects of success (or survival) was extremely slender, and the maximal loss was death. In general, any degree of risk can be reasonable, if enough importance is attached to the rewards of winning.

On a Marxist view it would not be uniquely rational in the original position to commit oneself to help realize the difference principle. For one has real concern over the elimination of social positions of great wealth and power. Knowing what the needs and desires associated with various social positions are, but not knowing what society or what social position is his own, someone in the original position cannot, it would be claimed, 'care very little . . . for what he might gain above the . . . stipend he can . . . be sure of by following the maximin rule'. For he must take into account the possibility that he is in fact a member of the best-off group in a highly nonegalitarian society. And in terms of the needs and desires characteristic of such groups it matters a great deal whether one has wealth, power, and status, or, on the other hand, one enjoys the comfortable but moderate standard of living that, at least on a Marxist view, would be found at all levels of a society satisfying the difference principle. Given the existence of typical upper-class needs for wealth, power, and status, it might be reasonable to choose a less egalitarian contract than one embodying the difference principle, even in light of the risks involved. To put it roughly, someone in the original position might argue: 'Perhaps I have a Rockefeller's need or a feudal baron's. In that case, life will be hell for me at any position in a difference-principle society, or as a worker or a serf in a non-difference-principle society. Given the possibility that I have luxurious needs,

I should not exclude the latter sort of society, where, if my needs turn out to be luxurious, I at least have a chance of enjoying the only sort of success I will find it worthwhile to have.'

In at least one respect, the hypothesis of special upper-class needs requires more explanation at this point. For if it is understood as a claim about mere felt needs, it cannot serve as an obstacle to Rawls' rationality argument. The difference principle might still emerge from the original position if a typical member of certain upper classes was strongly disinclined to give up his special status, felt (while he enjoyed it) that he could not do without it, strongly resisted losing special wealth and power, and was initially quite miserable after his class was deposed. For the rebellion or despondency of a representative member of a deposed upper class might be a passing phase; he might grow out of his old felt needs and become more or less satisfied with what the difference principle allots him. As a consequence, someone in the original position might discount the possible costs to himself of realizing the difference principle on the grounds that these costs will be transitory, in contrast to the permanent costs to the worst-off of violating the difference principle.

In speaking of the specially luxurious needs of at least some upper classes, I have meant relatively permanent and deep-seated needs, not mere felt needs. In particular, the following must be true of at least some best-off classes if Rawls' argument from rationality is to be blocked. If the difference principle is realized, the resulting unhappiness measured *over the course of the lifetime* of a typical member of the deposed best-off class must be such as to make some less egalitarian social principle at least as attractive from the standpoint of the original position.

That class differences in psychology are as permanent as my argument requires is supposed at a number of points in Marx's sociology. The most famous outcome of this view of class difference is Marx's theory of the dictatorship of the proletariat, a theory which makes no sense (taken as a whole) if deposed members of the old ruling class will grow out of their special needs in a short space of time. The history of upper-class responses to social change makes it plausible that there have been at least some classes with permanent and deep-seated luxurious needs. For a number of deposed upper classes (e.g., the landed aristocracy after the French Revolution) responded to reductions in status much smaller than the difference principle requires with violent and risky counter-activity, for a generation or more.

A need for great wealth, power, and status of the sort we are considering is not to be found throughout the population of any society. And it might be suggested that this consideration is all Rawls requires. Given the possibility that one would not find the best-off situations in a society violating the difference principle much more rewarding than the worst-off situations in a society satisfying that standard, it might be claimed that one should choose Rawls' principle in light of the risk involved in other arrangements. But in fact, it does not seem that this extremely conservative choice would be the only reasonable one. There are a number of other reasonable choices which give at least a certain amount of weight to the possibility that success excluded by the difference principle would be especially rewarding. (Note that rational gamblers like Orville Wright act as if their success will certainly matter even when they are not certain it will. When he took off, Orville may have been aware that other experimenters might have achieved self-sustained flight several hours before). In particular, if one is worried about admitting social systems in which one may turn out to be absolutely miserable, one can deal with this possibility by measures much less extreme than adopting the situation of the worst-off as one's basic standard for judging societies. For example, one might hedge one's bets by defining a certain social minimum, say one in which everybody who will work forty hours a week can lead a life free from acute hunger, sheltered from the elements, and safe from acute and constant dangers to his life. One might specify that any society must meet this minimum, or approach as close as possible to it, but that once that minimum is realized, the social arrangement is to be favored in which the happiness of the average person is greatest. Or one might hedge the principle of average utility not just with a social minimum, but with the requirement that the situation of the average person be no better than the situation of the median person. Supposing that one's attitude toward risks represents a reasonable mixture of the conservative and the daring, it would seem that either of these standards might be preferable to the difference principle. Yet each of them allows for extremely exploitive societies by the standard of the difference principle.[14]

[14] Or, in any case, their admitting such societies is a relevant possibility if Rawls is right in supposing that utilitarian principles are less egalitarian than his own, in ways that are relevant to the original position (cf. David Lyons, 'Rawls *versus* Utilitarianism', *Journal of Philosophy*, LXIX, 18 [1972]). As

Since hedged utilitarianism is not one of the conceptions of social justice Rawls considers, it is worth noting that this alternative is not at all artificial. A great many people feel that where justice is concerned the interests of the worst-off should be given special weight only in so far as a certain low social minimum ought to be guaranteed; thereafter, everyone's interest should be given equal weight, with no special weight attached to the welfare of the worst-off. A hedged utilitarianism is a plausible expression of this viewpoint. If a commitment to uphold such a principle is as attractive, in the original position, as a commitment to uphold the difference principle, then, given Rawls' assumption of a special obligation to uphold the latter standard, there is a live issue of social morality that ideal contractualism cannot resolve.

So far, I have left out one assumption Rawls makes in response to the claim that relatively daring attitudes toward risk might exist in the original position, attitudes which favor admitting increased life prospects for the best-off, even at the expense of the worst-off. Rawls seeks to justify a conservative attitude toward risk taking by supposing that people in the original position care about their immediate descendants, and want to be able to justify their choices to them. He points out that it is especially hard to defend a choice of principles of justice to one's children if it has a disastrous outcome for them.

It seems to me that Rawls' additional assumption does not in fact justify his giving unique weight to conservative attitudes toward risks. No doubt, if a person chooses, in the original position, a principle of justice that admits slave societies, and turns out himself to be a slave, his children will be bitter about his choice (especially if they are inclined to conservative gambles). But suppose he chooses a principle that excludes extremes of wealth and power, and turns out himself to have been at the top of an extreme and inequitable hierarchy. Again, his children may be bitter at his choice, since they have been deprived of the advantages they would have enjoyed, and cut off from the opportunity to pursue such advantages. In particular, the choice will appear indefensible to them if they are inclined to accept great risks for

far as the conflict between Rawls and utilitarianism is concerned, the outcome of my arguments will be that on the basis of certain social hypotheses the difference principle is not superior to a hedged utilitarianism such as the above, from the standpoint of the original position, even if the latter outcome does, in relevant ways, impose weaker constraints on inequality. I shall return to this topic at the end of this paper.

possible great rewards. In sum, since a person does not know whether his choice will deprive his children of great advantages, what needs his children have, or what their attitudes toward risk are, the desire to make choices he can defend to them does not entail a more conservative attitude toward risk on his own part. 'I wanted to give you a chance to enjoy the way of life you might require, on account of your needs', appears, in the original position, to be a reasonable consideration to advance in defending a choice to one's descendants. Certainly, the need to justify a choice to one's descendants does not increase the weight of extremely conservative choices sufficiently to exclude the sort of alternatives to the difference principle described above, e.g., utilitarianism hedged with a low, but livable social minimum.

If a commitment to help in the immediate realization of the difference principle would be no more rational than certain alternatives in the original position, might a commitment to the gradual realization of the difference principle be uniquely favored? In light of our previous arguments and given our assumption about the needs of the best-off, a gradualist commitment does not seem to be uniquely rational. A proposed commitment to improve the life prospects of the worst-off until they are maximized would be competing with a proposed commitment to improve their life prospects until some social minimum is reached, and then, let us say, to help maximize average utility. Assuming that one may turn out to have luxurious needs, and, hence, to care a great deal about nonegalitarian success, the choice of the latter social commitment, in the original position, would be no less rational than the choice of the former.

As we have seen, a general commitment to uphold the difference principle would emerge only if no 'representative individual' has sufficient need for advantages beyond what the difference principle guarantees to justify one in committing oneself to a less egalitarian arrangement, behind the veil of ignorance. I would like to conclude this section by indicating what is right and what is wrong about this Rawlsian assumption of moderate needs, from the Marxist standpoint we are examining. There is, to begin with, much truth to the requisite assumption, from a Marxist point of view, in that most people have the attitudes Rawls requires. If they knew their own needs, but were otherwise operating behind the veil of ignorance, then, seeing the existence guaranteed in a society governed by the difference principle, they would not take great risks to leave open the highest possibilities of success in a highly nonegalitarian society. This choice would result in part from a relatively small need for great wealth, power, and status, in part from a disinclination toward the sorts of existence characteristically associated with the highest positions in a society violating the difference principle. Thus, most people, on a Marxist view, would not care enough about the luxuries of a large slave holder to make the riskier choice, both because they do not care that much about luxuries, and because they would not want to be surrounded by those who hate them, or who would if they knew their role. Similarly, someone with the needs and desires typical of mankind would not care enough about the material standards of a top corporate executive, at least after perceiving their association with such corporate phenomena as speeding workers up and laying them off, making goods seem worth more than they are, and trying to defeat other executives in corporate politics.

If the psychological assumption Rawls requires is, in fact, true of most people, only the following premise is needed to apply it to the original position: Morally relevant facts of psychology that are true of most people are true of typical representatives of all major income groups. In other words, the requisite assumption about attitudes toward highly nonegalitarian success in the original position depends, in effect, on the idea that morally important aspects of human nature are not too sharply differentiated along class lines. I have argued that on a Marxist view such a split in human nature does obtain, in at least the following respect: The existence of a special need for wealth, power, and status typical of an upper class, or, more precisely, a dominant exploitive class, might lead one to commit oneself to a weaker constraint on inequality than Rawls' in a situation like that of the original position, in which one does not know one's actual place in society. Thus, the outcome of the original position (indeed, the very existence of some outcome) would be crucially affected by a split between what most people are like and what people in a certain income group are like.

I should now like to consider, very briefly, two other arguments suggested by remarks of Rawls', for commitment to the difference principle. For one thing, Rawls claims that his standard of social justice is especially stable in that when a system is known to realize it, people will come to commit themselves to upholding this system (p. 177). Since the best-off would, of course, prefer stability, this looks like a reason for a commitment to further

Rawls' principles, even though it may require one to give up certain advantages.

Obviously, any special stability associated with Rawls' favored social rules can, in principle, be outweighed by losses in other dimensions. Stability is just one consideration to be taken into account in the rational, self-interested choice of a social contract; otherwise, universal sedation would be an attractive choice. In particular, assuming that one might turn out to be a member of an exploitive ruling class, one must consider to what extent gains from increased stability following the realization of the difference principles would compensate for possible losses in one's status, compared with the best-off status that other, less egalitarian principles might allow. Suppose, for example, that the principle of average utility were chosen instead of the difference principle and that certain societies, when transformed to accord with the former principle, are significantly less egalitarian than those governed by the difference principle. A society of this relatively nonegalitarian kind might last for many consecutive generations. After all, extremely nonegalitarian social arrangements have lasted for hundreds, even thousands of years. In such a non-Rawlsian society, educational institutions (taking the phrase in its broadest sense) would teach everyone, in effect, that more egalitarian arrangements such as those dictated by the difference principle, are 'more than justice requires'. And the state, through its police function (again understood in the broadest sense), would defend the institutions giving rise to a non-Rawlsian distribution. The costs of instability—and the alternative costs of coercion—in such a society may be greater than those in a society governed by the difference principle. But there is no reason to suppose that the relatively high status of the best-off would not more than compensate for the relative costs to the best-off of instability. From the standpoint of a best-off class with specially luxurious needs the non-Rawlsian society could be vastly preferable. Thus, given the hypothesis of special upper-class needs (and perhaps even without it), considerations of stability would not give crucial weight to a commitment to realize the difference principle, if the interests of every class in the circumstances of justice are to be taken into account.

There is one more argument that Rawls presents for the difference principle that does not explicitly appeal to the idea that the burdens of upholding this principle are uniquely tolerable. Rawls argues that by committing themselves to exclude inequalities that are not to everyone's advantage, the bargainers in the original position agree to a social arrangement in which everyone can have self-respect (pp. 178–9). If this argument were used to support a commitment to uphold the difference principle throughout the circumstances of justice, I would propose that the bases of self-respect may be relevantly different from class to class. In particular, on the view of class conflict we are examining, most members of some exploitive classes have a need for great wealth and power, and for much more of them than most other people, a need which typical members of exploited classes do not have to anything like the same degree. This need must typically be satisfied, if a member of the exploitive class is to respect himself. Indeed, it seems plausible that many a feudal lord, a late nineteenth-century tycoon, or even a member of some contemporary ruling class would rather kill himself than accept the loss of his special position in society. Thus, in a society conforming to the difference principle, in which, no doubt, the self-esteem of the majority would be increased, the self-esteem of former exploiters might be destroyed. More generally, on the Marxist view under consideration, the needs for self-respect of the typical members of the different classes conflict in a way that Rawls' principle cannot resolve.

If the above arguments are correct, it seems highly unlikely that a commitment to accept advances toward the realization of the difference principle would emerge from an original position in which the following propositions are known to hold true in some societies:

(1) The reduction of inequalities requires class struggle. Any overall social arrangement that the best-off class can accept has unacceptable consequences for most people affected by it. This hypothesis was seen to be incompatible with arguments concerning the strains of commitment, in support of a commitment to uphold the difference principle.
(2) The best-off class in society is a ruling class. Political and ideological institutions work exclusively or almost exclusively in its interests. The argument that social cooperation requires realizing the difference principle depended on the rejection of the ruling class hypothesis.
(3) Typical members of the ruling class have an especially great need for wealth, power, and status. This hypothesis was seen to conflict with arguments for commitment to Rawls' principle

as uniquely rational in the original position. The corresponding hypothesis about class differences in the sources of self-respect was in conflict with the argument that one should commit oneself to upholding the difference principle because it is a universally effective basis for self-respect.

Very roughly, then, the derivation of a commitment to accept and further the difference principle throughout the circumstances of justice would depend, for its validity, on the following view of what societies in those circumstances are like: class conflict is moderate; there is no ruling class; the basic desires characteristic of one class are not drastically different from those characteristic of another.

Our discussion of Rawls' arguments from the point of view of Marxist social theory naturally gives rise to the following question: 'Given that a commitment to uphold the difference principle does not emerge from the original position if the Marxist hypotheses we have considered are correct, does any agreement concerning the reduction of economic inequality emerge?' The answer, I think, is 'no'. The specific form of this negative result depends on whether one accepts or rejects Rawls' exclusion of statistical arguments from the deliberations in the original position. If estimates of probabilities are excluded, then no honest commitment concerning the promotion of social justice is possible. If probabilistic reasoning is admitted, then too many agreements are equally acceptable, and the outcome of the original position is indeterminate.

First of all, if we follow Rawls in excluding probabilistic reasoning, no one in the original position will be able to make the required honest commitment to a social contract, since one cannot be sufficiently certain of being able to accept the social changes the contract requires. If, when the veil of ignorance is lifted, one turns out to be a typical member of an exploitive ruling class, one may be unable to accept the consequences of the difference principle or of any constraint or inequality of similar strength. But, as Marx and Rawls would agree, any weaker constraint on inequality would be intolerable to the working class. Thus it would seem that no commitment concerning the reduction of economic inequality can emerge from the original position.

The situation may be somewhat different if probabilistic reasoning is admitted. Perhaps in that case one could honestly agree to support the difference principle, on the grounds that one is very unlikely to be a member of an exploitive ruling class, unable to accept its consequences. I say 'perhaps', because even here the possibility of honest commitment is not obvious. After all, one has reason to believe one may actually be a person who is incapable of accepting what the difference principle requires. It is as if, at a party, one knows that LSD has been dropped in someone's drink, and one is asked to run an errand within the next hour. In such a case, it is by no means obvious one can honestly make the relevant promise.

Still, suppose we grant for the sake of argument that an honest commitment to the difference principle might be based on its probable tolerability. At this point, the following dilemma arises: commitment to the difference principle is but one of several possible outcomes of the search for a commitment to social justice in the original position, and probabilistic considerations (such as must, in all consistency, now be admitted) do not decisively favor one outcome over the others. In addition to the difference principle, it can be foreseen that utilitarian standards are very likely to be found tolerable, as are utilitarian standards with some requirement of a low social minimum added on. But, especially as between the latter, mixed standards and the difference principle, there seem to be no decisive arguments either way. One might adopt Rawls' principle, as we saw, on the grounds that it makes the most sense in terms of the needs and desires one is likely to have. On the other hand, one might adopt a hedged utilitarian principle on the grounds that not knowing what one's actual desires are, one wants to maximize the extent to which they will probably be fulfilled, whatever they turn out to be, while excluding outcomes that are disastrous from all points of view, and uniquely disastrous from most. Similarly, each conception has its drawbacks. In the Rawlsian solution, one completely dismisses needs one may actually have. In the hedged utilitarian solution, one assumes a greater risk of nonegalitarian outcomes that represent irrational bets from the standpoint of the needs and desires one probably has.[15] In sum, even if one accepts

[15] In any case, Rawls would characterize the more utilitarian contract in this way. This characterization seems reasonable at least on the following basis: more or stronger factual assumptions would be required to preclude the highly nonegalitarian outcomes in question as consequences of the hedged utilitarian contract, than would be required for Rawls' principles of justice. The participants in the original position are plausibly assumed to be reluctant to bet on the truth of social facts (which they do, in fact, know) where this is avoidable. Thus, they regard the risk of a highly nonegali-

the need to hedge one's bets in the original position, it becomes quite indeterminate what hedging one's bets requires, once one sees that the availability of highly nonegalitarian success has a different importance with respect to sets of needs and desires typical of different income groups.[16]

10 Rawls on Liberty and its Priority

H. L. A. HART*

1. *Introductory*

No book of political philosophy since I read the great classics of the subject has stirred my thoughts as deeply as John Rawls' *A Theory of Justice*. But I shall not in this article offer a general assessment of this important and most interesting work. I shall be concerned with only one of its themes, namely, Rawls' account of the relationship between justice and liberty, and in particular with his conception that justice requires that liberty may only be limited for the sake of liberty and not for the sake of other social and economic advantages. I have chosen this theme partly because of its obvious importance to lawyers who are, as it were, professionally concerned with limitations of liberty and with the justice or injustice of such limitations. I choose this theme also because this part of Rawls' book has not, I think, so far received in any of the vast number of articles on and reviews of the book which have been published, the detailed attention which it deserves. Yet, as Sidgwick found when he considered a somewhat similar doctrine ascribing priority to liberty over other values, such a conception of liberty, though undoubtedly striking a responsive chord in the heart of any liberal, has its baffling as well as its attractive aspect,[1] which becomes apparent when we consider,

tarian outcome as relevantly greater for the more utilitarian contract (cf. p. 160).

[16] I am indebted to Richard Boyd, Gerald Dworkin, David Lyons, and Hilary Putnam for suggestions and criticisms that helped me in writing this essay.

* Research Fellow of University College, Oxford.

[1] H. Sidgwick, *The Methods of Ethics* (7th ed. 1907) Book III, Ch. V, § 4. 'I admit that it commends itself much to my mind. . . . But when I endeavour

as Rawls intends that we should, what the application of this doctrine would require in practice.

Part of what follows is concerned with a major question of interpretation of Rawls' doctrine and the rest is critical. But I am very conscious that I may have failed to keep constantly in view or in proper perspective all the arguments which Rawls, at different places in this long and complex work, concentrates on the points which I find unconvincing. I would not therefore be surprised if my interpretation could be corrected and my criticisms answered by some further explanation which the author could supply. Indeed I do not write to confute, but mainly in the hope that in some of the innumerable future editions of this book Rawls may be induced to add explanation of these points.

I hope that I can assume that by now the main features of Rawls' *A Theory of Justice* are familiar to most readers, but for those to whom it is not, the following is a minimum account required to make this article intelligible.

First, there is what Rawls terms the 'Main Idea'. This is the striking claim that principles of justice do not rest on mere intuition yet are not to be derived from utilitarian principles or any other teleological theory holding that there is some form of good to be sought and maximized. Instead, the principles of justice are to be conceived as those that free and rational persons concerned to further their own interests would agree should govern their forms of social life and institutions if they had to choose such principles from behind a 'veil of ignorance'—that is, in ignorance of their own abilities, of their psychological propensities and conception of the good, and of their status and position in society and the level of development of the society of which they are to be members. The position of these choosing parties is called 'the original position'. Many discussions of the validity of this Main Idea have already appeared and it will continue to be much debated by philosophers, but for the purposes of this article I shall assume that if it could be shown that the parties in the original position would choose the principles which Rawls identifies as principles of justice, that would be a strong argument in their favour. From the Main Idea Rawls makes a transition to a general form or 'general conception' of the principles that the parties in the original position would choose. This

to bring it into closer relation to the actual circumstances of human society it soon comes to wear a different aspect'.

general conception of justice is as follows:

> All social values—liberty and opportunity, income and wealth, and the bases of self-respect—are to be distributed equally unless an unequal distribution of any, or all, of these values is to everyone's advantage [p. 62].

This general conception of justice, it should be observed, refers to the equal distribution of liberty but not to its maximization or extent. However, most of the book is concerned with a special interpretation of this general conception which refers both to the maximization and the equality of liberty. The principal features of this special conception of justice are as follows:

> *First Principle* ['the principle of greatest equal liberty' (p. 124)]
> Each person is to have an equal right to the most extensive total system of equal basic liberties compatible with a similar system of liberty for all.
>
> *Second Principle*
> Social and economic inequalities are to be arranged so that they are . . . to the greatest benefit of the least advantaged . . . [p. 302].[2]

To these two principles there are attached certain priority rules of which the most important is that liberty is given a priority over all other advantages, so that it may be restricted or unequally distributed only for the sake of liberty and not for any other form of social or economic advantage.

To this account there must be added two points specially relevant to this article. First, Rawls regards his two principles as established or justified not simply by the fact that they would be chosen, as he claims they would, by the parties in the original position, but also by the general harmony of these principles with ordinary 'considered judgments duly pruned and adjusted' (p. 20).[3] The test of his theory, therefore, is in part whether the principles he identifies illuminate our ordinary judgments and help to reveal a basic structure and coherence underlying them.

[2] I have here omitted the provisions for a just savings principle and for equality of opportunity, which Rawls includes in this formulation of his second principle, since they are not relevant to the present discussion.

[3] Rawls, in fact, speaks of a 'reflective equilibrium' between principles and ordinary judgments, since he envisages that where there are initial discrepancies between these we have a choice of modifying the conditions of the initial position in which principles are chosen or modifying in detail the judgments (pp. 20 ff.).

Secondly, it is an important and interesting feature of Rawls' theory that once the principles of justice have been chosen we come to understand what their implementation would require by imagining a four-stage process. Thus, we are to suppose that after the first stage, when the parties in the original position have chosen the principles of justice, they move to a constitutional convention. There, in accordance with the chosen principles, they choose a constitution and establish the basic rights or liberties of citizens. The third stage is that of legislation, where the justice of laws and policies is considered; enacted statutes, if they are to be just, must satisfy both the limits laid down in the constitution and the originally chosen principles of justice. The fourth and last stage is that of the application of rules by judges and other officials to particular cases.

11. *Liberty and Basic Liberties*

Throughout his book Rawls emphasizes the distinction between liberty and other social goods, and his principle of greatest equal liberty is, as I have said, accompanied—in his special conception of justice as distinct from his general conception—by a priority rule which ssigns to liberty, or at least to certain forms of liberty institutionally defined and protected, a priority which forbids the restriction of liberty for the sake of other benefits: liberty is only to be restricted for the sake of liberty itself. In the general conception of justice there is no such priority rule and no requirement that liberty must be as extensive as possible, though it is to be equally distributed unless an unequal distribution of it is justified as being to everyone's advantage (p. 62). The special conception is to govern societies which have been developed to the point when, as Rawls says, 'the basic wants of individuals can be fulfilled' (p. 543) and social conditions allow 'the effective establishment of fundamental rights' (pp. 152, 542). If these favourable conditions do not obtain, equal liberty may be denied, if this is required to 'raise the level of civilization so that in due course these freedoms can be enjoyed' (p. 152).

I find it no easy matter, on some quite crucial points, to interpret Rawls' complex doctrine, and there is one initial question of interpretation which I discuss here at some length. But it is perhaps worth saying that to do justice to Rawls' principle of greatest equal liberty it is necessary to take into account not only what he says when expressly formulating, expounding, and illustrating this principle, but also what he says about some other apparently separate issues—in particular, natural duties (pp. 114 ff., 333 ff.), obligations arising from the principle of fairness (pp. 108 ff.), permissions (pp. 116 ff.), and the common good or common interest (pp. 97, 213, 246), for these may apparently supplement the rather exiguous provision for restrictions on liberty which are all that, at first sight, his principle of greatest equal liberty seems to allow.

The initial question of interpretation arises from the following circumstances. Rawls in his book often refers in broad terms to his first principle of justice as 'the principle of greatest equal liberty' (e.g., p. 124), and in similarly broad terms to its associated priority rule as the rule that 'liberty can be restricted only for the sake of liberty' (pp. 250, 302). These references to liberty in quite general terms, and also Rawls' previous formulation in his articles of this first principle as the principle that everyone has 'an equal right to the most extensive liberty compatible with a like liberty for all',[4] suggest that his doctrine is similar to that criticized by Sidgwick.[5] It is probable that Sidgwick had chiefly in mind a formulation of a principle of greatest equal liberty urged by Herbert Spencer in his long forgotten *Social Statistics*.[6] This was effectively criticized by Sidgwick as failing to account for some of the most obvious restrictions on liberty required to protect individuals from harms other than constraint or deprivation of liberty, and indeed as forbidding the institution of private property, since to own anything privately is to have liberty to use it in ways denied to others. Spencer attempted to get out of this difficulty (or rather outside it) by simply swallowing it, and reached the conclusion that, at least in the case of land, only

[4] Rawls, 'Justice as Fairness', *Philosophical Review* (1958), Vol 67, pp. 164, 165; *see* Rawls, 'The Sense of Justice', *Philosophical Review* (1963), Vol. 72; J. Rawls, 'Distributive Justice', in *Politics, Philosophy and Society* (3rd Series, Oxford, 1967). This formulation in these articles should not be confused with the formulation of the 'general conception' of justice in the book. See pp. 3 ff.

[5] H. Sidgwick, *supra* note 1, Book III, Ch. V., §§ 4–5 and Ch. XI, § 5.

[6] *See* H. Spencer, *Social Statistics* (1850). Criticisms of Spencer's theory in terms very similar to Sidgwick's criticisms were made by F. W. Maitland in 1 *Collected Papers* 247 (H. Fisher ed. 1911). Maitland treated Spencer's doctrine of equal liberty as virtually identical with Kant's notion of mutual freedom under universal law expounded in the latter's *Rechtslehre*. I am grateful to Professor B. J. Diggs for pointing out to me important differences between Rawls' doctrine of liberty and Kant's conception of mutual freedom under universal law.

property held in common by a community would be consistent with 'equal liberty'[7] and hence legitimate. Rawls in his book simply lists without argument the right to hold personal property, but not property in the means of production, as one of the basic liberties (p. 61), though, as I shall argue later, he does this at some cost to the coherence of his theory.

Rawls' previous formulation of his general principle of greatest equal liberty—'everyone has an equal right to the most extensive liberty compatible with the like liberty for all'—was then very similar to the doctrine criticized by Sidgwick. But Rawls' explicit formulation of it in his book is no longer in these general terms. It refers not to 'liberty' but to basic or fundamental *liberties*, which are understood to be legally recognized and protected from interference. This, with its priority rule, as finally formulated, now runs as follows:

> Each person is to have an equal right to the most extensive total system of equal basic liberties compatible with a similar system of liberty for all. . . .
> [L]iberty can be restricted only for the sake of liberty. There are two cases: (a) a less extensive liberty must strengthen the total system of liberty shared by all; (b) a less than equal liberty must be acceptable to those with the lesser liberty [p. 302].

Even to this, however, for complete accuracy a gloss on the last sentence is needed because Rawls also insists that 'acceptable to those with the lesser liberty' means not acceptable just on any grounds, but only acceptable because affording a greater protection of their other liberties (p. 233).

The basic liberties to which Rawls' principle thus refers are identified by the parties in the original position[8] from behind the veil of ignorance as essential for the pursuit of their ends, whatever those ends turn out to be, and so as determining the form of their society. Not surprisingly, therefore, the basic liberties are rather few in number and Rawls gives a short list of them which he describes in the index as an 'enumeration' (p. 540), though he warns us that these are what they are only 'roughly speaking' (p. 61). They comprise political liberty, that is, the right to vote and be eligible for public office, freedom of speech and of

[7] H. Spencer, *supra* note 6.

[8] E.g., 'equal liberty of conscience is the only principle that parties in the original position can acknowledge' (p. 207).

assembly; liberty of conscience and freedom of thought; freedom of the person, along with the right to hold personal property; and freedom from arbitrary arrest and seizure.

Now the question of interpretation is whether Rawls' change of language from a principle of greatest equal liberty couched in quite general terms ('everyone has an equal right to the most extensive *liberty*'), to one referring only to specific basic *liberties*, indicates a change in his theory. Is the principle of liberty in the book still this quite general principle so that under the priority rule now attached to it no form of liberty may be restricted except for the sake of liberty? It is difficult to be sure but my own view on this important point is that Rawls no longer holds the quite general theory which appeared in his articles perhaps because he had met the difficulties pointed out by Sidgwick and others. There are, I think, several indications, besides the striking change in language, that Rawls' principle is now limited to the list of basic liberties allowing of course for his statement that the actual list he gives is only rough. The first indication is the fact that Rawls does not find it necessary to reconcile the admission of private property as a liberty with any general principle of *maximum* equal liberty, or of 'an equal right to the most extensive liberty', and he avoids the difficulties found in Herbert Spencer's doctrine by giving a new sense to the requirement that the right to hold property must be equal. This sense of equality turns on Rawls' distinction between liberty and the value or worth of liberty (pp. 204, 225 ff.). Rawls does not require, except in the case of the *political liberties* (the right to participate in government and freedom of speech), that basic liberties be equal in value, or substantially equal, so he does not require, in admitting the right to property as a basic equal liberty, either that property should be held in common so that everyone can enjoy the same property or that separately owned property should be equal in amount. That would be to insist that the value of the right to property should be equal. What is required is the merely formal condition that the *rules* (pp. 63–4) governing the acquisition, disposition, and scope of property rights should be the same for all. Rawls' reply to the familiar Marxist criticism that in this case we shall have to say that the beggar and the millionaire have equal property rights would be to admit the charge, but to point out that, in his system, the unequal value of these equal property rights would be cut down to the point where inequality would be justified by the working of the difference principle, according to which economic inequalities are justified only if they are for the benefit of the least advantaged (p. 204).

The second indication that Rawls' principle of greatest equal liberty and its priority rule 'liberty can be restricted only for the sake of liberty' (p. 302) is now limited to the basic liberties is his careful and repeated explanation that, though the right to hold property is for him a 'liberty', the choice between private capitalism and state ownership of the means of production is left quite open by the principles of justice (pp. 66, 273–4). Whether or not the means of production are to be privately owned is something which a society must decide in the light of the knowledge of its actual circumstances and the demands of social and economic efficiency. But, of course, a decision to limit private ownership to consumer goods made on such grounds would result in a less extensive form of liberty than would obtain if private ownership could be exercised over all forms of property. Rawls' admission of this restriction as allowable so far as justice is concerned would be a glaring inconsistency if he was still advancing the general principle that there must be 'an equal right to the most extensive liberty', for that, under the priority rule, would entail that *no form* of liberty may be narrowed or limited for the sake of economic benefits, but only for the sake of liberty itself.

These considerations support very strongly the interpretation that Rawls' principle of greatest equal liberty, as it is developed in this book, is concerned only with the enumerated basic liberties, though of course these are specified by him only in broad terms. But I confess that there are also difficulties in this interpretation which suggest that Rawls has not eliminated altogether the earlier general doctrine of liberty, even though that earlier doctrine is not, as I have explained above, really consistent with Rawls' treatment of the admissible limitations of the right of property. For it seems obvious that there are important forms of liberty—sexual freedom and the liberty to use alcohol or drugs among them—which apparently do not fall within any of the roughly described basic liberties;[9] yet it would be very

[9] It has been suggested to me that Rawls would regard these freedoms as basic liberties falling under his broad category of liberty of conscience, which is concerned not only with religious but with moral freedom. But Rawls' discussion of this (pp. 205 ff.) seems to envisage only a man's freedom to fulfil moral *obligations* as he interprets them, and sexual freedom would therefore only fall under this category for those to whom the promptings of

surprising if principles of justice were silent about their restriction. Since John Stuart Mills' essay *On Liberty*, such liberties have been the storm centre of discussions of the proper scope of the criminal law and other forms of social coercion, and there is, in fact, just one passage in this book from which it is clear that Rawls thinks that his principles of justice are not silent as to the justice of restricting such liberties (p. 331). For in arguing against the view that certain forms of sexual relationship should be prohibited simply as degrading or shameful, and so as falling short of some 'perfectionist' ideal, Rawls says that we should rely not on such perfectionist criteria but on the principles of justice and that according to these no reasonable case for restriction can be made out.

There is much that I do not understand in this short passage. Rawls says here that justice requires us to show, before restricting such modes of conduct, either that they interfere with the basic liberties of others or that 'they violate some natural duty or some obligation'. This seems an unexplained departure from the strict line so often emphasized in the case of basic liberties, that liberty may be restricted only for the sake of liberty. Is there then a secondary set of principles for nonbasic liberties? This solution would have its own difficulties. The natural duties to which Rawls refers here, and the principle from which obligations, such as the obligation to keep a promise, derive, are, according to Rawls, standards of conduct for *individuals* which the parties in the original position have gone on to choose after they have chosen the principles of justice as standards for *institutions*, which I take it include the law. If liberty may be restricted to prevent violation of any such natural duties or obligations, this may rather severely narrow the area of liberty, for the natural duties include the duty to assist others when this can be done at small cost and the duty to show respect and courtesy, as well as duties to support just institutions, not to harm the innocent, and not to cause unnecessary suffering. Further, since the parties in the original position are said to choose the principles of justice as standards for institutions

passion presented themselves as calls of moral duty. Others have suggested that these freedoms would fall under Rawls' category of freedom of the person; but this seems most unlikely to me in view of his collocation of it with property ('freedom of the person along with the right to hold personal property'). It is to be noted also that sexual freedom is spoken of as a 'mode of conduct' (p. 331) and the possibility of its interference with 'basic liberties' (not '*other*' basic liberties) is mentioned.

before they choose the natural duties for individuals, it is not clear how the former can incorporate the latter, as Rawls suggests they do when he says that principles of justice require us to show, before we restrict conduct, that it violates either basic liberties or natural duties or obligations.

I hope that I have not made too much of what is a mere passing reference by Rawls to liberties which do not appear to fall within his categories of basic liberties, but have been at the centre of some famous discussions of freedom. I cannot, however, from this book see quite how Rawls would resolve the difficulties I have mentioned, and I raise below the related question whether liberties which are plainly 'basic' may also be restricted if their exercise involves violation of natural duties or obligations.

III. *Limiting Liberty for the Sake of Liberty*

I turn now to consider the principle that basic liberties may be limited only for the sake of liberty. Rawls expresses this principle in several different ways. He says that basic liberties may be restricted or unequally distributed only for the sake of a greater 'system of liberty as a whole' (p. 203); that the restriction must yield 'a greater equal liberty' (p. 229) or 'the best total system of equal liberty' (p. 203) or 'strengthen' that system (p. 250) or be 'a gain for . . . freedom on balance' (p. 244).

What, then, is to limit liberty for the sake of liberty? Rawls gives a number of examples which his principle would permit. The simplest case is the introduction of rules of order in debate (p. 203), which restrict the liberty to speak when we please. Without this restriction the liberty to say and advocate what we please would be grossly hampered and made less valuable to us. As Rawls says, such rules are necessary for 'profitable' (p. 203) discussion, and plainly when such rules are introduced a balance is struck and the liberty judged less important or less valuable is subordinated to the other. In this very simple case there seems to be a quite obvious answer to the question as to which of the two liberties here conflicting is more valuable since, whatever ends we are pursuing in debate, the liberty to communicate our thought in speech must contribute more to their advancement than the liberty to interrupt communication. It seems to me, however misleading, to describe even the resolution of the conflicting liberties in this very simple case as yielding a 'greater' or 'stronger' total system of liberty, for these phrases suggest that no values other than liberty and dimensions of it, like extent, size, or strength, are involved. Plainly what such rules of debate help to secure is not a *greater* or more extensive liberty, but a liberty to do something which is more valuable for any rational person than the activities forbidden by the rules, or, as Rawls himself says, something more 'profitable'. So some criterion of the value of different liberties must be involved in the resolution of conflicts between them; yet Rawls speaks as if the system 'of basic liberties' were self-contained, and conflicts within it were adjusted without appeal to any other value besides liberty and its extent.

In some cases, it is true, Rawls' conception of a greater or more extensive liberty resulting from a more satisfactory resolution of conflicts between liberties may have application. One fairly clear example is provided by Rawls when he says that the principle of limiting liberty only for the sake of liberty would allow conscription for military service in a war genuinely undertaken to defend free institutions either at home or abroad (p. 350). In that case it might plausibly be said that only the quantum or extent of liberty was at stake; the temporary restriction of liberty involved in military conscription might be allowed to prevent or remove much greater inroads on liberty. Similarly, the restriction imposed in the name of public order and security, to which Rawls often refers (pp. 97, 212–13), may be justified simply as hindering greater or more extensive hindrances to liberty of action. But there certainly are important cases of conflict between basic liberties where, as in the simple rules of debate case, the resolution of conflict must involve consideration of the relative value of different modes of conduct, and not merely the extent or amount of freedom. One such conflict, which, according to Rawls' four-stage sequence, will have to be settled at a stage analogous to a constitutional convention, is the conflict between freedom of speech and of the person, and freedom to participate in government through a democratically elected legislature (pp. 228–30). Rawls discusses this conflict on the footing that the freedom to participate in government is to be considered as restricted if there is a Bill of Rights protecting the individual's freedom of speech or of the person from regulation by an ordinary majority vote of the legislature. He says that the kind of argument to support such a restriction, which his principles of justice require, is 'a justification which appeals only to a greater equal liberty' (p. 229). He admits that different opinions about the value of the conflicting liberties will affect the way in which different persons view this conflict. Nonetheless, he insists that to arrive at a just resolution of the conflict we must try to find the point at which 'the danger to liberty from the marginal loss in control over those holding political power just balances the security of liberty gained by the greater use of constitutional devices' (p. 230). I cannot myself understand, however, how such weighing or striking of a balance is conceivable if the only appeal is, as Rawls says, to 'a greater liberty'.

These difficulties in the notion of a greater total liberty, or system of liberty, resulting from the just resolution of conflict between liberties, are made more acute for me by Rawls' description of the point of view from which he says all such conflicts between liberties are to be settled—whether they occur at the constitution making stage of the four-stage sequence, as in the case last considered, or at the stage of legislation in relation to other matters.

Rawls says that when liberties conflict the adjustment which is to secure 'the best total system' is to be settled from the standpoint of 'the representative equal citizen', and we are to ask which adjustment 'it would be rational for him to prefer' (p. 204). This, he says, involves the application of the principle of the common interest or common good which selects those conditions which are necessary for 'all to equally further their aims' or which will 'advance shared ends' (p. 97). It is, of course, easy to see that very simple conflicts between liberties, such as the debating rules case, may intelligibly be said to be settled by reference to this point of view. For in such simple cases it is certainly arguable that, whatever ends a man may have, he will see as a rational being that the restrictions are required if he is to pursue his ends successfully, and this can be expressed in terms of 'the common good' on the footing that such restrictions are necessary for all alike. But it would be quite wrong to generalize from this simple case; other conflicts between basic liberties will be such that different resolutions of the conflict will correspond to the interests of different people who will diverge over the relative value they set on the conflicting liberties. In such cases, there will be no resolution which will be uniquely selected by reference to the common good. So, in the constitutional case discussed above, it seems difficult to understand how the conflict can be resolved by reference to the representative equal citizen, and without appeal to utilitarian considerations or to some conception of what all individuals are morally entitled to have as a matter of human dignity or moral right. In particular, the general strategy which Rawls ascribes to the parties in the original position of choosing the alternative that yields the best worst position is no help except in obvious cases like the debating rules case. There, of course, it can be argued that it is better to be restricted by reasonable rules than to be exposed to unregulated interruption, so that it is rational to trade off the liberty to speak when you please for the more valuable benefit of being able to communicate more or less effectively what you please. Or, to put the same exceedingly simple point in the 'maximin' terms which Rawls often illuminatingly uses, the worst position under the rule (being restrained from interruption but given time to speak free from interruption) is better than the wost position without the rule (being constantly exposed to interruption though free to interrupt).

Such simple cases, indeed, exist where it can be said that all 'equal citizens', however divergent their individual tastes or desires, would, if rational, prefer one alternative where liberties conflict. But I do not understand how the notion of the rational preference of the representative equal citizen can assist in the resolution of conflicts where reasonable men may differ as to the value of conflicting liberties, and there is no obviously best worst position which a rational man would prefer. It is true that at the stages in the four-stage sequence where such conflicts have to be resolved there is no veil of ignorance to prevent those who have to take decisions knowing what proportions of the population favour which alternatives. But I do not think Rawls would regard such knowledge as relevant in arguments about what it would be rational for the representative equal citizen to prefer; for it would only be relevant if we conceive that this representative figure in some way reflects (perhaps in the relative strength or intensity of his conflicting desires) the distribution of different preferences in the population. This, however, would be virtually equivalent to a utilitarian criterion and one that I am sure is far from Rawls' thoughts. I would stress here that I am not complaining that Rawls' invocation of 'the rational preference of the representative equal citizen' fails to provide a decision procedure yielding a determinate answer in all cases. Rather, I do not understand, except in the very simple cases, what sort of argument is to be used to show what the representative's rational preference would be and in what sense it results in 'a greater liberty'.

Of course, it is open to Rawls to say, as he does, that arguments concerning the representative's rational preference will often be equally balanced, and in such cases justice will be indeterminate. But I do not think that he can mean justice is to be indeterminate whenever different people value alternatives differently. Indeed, he is quite clear that, in spite of such difference in valuation, justice does require that there be some constitutional protections for individual freedom, though these will limit the freedom to participate in government;[10] the only indeterminacy he contemplates here is as to the particular form of constitutional protection to be selected from a range of alternatives all of which may be permitted by principles of justice. Yet, if opinion is divided on the main issue (that is, whether there should be any or no restrictions on legislative power to protect individual freedom), I do not understand what sort of argument it is that is supposed to show that the representative equal citizen would prefer an affirmative answer on this main issue as securing 'the greater liberty'.

This difficulty still plagues me even in relatively minor cases where one might well accept a conclusion that principles of justice are indeterminate. Thus, suppose the legislator has to determine the scope of the rights of exclusion comprised in the private ownership of land, which is for Rawls a basic liberty,[11] when this basic liberty conflicts with others. Some people may prefer freedom of movement not to be limited by the rights of landowners supported by laws about trespass; others, whether they are landowners or not, may prefer that there be some limitations. If justice is indeterminate in this minor case of conflicting

[10] 'The liberties of equal citizenship must be incorporated into and protected by the constitution' (p. 197). 'If a bill of rights guaranteeing liberty of conscience and freedom of thought and assembly would be effective then it should be adopted' (p. 231).

[11] It has been suggested to me by Mr. Michael Lesnoff that Rawls might not consider the private ownership of land to be a basic liberty since, as noted above, justice according to Rawls leaves open the question whether there is to be private ownership of the means of production. I am not, however, clear what is included in the scope of the basic liberty which Rawls described as 'the right to hold [personal] property' (p. 61). Would it comprise ownership or (in a socialized economy) a tenancy from the state in land to be used as a garden? If not, the example in the text might be changed to that of a conflict between pedestrians' freedom of movement and the rights of automobiles.

liberties, then no doubt we would fall back on what Rawls terms procedural justice, and accept the majority vote of a legislature operating under a just constitution and a fair procedure, even if we cannot say of the outcome that it is in itself a just one. But, presumably, in considering what measures to promote and how to vote, the legislators must, since this is a case, though a minor one, of conflicting basic liberties, begin by asking which of the alternatives a representative equal citizen would, if rational, prefer, even if they are doomed to discover that this question has no determinate answer. But indeterminacy and unintelligibility are different things, and it is the intelligibility of the question with which I am concerned. What do the legislators mean in such cases when they ask which alternative it would be rational for the representative equal citizen to prefer as securing the greater liberty, when they know that some men may value privacy of property more than freedom of movement, and others not? If the question is rephrased, as Rawls says it can be, as a question involving the principle of the common good, then it will presumably appear as the question which alternative will in the long run most advance the good of all, or ends that all share. This might be an answerable question in principle if it could be taken simply as the question which alternative is likely most to advance everyone's general welfare, where this is taken to include economic and other advantages besides liberty. If, for example, it could be shown that unrestricted freedom of movement over land would tend to reduce everyone's food supply, whereas no bad consequences likely to affect everyone would result from the other alternative, then the conflict should be resolved in favour of restriction of movement. But this interpretation of the question in terms of welfare seems ruled out by the principle that liberty may only be limited for the sake of liberty, and not for social or economic advantages. So, I think, that the conception of the rational choice of the representative equal citizen needs further clarification.

IV. *Limiting Liberty to Prevent Harm or Suffering*

I now turn to the question whether the principle of limiting liberty only for the sake of liberty provides adequately for restrictions on conduct which causes pain or unhappiness to others otherwise than by constraining liberty of action. Such harmful conduct in some cases would be an exercise of the basic liberties, such as freedom of speech, for example, or the use of property, though in other cases it may be the exercise of a liberty not classed by Rawls as basic. It would be extraordinary if principles of justice which Rawls claims are in general in harmony with ordinary considered judgments were actually to exclude (because they limited liberty otherwise than for the sake of liberty) laws restraining libel or slander, or publications grossly infringing privacy, or restrictions on the use of private property (e.g. automobiles) designed to protect the environment and general social amenities. These restrictions on the basic liberties of speech and private property are commonly accepted as trade-offs not of liberty for liberty, but of liberty for protection from harm or loss of amenities or other elements of real utility.

There are two ways in which perhaps Rawls' principles can at least partly fill this gap.[12] In some cases more plausibly than others he might argue that an unrestricted liberty to inflict what we call harm or suffering on others would in fact restrict the victim's liberty of action in either or both of two ways. The physical injury inflicted might actually impair the capacity for action, or the knowledge that such harmful actions were not prohibited might create conditions of apprehension and uncertainty among potential victims which would grossly inhibit their actions. But such arguments seem quite unplausible except in cases of conduct inflicting serious physical harm on individuals, and even there, when such restrictions are accepted as a reasonable sacrifice of liberty, it seems clear that if pain and suffering and distress were not given a weight independent of the tendency of harmful conduct to inhibit the victim's actions or incapacitate him from action, the balance would often, in fact, not be struck as it is.

It is, however, necessary at this point again to take into

[12] Professor Dworkin and Mr. Michael Lesnoff have suggested to me that what I describe here as a 'gap' may not in fact exist, since Rawls' basic liberties may be conceived by him as limited *ab initio* so that they do not include the liberty to act in a way damaging to the interests or liberties of others. But though it is certainly consistent with much of Rawls' discussion of basic liberties to treat his admittedly rough description of them as simply indicating areas of conduct within which the parties in the original position identify specific rights *after* resolving conflicts between the several liberties and the interests or liberties of others, this does not fit with Rawls' account of the basic liberties as liable to conflict, nor with his account of the conflicts as resolved not by the parties in the original position but by constitutional convention or body of legislators adopting the point of view of the representative equal citizen.

account those natural duties which are standards of individual conduct, as distinct from principles of justice, which are standards for institutions. These duties include the duty not to harm others or cause 'unnecessary suffering' and also the duty to come to the assistance of others. In discussing the acceptance of such duties by the parties in the original position, Rawls represents them as calculating that the burdens of such duties will be outweighed by the benefits (p. 388); so natural duties represent cases where, like the simple rules of debate case, the best worst position for all rational men can be identified, and in these cases even from behind the veil of ignorance. Even there it will appear to the parties as rational self-interested persons that it is, for example, better to be restrained from practising cruelty to others while protected from them than to be exposed to others' cruelty while free to practise it, and better to have to provide modest assistance to others in need than never to be able to rely on such assistance being forthcoming. So it is plain that these natural duties might fill part of the gap left open by the principle that liberty may only be limited for the sake of liberty, if Rawls means (though he does not explicitly say it) that even the basic liberties may be restricted if their exercise would infringe any natural duty. But again, these natural duties chosen from behind the veil of ignorance would only account for very obvious cases where the benefits of the restrictions would, for all rational men, plainly outweigh the burdens. This will not help where divergent choices could reasonably be made by different individuals in the light of their different interests, and it seems to me that this will very often be the case. Some persons, given their general temperament, might reasonably prefer to be free to libel others or to invade their privacy, or to make use of their own property in whatever style they like, and might gladly take the risk of being exposed to these practices on the part of others and to the consequences of such practices for themselves and the general social and physical environment. Other persons would not pay this price for unrestricted liberty in these matters, since, given their temperament, they would value the protections afforded by the restrictions higher than the unrestricted liberty. In such cases restrictions on the basic liberties of speech or private property cannot be represented as a matter of natural duty on the footing that rational men, whatever their particular temperament, would opt for the restrictions just as they might opt for general restrictions on killing or the use of violence.

Of course, it is certainly to be remembered that justice for Rawls does not exhaust morality; there are, as he tells us, requirements, indeed duties, in relation to animals and even in relation to the rest of nature which are outside the scope of a theory of justice as a theory of what is owed to rational individuals (p. 512). But even if there are such moral duties, regarding even rational beings, I do not think that Rawls would consider them as supplementing principles of justice which apply to institutions. I take it, therefore, that restrictions on the basic liberties excluded by the principles of justice because they are not restrictions of liberty for the sake of liberty could not be independently supported as just by appeal to other principles of morality. The point here is not that Rawlsian justice will be shown to be indeterminate at certain points as to the propriety of certain restrictions on liberty; it is, on the contrary, all too determinate since they seem to exclude such restrictions as actually unjust because they do not limit liberty only for the sake of liberty. I take it Rawls would not wish to meet this point by simply adding to his principles of justice a further supplement permitting liberty to be restricted if its exercise violated not only the natural duties but any requirements of morality, for this would, it seems to me, run counter to the general liberal tenor of his theory.

V. *The Choice of Basic Liberties*

I think the most important general point which emerges from these separate criticisms is as follows. Any scheme providing for the general distribution in society of liberty of action necessarily does two things: first, it confers on individuals the advantage of that liberty, but secondly, it exposes them to whatever disadvantages the practices of that liberty by others may entail for them. These disadvantages include not only the case on which Rawls concentrates, namely interference with another individual's basic liberties, but also the various forms of harm, pain, and suffering against which legal systems usually provide by restrictive rules. Such harm may also include the destruction of forms of social life or amenities which otherwise would have been available to the individual. So whether or not it is in any man's interest to choose that any specific liberty should be generally distributed depends on whether the advantages for him of the exercise of that liberty outweigh the various disadvantages for him of its general practice by others. I do not think Rawls recognizes this adequately in his discussion of conflicting liberties and his theory of natural duties. His recognition is inadequate, I think, because his doctrine insists that liberty can only be limited for the sake of liberty, and that when we resolve conflicts we must be concerned only with the extent or amount of liberty. This conceals the character of the advantages and disadvantages of different sorts which must be involved in the resolution of such conflicts; and his doctrine also leads him to misrepresent the character of all except those most simple conflicts between liberty and other benefits which are resolved by the parties in the original position when they choose the natural duties. Throughout, I think, Rawls fails to recognize sufficiently that a weighing of advantage and disadvantage must always be required to determine whether the general distribution of any specific liberty is in a man's interest, since the exercise of that liberty by others may outweigh the advantages to him of his own exercise of it. A rather startling sign that this is ignored appears in Rawls' remark that 'from the standpoint of the original position, it is rational' for men to want as large a share as possible of liberty, since 'they are not compelled to accept more if they do not wish to, nor does a person suffer from a greater liberty' (p. 143). This I find misleading because it seems to miss the vital point that, whatever advantage for any individual there may be in the exercise of some liberty taken in itself, this may be outweighed by the disadvantages for him involved in the general distribution of that liberty in the society of which he is a member.

The detailed criticisms which I have made so far concern the *application* of Rawls' principle of greatest equal liberty. But the general point made in the last paragraph, if it is valid, affects not merely the application of the principles of justice once they have been chosen but also the argument which is designed to show that the parties would in the conditions of the original position, as rational self-interested persons, choose the basic liberties which Rawls enumerates. Even if we assume with Rawls that every rational person would prefer as much liberty as he can get if no price is to be paid for it, so that in that sense it is true that no one 'suffers from a greater liberty', it does not follow that a liberty which can only be obtained by an individual at the price of its general distribution through society is one that a rational person would still want. Of course, Rawls' natural duties represent some obvious cases where it can fairly be said that any rational person would prefer certain restrictions to a generalized liberty. In other, less simple cases, whether it would be rational to prefer liberty at the cost of others having it too must depend on one's temperament and desires. But these are hidden from the parties in the original position and, this being so, I do not understand how they can make a rational decision, in terms of self-interest, to have the various liberties at the cost of their general distribution. Opting for the most extensive liberty for all cannot, I think, be presented as always being the best insurance against the worst in conditions of uncertainty about one's own temperament and desires.

VI. *The Argument for the Priority of Liberty*

I will end by explaining a difficulty which I find in the main argument which Rawls uses to show that the priority of liberty prohibiting exchanges of liberty for economic or other social advantages must be included among the requirements of justice. According to Rawls' theory, the rational, self-interested parties in the original position choose this priority from behind the veil of ignorance as part of the special conception of justice, but they choose it on the footing that the rule is not to come into play unless or until certain favourable social and economic conditions have actually been reached in the society of which they will be members. These favourable conditions are identified as those which allow the effective establishment and exercise of the basic liberties (p. 152), and when basic wants can be fulfilled (pp. 542–3). Until this point is reached the general conception of justice is to govern the society, and men may give up liberties for social and economic gains if they wish.

I do not think that Rawls conceives of the conditions which bring the priority rule into play as a stage of great prosperity.[18]

[18] It is plain that under this identification the conditions for the application of the special conception of justice may be reached at very different levels of material prosperity in different societies. Thus, in a small agrarian society or in a society long used to hard conditions, men might be capable of establishing and exercising political liberties at a much lower standard of living than would be possible for inhabitants of a large, modern industrial society. But in view of the fact that Rawls describes the relevant stage as one where conditions merely 'allow' or 'admit' the effective establishment and realization of basic liberties, it is not clear to me whether he would consider the special conception of justice applicable to a very wealthy society where, owing to the unequal distribution of wealth, poverty prevented considerable

At any rate, it is quite clear that when this stage is reached there may still be in any society people who want more material goods and would be willing to surrender some of their basic liberties to get them. If material prosperity at this stage were so great that there could then be no such people, the priority rule then brought into operation could not function as a prohibitory rule, for there would be nothing for it to rule out. As Rawls says, we need not think of the surrender of liberties which men might still be willing to make for greater economic welfare in very extreme terms, such as the adoption of slavery (p. 61). It might be merely that some men, perhaps a majority, perhaps even all, in a society might wish to surrender certain political rights the exercise of which does not appear to them to bring great benefits, and would be willing to let government be carried on in some authoritarian form if there were good reasons for believing that this would bring a great advance in material prosperity. It is this kind of exchange which men might wish to make that the priority rule forbids once a society has reached the quite modest stage where the basic liberties can be effectively established and the basic wants satisfied.

Why then should this restrictive priority rule be accepted as among the requirements of justice? Rawls' main answer seems to be that, as the conditions of civilization improve, a point will be reached when *from the standpoint of the original position*, 'it becomes and then remains . . . irrational to acknowledge a lesser liberty for the sake of greater material means . . .' because 'as the general level of well-being rises, only the less urgent material wants remain' (p. 542) to be satisfied and men come increasingly to prize liberty. 'The fundamental interest in determining our plan of life *eventually* assumes a prior place' and 'the desire for liberty is the chief regulative interest that the parties [in the original position] must suppose they all will have in common *in due course*' (p. 543, emphasis added). These considerations are taken to show the rationality, from the standpoint of the parties in the original position, of ranking liberty over material goods, represented by the priority rule.

The core of this argument seems to be that it is rational for the parties in the original position, ignorant as they are of their own temperaments and desires and the conditions of the society

numbers from actually exercising the basic liberties. Would it be unjust for the poor in such a society to support an authoritarian form of government to advance their material conditions?

of which they are to be members, to impose this restriction on themselves prohibiting exchanges of liberty for other goods because 'eventually' or 'in due course' in the development of that society the desire for liberty will actually come to have a greater attraction for them. But it is not obvious to me why it is rational for men to impose on themselves a restriction against doing something they may want to do at some stage in the development of their society because at a later stage ('eventually' or 'in due course') they would not want to do it. There seems no reason why a surrender of political liberties which men might want to make purely for a large increase in material welfare, which would be forbidden by the priority rule, should be permanent so as to prevent men, when great affluence is reached, restoring the liberties if they wished to do so; it is not as if men would run the risk, if there were no priority rule, of permanently losing liberties which later they might wish to have. I think, however, that probably Rawls' argument is really of the following form, which makes use again of the idea that under certain conditions of uncertainty rational beings would opt for the alternative whose worst consequences would be least damaging to one's interests than the worst consequences of other alternatives. Since the parties in the original position do not know the stage of development of their society, they must, in considering whether to institute a priority rule prohibiting exchanges of liberty for economic goods, ask themselves which of the following alternatives, A or B, is least bad:

A. If there is no priority rule and political liberties have been surrendered in order to gain an increase in wealth, the worst position is that of a man anxious to exercise the lost liberties and who cares nothing for the extra wealth brought him by surrender.

B. If there is a priority rule, the worst position will be that of a person living at the bottom economic level of society, just prosperous enough to bring the priority rule into operation, and who would gladly surrender the political liberties for a greater advance in material prosperity.

It must, I think, be part of Rawls' argument that for any rational self-interested person B is the best worst position and for that reason the parties in the original position would choose it. I am not sure that this is Rawls' argument, but if it is, I do not find it convincing. For it seems to me that here again the parties in the original position, ignorant as they are of the character and strength of their desires, just cannot give any determinate answer if they ask which of the positions, A or B, it is then, in their condition of ignorance, most in their interests to choose. When the veil of ignorance is lifted some will prefer A to B and others B to A.

It may be that a better case along the line of argument just considered could be made out for some of the basic liberties, for example, religious freedom, than for others. It might be said that any rational person who understood what it is to have a religious faith and to wish and practise it would agree that for any such person to be prevented by law from practising his religion must be worse than for a relatively poor man to be prevented from gaining a great advance in material goods through the surrender of a religious liberty which meant little or nothing to him. But even if this is so, it seems to me that no *general* priority rule forbidding the exchange, even for a limited period, of any basic liberty which men might wish to make in order to gain an advance in material prosperity, can be supported by this argument which I have ascribed, possibly mistakenly, to Rawls.

I think the apparently dogmatic course of Rawls' argument for the priority of liberty may be explained by the fact that, though he is not offering it merely as an ideal, he does harbour a latent ideal of his own, on which he tacitly draws when he represents the priority of liberty as a choice which the parties in the original position must, in their own interest, make as rational agents choosing from behind the veil of ignorance. The ideal is that of a public-spirited citizen who prizes political activity and service to others as among the chief goods of life and could not contemplate as tolerable an exchange of the opportunities for such activity for mere material goods or contentment. This ideal powerfully impregnates Rawls' book at many points which I have been unable to discuss here. It is, of course, among the chief ideals of Liberalism, but Rawls' argument for the priority of liberty purports to rest on interests, not on ideals, and to demonstrate that the general priority of liberty reflects a preference for liberty over other goods which every self-interested person who is rational would have. Though his argument throws much incidental light on the relationship between liberty and other values, I do not think that it succeeds in demonstrating its priority.

1 Equal Liberty and Unequal Worth of Liberty

NORMAN DANIELS*

I

Liberal political theory has traditionally attempted to provide a two-fold justification. On the one hand, liberal theorists have argued for the equality of various political liberties. Of course, different theorists were concerned with different sets of equal basic liberties. Hobbes justified only a narrow set of equal liberties of the person, for example, the liberty to refuse to testify against oneself.[1] Locke argued for a broader set of equal liberties of political participation, and Mill tried to defend broad, equal liberties of thought and expression. On the other hand, while justifying some degree of equality in the political sphere, these liberal theorists at the same time accepted and justified significant inequalities in income, wealth, powers, and authority between both individuals and classes.[2] Usually, they viewed these inequalities as the necessary or fair outcomes of differences in skill, intelligence, or industriousness operating within the framework of a competitive market. Despite the highly divergent theoretical frameworks used to justify these political equalities and socio-economic inequalities, including appeals to natural rights, to social contracts, and to different forms of utilitarianism, there was always a shared assumption. Liberal theorists uniformly assumed that political equality is compatible with significant social and economic inequalities, that they can exist together.

A similar assumption is implicit in John Rawls' 'special conception'[3] of justice as fairness and in his powerful and sophisticated

* Assistant Professor, Tufts University.

[1] Thomas Hobbes, *Leviathan: or the Matter, Form, and Power of a Commonwealth Ecclesiastical and Civil* (New York: E. P. Dutton & Co., 1950), Everyman Edition, Chapter XXI.

[2] Rawls tends to define classes as in contemporary stratification theory, in terms of weighted parameters of income, prestige, etc. Though I prefer the Marxist treatment, in which classes are defined in terms of relations to production, I shall here follow Rawls' usage unless otherwise indicated.

[3] The *Special* conception of justice as fairness, which gives priority to

arguments for it. Rawls' First Principle requires broad equality in the political sphere, stipulating a maximally extensive system of equal basic liberties (p. 302). Moreover, this liberty is given 'priority' and 'can be restricted only for the sake of liberty' (p. 302). That is, neither the extent nor the equality of liberty can be traded away for other social goods. Rawls' Second Principle, however, permits inequalities in the social and economic sphere. Specifically, inequalities in income, wealth, powers and authority between individuals or classes are permitted if such inequalities maximize a suitable index of the 'primary social goods' enjoyed by the worst-off representative members of society.

The assumption that First Principle equalities and Second Principle inequalities are compatible is likely to be weak and unproblematic if the inequalities allowed by the Second Principle are very small. Rawls certainly hopes they will not be large and that most of the inequalities we judge unjust or unfair today would be ruled out by the Second Principle. But he offers no argument from the social sciences capable of demonstrating, or even designed to demonstrate, the impossibility or improbability that large or significant inequalities will always fail to satisfy the Second Principle. Such an argument would have to show that inequalities greater than a certain magnitude always fail to benefit the worst-off maximally.[4] Moreover, Rawls sets no moral restriction on the absolute size of 'fair' inequalities, perhaps because he thinks that envy, which is ruled out of the original position, would be the only basis for opposing large inequalities. But lacking assurance from either social theory or special moral principles, we are forced to assess the compatibility of First and Second Principle equalities and inequalities in the least favorable case, where social and economic inequalities may prove to be large.

The incompatibility that may obtain between First Principle and significant Second Principle inequalities is not, of course, logical incompatibility. There is no logical contradiction involved. But it is not enough that there be no logical contradiction here. When Rawls proposes his principles as a model for 'ideal theory'

equality of liberty, applies only when a certain level of material well-being has been achieved in a society. Otherwise, the *general* conception applies and the difference principle governs all social goods, liberty included. Cf. Sect. 11, 26, 82.

[4] There may be the basis for just such an argument in Marxist theory, but there seems to be no theoretical basis for it in non-Marxist social sciences.

(see Sect. 39), he is not concerned with mere logical possibility. Rather, he requires that his ideal must be socially possible (p. 138). It must comprise a workable conception of justice in light of what we know from general social theory, including psychology, sociology, history, economics, and political science (though, of course, we are not told how to achieve the workable ideal). Thus Rawls argues at length that the special conception of justice as fairness is consonant with 'the principles of moral psychology'. Indeed, one of Rawls' arguments for justice as fairness is that it is more stable than other theories because it is more in accord with the principles of moral psychology (cf. Sect. 76). 'Stability' in this sense is one determinant of social possibility and is an important empirical constraint on the content of an adequate moral theory.

There are other determinants of social possibility, however. If the degree to which a moral theory accords with principles of moral psychology provides one measure of stability, then the degree to which a moral theory is in line with principles from other areas of the social sciences presumably would provide other measures of stability and, thus, other determinants of social possibility. For example, if we have good reason to believe that the arrangements authorized by a conception of justice are not in line with the principles of political science or with what we know from history, then that conception of justice is to a certain degree unstable and perhaps not socially possible. Specifically, if social theory gives us good reason to believe that significant individual and class inequalities in wealth and powers cause or produce inequalities of liberty, then this fact becomes relevant to assessing Rawls' ideal.[5] It would indicate a serious form of instability, though not one Rawls discusses when he talks about the stability of conceptions of justice. More generally, this fact, if established, would raise a serious question whether Rawls' ideal is realizable, that is, whether or not Rawls' two principles of justice describe a socially possible or consistent system of justice

My concern in this paper is to explore the consistency of the two principles of justice. In Section II, I argue that it is a serious question in social theory, one not to be ignored, whether or not

[5] The argument for class inequalities causing inequalities of liberty is both clearer and stronger if we restrict attention to the dominant economic class in any given period, defined by its control over the means of production, and consider the various ways in which it has controlled the major institutions, notably the state.

the kinds of inequalities allowable by Rawls' Second Principle are compatible with the demands of his First Principle. Further, I will show that Rawls' steps to accommodate this possibility are inadequate. In Section III, I discuss Rawls' distinction between liberty and worth of liberty, arguing that it is arbitrary. In Section IV, using considerations internal to Rawls' theory, I show that the appeal to worth of liberty cannot reconcile the First and Second Principles. If I am right, then these points raise a serious problem, not just for Rawls, but for this central assumption about compatibility which he shares with earlier theories. As I argue in Section V, Rawls' principles may drive him and other liberal theorists toward far greater egalitarianism than was expected.

II

Our historical experience, as Rawls acknowledges (p. 226), is that inequalities of wealth and accompanying inequalities in powers tend to produce inequalities of liberty. For example, universal suffrage grants the wealthy and the poor identical voting rights. But the wealthy have more ability than the poor to select candidates, to influence public opinion, and to influence elected officials.[6] Consequently, a clear inequality in the liberty to participate in the political process emerges. Similarly, though wealthy and poor are equally entitled to a fair trial, are 'equal before the law', the wealthy have access to better legal counsel, have more opportunity to influence the administration of justice both in specific cases and in determining what crimes will be prosecuted, and have greater ability to secure laws that favor their interests. Again, the wealthy and the poor are equally free to express (non-libelous) opinions in the appropriate circumstances. Yet, the wealthy have more access to and control over the media and so are freer to have their opinions advanced. This inequality in 'free-

[6] To the extent that elected (and appointed) officials are drawn in disproportionate numbers from the better-off and best-off classes, 'influence' and 'control' by the wealthy over public officials becomes less the issue than the awareness of common class interests. To the extent that elected officials initially drawn from the worse-off classes begin to identify their interests with the interests of better-off classes, perhaps even as a result of the kinds of 'incentives' provided for by the Second Principle, then the perception of common interests again supersedes direct 'influence' and 'control'. These processes are accelerated by various ideological factors, like professionalism, elitism, and racism.

dom of speech' is one of the greatest importance since it means views which represent particular interests, that is, those of the best-off classes, are most likely to get advanced. Indeed, whole ideologies reflecting those interests may be promoted in this manner. What is worse, even greater inequalities in liberty emerge when we note that there are combined effects. For example, if the wealthy have greater liberty to affect the political process, then they may also acquire greater influence over the schools and what is taught in them. But the combined effects of control over the schools and the media give the wealthy vastly greater 'freedom of expression' than those less well-off. In turn, their resulting influence over public information and training produces further increases in their political effectiveness. The inequalities in liberty compound each other.

In these generalized examples, the inequality of liberty does not result primarily from abuses, like bribery. Instead, the inequality derives from the (usually) legal exercise of abilities, authority, and powers that come with wealth. Moreover, the examples do not seem to depend primarily, if at all, on violations of what Rawls calls 'fair equality of opportunity' to achieve offices to which advantages are attached (cf. Sect. 14). To the extent that some inequality of opportunity is operative in these examples, it, too, may be an effect of inequalities in wealth and concomitant powers. Indeed, the resulting inequality of opportunity may be as hard to avoid as the inequalities of liberty which here concern us.[7]

Examples such as these do not, of course, prove that inequalities in wealth and power cause inequalities in basic liberties in all workable political systems. If one thought that the mechanisms through which unequal wealth operates to destroy equal liberty were simple and insolatable, then perhaps constitutional provisions could be devised to solve the problem. Rawls, for example, suggests constitutional provisions for the public funding of political parties and for the subsidy of public debate (pp. 225–6). But there is little reason to believe that the mechanisms are so simple and that such safeguards would work. The current United States tax deduction for subsidizing election campaigns, for example, is unlikely to wrest control of the major parties from the hands of the wealthy. From what we do know about cases

[7] Rawls admits practical limitations, such as the continued existence of the family as a social unit, to securing fair equality of opportunity in a context of social and economic inequalities.

of class divided societies, the process of political control by the dominant economic class is highly complicated, and, much more than the direct 'buying' of influence, it involves the combined effect of vast economic powers and control over ideological institutions. At any rate, in the absence of a comprehensive 'political sociology', as Rawls calls it (p. 227), it is safe to say that we fail to know what all, or even the main, causal mechanisms are. Therefore, we fail to know if constitutional safeguards could satisfactorily interfere with them.

Actually, the situation is a bit worse than is indicated by our claiming lack of comprehensive knowledge. Rawls himself admits that it is an open question, even in theory, whether or not we could eliminate the relevant mechanisms. As he suggests, 'The democratic political process is at best regulated rivalry, it does not *even in theory* have the desirable properties that price theory ascribes to truly competitive markets' (p. 226, my emphasis). Part of the problem, of course, is that the 'regulating' becomes the task of those needing regulating and there is no equivalent of market forces to redistribute imbalances.[8]

From the point of view of the original position, then, reliance on reassurances about the possibility of constitutional safeguards seems highly risky. Persons in the original position are aware how little is really known about the relevant mechanisms. They also know that there even are theoretical reasons for skepticism that an adequate set of constitutional safeguards could be devised. Assuming these rational agents value equal liberty as strongly as Rawls says they do, they would not want to risk losing that equal liberty. But if such agents do not know if it is socially possible to prevent unequal wealth and powers from destroying equal liberty, then they would not want to take a chance on an untested constitutional blueprint. Accordingly, they might not be able to accept the conjunction of the First and Second Principles.

[8] The point is put more strongly in Marxist theory of the state. The state cannot exist to 'reconcile' and regulate class conflicts. Rather, both in theory and in practice, it always functions as an instrument of class conflict, controlled by the dominant class. My own view is that formal, procedural guarantees are never sufficient to make sure small groups or classes cannot gain significant advantages in political liberty and power. For example, if a broad class, like the working class, controlled the political apparatus, then only its own developed class consciousness and political understanding, not constitutional guarantees, could protect it against attempts by smaller classes to reinstate a condition of less political equality. Equalizing liberty between antagonistic classes by devising the proper form of government is an impossible dream.

III

Fortunately, Rawls does not rely on the unsupported hope that we can find a constitutional blueprint for eliminating the effects of unequal wealth and powers on liberty. Instead, he tries to circumvent the problem by introducing a distinction between *liberty* and *the worth of liberty*. *Liberty*, 'represented by the complete system of the liberties of equal citizenship' (p. 204), continues to be distributed in accordance with the First Principle. But the new social good, *the worth of liberty* to persons or groups, 'is proportioned to their capacity to advance their ends within the framework the system defines' (p. 204). Apparently, then, it is distributed in accordance with the Second Principle. As a consequence, the incompatibility between equal liberty and unequal wealth and power, between the First and Second Principles, seems to disappear. Unequal wealth and unequal powers no longer cause inequality of liberty itself, only inequality in the worth of liberty:

> Freedom as equal liberty is the same for all; the question of compensation for a lesser than equal liberty does not arise. But the worth of liberty is not the same for everyone. Some have greater authority and wealth and therefore greater means to achieve their aims [p. 204].

We shall consider in turn two questions. In this section we shall ask if Rawls' distinction between 'liberty' and 'worth of liberty' is an arbitrary one, that is, if there is good reason for Rawls to make it. The point here is whether it is useful to talk about something as a 'liberty' when we can not effectively exercise it. Is it useful to be able to say, 'my liberty is equal to Rockefeller's, but I can not exercise "it" equally'? In the next section we will grant Rawls his use of the distinction, and inquire if it really helps him to reconcile the First and Second Principles.

To see if the distinction between liberty and worth of liberty is arbitrary, we must first see how Rawls analyzes liberty. Basically, he follows Felix Oppenheim and Gerald MacCallum in treating liberty as a triadic relation holding between agents, constraints, and acts, and having the general form: 'This or that person (or persons) is free (or not free) from this or that constraint (or set of constraints) to do (or not to do) so and so' (p. 202).

Agents include persons and associations, such as states and classes. Constraints 'range from duties and prohibitions defined by law to coercive influences arising from public opinion and social pressure' (p. 202), though Rawls is mainly concerned with legal restrictions. The crucial point for our discussion is that economic factors, and perhaps other factors (like ideology) are explicitly excluded from among the constraints *definitive* of liberty.

> The inability to take advantage of one's rights and opportunities as a result of poverty and ignorance, and a lack of means generally, is sometimes counted among the constraints definitive of liberty. I shall not, however, say this, but rather I shall think of these things as affecting the worth of liberty, the value to individuals of the rights that the First Principle defines [p. 204].

Rawls also remarks that 'greater authority and wealth' implies having 'greater means to achieve . . . aims' and thus greater worth of liberty (p. 204). Thus the full range of economic factors, and not just abject poverty, are excluded from the category of constraints defining liberty.

The question whether Rawls' distinction between liberty and worth of liberty is arbitrary reduces, then, to the question whether it is arbitrary to exclude economic factors from the category of constraints defining liberty. It is often assumed that constraints defining liberty must be legal restrictions, but, as we have seen, Rawls agrees with Mill and does not view being a legal restriction as a necessary condition for being a defining constraint. Non-legal, but legally permissible coercions, like public opinion and social pressure, are counted among the defining constraints. But, assuming for the moment that economic factors are not reflected in legal restrictions, why distinguish them from other non-legal coercions? Presumably, social pressure and public opinion act as defining constraints because they create obstacles for agents who might desire to perform certain acts. In exactly the same way, however, economic factors also act as systematic, socially produced obstacles (or aids) for hindering agents desiring to perform certain acts.

It might be thought that we should treat economic factors differently from other non-legal coercions if we can find a relevant difference in the way the obstacles work. Legal constraints and pressures of public opinion appear as socially produced obstacles acting outside the agent whose liberty is in question. They appear to be the results of other agents' activities. In contrast, lack of money might appear to function more like the lack of a capacity or an ability than like the effect of another agents' activities. That is, it might appear more like something internal to the agent whose liberty is in question. Since there is some reason to exclude certain 'internal' abilities from among liberty-defining constraints, there might seem to be reason to treat economic factors in the same ways.

This defense of a difference is not quite adequate. However much one might want to exclude obviously psychological abilities or capacities from among the liberty-defining constraints,[9] it ought to be clear that the institutions which define and determine income and wealth, including the rights of transfer, exchange and protection of property, cannot be assimilated to them. There is a much more direct parallel here between public pressure and economic factors than there is between economic factors and capacities. If social pressure prohibits me from sending my children to private school, then I am not at liberty to do so. But, if I cannot afford to send my children, I am not at liberty to do so either. If I send them, they'll be sent home by forces as external as those involved in social pressure. Similarly, if it makes sense to claim that economic factors define only worth of liberty but not liberty itself, then it makes equally good sense to say that other non-legal coercions also define only worth of liberty. The special exclusion of economic factors seems arbitrary.

One way around the charge of arbitrariness might be to exclude all non-legal coercions from the category of defining constraints. That is, since economic factors are to be excluded, then let us exclude social pressures and public opinion as well. Unfortunately, this maneuver will not solve the problem, either. Requiring that constraints definitive of liberty must be legal restrictions will not entail excluding economic factors since economic factors are always reflected in the laws and constitution. In fact, economic factors are constituted by laws which recognize and enforce property rights, including rights for the transferral, exchange and protection of property. If I do not have the money

[9] More difficult questions arise when we try to assess the degree to which certain 'internal' capacities, like motivation, are social products, perhaps even developed in accordance with a particular plan for their distribution. Attitudes and beliefs that result from training and indoctrination may similarly function as obstacles to the performance of certain acts and also raise questions about what can count as constraints defining liberty.

to afford adequate legal counsel, then it is because there are laws which establish the rights of lawyers to refuse to counsel me and of police to arrest me if I create a commotion demanding the legal counsel I want. So, if economic factors are to be excluded from constraints defining liberty, Rawls owes us another, hopefully sufficient, criterion for distinguishing economic factors from other legal restrictions.[10] No other criterion is provided us, however.

We might sympathize with the desire to mark *some* distinction between having liberties and having the ability to exercise those liberties, otherwise defined. If I am unable to speak in public because I am shy, it would be a mistake to conclude that I have no liberty to speak.[11] Some kinds of obstacles are not the ones we want to include among liberty-defining constraints. But simply believing it important to make some distinction here does not by itself tell us to draw it where Rawls proposes, since economic factors share important features with both social pressures and legal retrictions.

There is, though, a history to, and perhaps therein an explanation, for the exclusion of economic factors from constraints defining liberty. For example, although MacCullum and Oppenheim say things that seem to allow economic factors to serve as defining constraints,[12] Oppenheim in particular insists that liberty-defining constraints always be identifiable as individuals or groups of individuals. According to this methodological individualism, legal restrictions are to be translatable into the actions and powers of various officers and agents of the state. In contrast, however, economic inequalities have traditionally, for example, in Hobbes and Locke, been construed not as the work of identifiable individuals, but, rather, as the effect of impersonal market forces

[10] Incidentally, as Hugo Adam Bedau has pointed out to me, being a legal restriction is not by itself sufficient condition for being a constraint on liberty since all legal systems, including Rawls' ideal one, leave room for discretion in the prosecution of laws; not all laws in fact, then, constrain anyone.

[11] Although even this obvious case is not entirely clear. Suppose, for example, that people with blue eyes and black skin are the object of an inferiority theory which claims they are always poor public speakers. Suppose this theory is widely believed and acted on, so that people with blue eyes and black skin are rarely listened to and may even be ridiculed. As a result, many such people become extremely shy. Suppose further that it is the shyness, and not the response of others, that acts as the obstacle to speaking. This situation might make it plausible to think that blue-eyed, black-skinned people actually had less liberty to speak.

[12] Felix Oppenheim, *Dimensions of Freedom* (New York: St. Martins, 1961), p. 123.

which operate vis-a-vis individuals or classes much like the laws of nature, i.e., anonymously. Since we do not, in political theory, treat anonymous natural laws as constraints definitive of liberty, then by analogy, market effects would seem equally excludable.[18] This historical explanation, however, cannot serve as justification for Rawls' distinction. If we view, as Rawls does, the market as an institution whose outcomes and processes we can deliberately manipulate, then market operations and outcomes are no longer anonymous, natural-law-like forces and effects to be distinguished from legal restrictions.

Perhaps we are barking up an unnecessary tree. Rawls' exclusion of economic factors from among the constraints defining liberty seems arbitrary because we have sought without success a special rationale for it. But, it might be argued, no rationale is really needed. If the special treatment of economic factors has systematic, beneficial ramifications in Rawls' overall theory, then that may be justification enough. And we did see that Rawls expects just such a systematic, beneficial effect, namely the reconciliation of the First Principle demand for equal liberty and the possible Second Principle effect of destroying equal liberty. Unfortunately, this final defense seems only to beg the question. Reconciling the two principles by appeal to the special definition of liberty is exactly what needs justification in the first place. Besides, there are far more interesting reasons, internal to Rawls' theory, which prevent Rawls from using the distinction between liberty and worth of liberty to reconcile the top two principles of justice. I shall turn to these now.

IV

Considerations internal to Rawls' own theory open him to the charge that equal liberty without equal worth of liberty is a worthless abstraction. No doubt, this charge could be explored directly by trying to discover what value there is to equality of liberty if the liberty cannot effectively be equally exercised. But it will be more illuminating of Rawls' theory to ask the question from the point of view of agents in the original position. Specifically, we shall want to know if it is rational to choose equal liberty without also choosing equal worth of liberty. If the

[18] I am obliged to my colleagues Hugo Bedau and David Israel for helpful discussion of this and other points.

answer is 'no', then Rawls cannot use worth of liberty to reconcile his First and Second Principles.

In answering this question, we will arrive at what might be called a 'relative rationality proof', analogous to relative consistency proofs in mathematics. Our argument will have the overall form: If it is rational to choose x for reasons R in the original position, then, if R constitutes equally good reasons for choosing y, then it is also rational to choose y in the original position. Specifically, if it is rational to choose equal liberty in the original position for the reasons Rawls gives, and the same reasons are equally good reasons for choosing equal worth of liberty, then it is equally rational to choose equal worth of liberty. In short, choosing equal worth of liberty is rational if choosing equal basic liberties is. Being concerned with the relative consistency and not the validity of Rawls' argument, I will not question whether or not Rawls is right in concluding that choosing equal basic liberties is rational in the original position. Accordingly, I shall use, rather than assess, Rawls' reasons for choosing equal liberty, and I shall use, rather than criticize, his contractarian method. Of course, I do have to paraphrase Rawls' argument in order to isolate his reasons for choosing equal liberty.

Rawls gives only two arguments in which he is explicitly concerned with showing that basic liberties must be distributed equally. The first such argument is for equal liberties of conscience. The second, potentially more general, is for equal liberties of political participation. Both arguments draw, as we shall see, on Rawls' general claim that liberty is to be given priority over other primary social goods. The general argument for the priority of liberty, however, leaves open the question how liberty is to be distributed. But the equal distribution of liberty is both what mainly concerns us here and what is special about these two arguments. Accordingly, I will not discuss the general priority argument, especially since it has been widely discussed elsewhere.[14]

The argument for equal liberty of conscience seems to rest on the special importance of moral and religious obligations. It is by appeal to this special importance that Rawls establishes, first, the priority of liberty of conscience and, then, its equal distri-

[14] Discussions of the priority of liberty can be found in several selections in this volume, in particular in the articles by Nagel, Hart, and Scanlon. Brian Barry also discusses the question in *The Liberal Theory of Justice: A Critical Examination of the Principal Doctrines in 'A Theory of Justice' by John Rawls* (Oxford: Clarendon Press, 1973), Chap. 7.

bution. These two phases of the argument are obvious in the following paraphrase (cf. pp. 205–8): (1) Persons in the original position 'have interests which they must protect as best they can'; (2) among these interests there *may be* self-imposed moral or religious obligations which are very important to chosen life-plans; (3) Persons in the original position know that at some level of material well-being, even the worst-off members of any society would prefer (increments in) the liberty to meet moral and religious obligations over any further increments in the index of other primary social goods; (4) accordingly, persons in the original position would choose to give liberty of conscience priority over other primary social goods; (5) liberty of conscience, even if given priority over other goods, could be distributed in accordance with (a) majority will, (b) the utilitarian principle, or (c) the principle of equality; (6) moral and religious obligations are so important, however, that agents in the original position cannot gamble on their being able to meet such obligations; but (7) because of the veil of ignorance, persons in the original position cannot know if they'll be in the majority; and (8) because the freedom to meet such obligations is so important to them, they cannot afford to gamble on being in the majority; so, (9) majority will would not be an acceptable principle for distributing liberty of conscience; similarly, (10) because such obligations are so important, agents in the original position would not gamble that the utilitarian principle will allow for meeting them; consequently, (11) only the principle of equality remains as a rational basis for distribution; and so, (12) rational agents in the original position would choose equal liberty of conscience and give it priority over other primary social goods. Rawls supplements this central argument by claiming that the equality principle, more than other principles, respects the interests of the next generation, which agents in the original position are obliged to consider. We shall ignore this supplementary argument.

Before showing that an analogous argument to (1)–(12) can be made for equal worth of liberty, a few points are worth noting about Rawls' argument itself. First, Rawls views the conclusion of the argument as 'settled', as 'one of the fixed points of our considered judgments of justice' (p. 206). He never considers any of the very serious questions that can be raised about self-imposed demands of conscience. It might be argued, for example, that some religious views tend to impose and emphasize divisions and barriers among people with a generally harmful effect. If people in the original position know of these effects, why would they want to risk being exposed to them? Or, it might be objected that it is hard to grant liberty of conscience to a parent without also granting the ability to indoctrinate children. From the point of view of the original position, would a rational agent want to risk being the child of a parent free to pursue fanatical religious views?

A more general problem also emerges. Does treating equal liberty of conscience as a 'fixed point' tend to imply, and therefore to impose, an unnecessary and possibly dangerous relativism on the assessment of the truth and effects of various religious, moral and philosophical views? Are not some such views false, dangerous, and immoral? Why should persons in the original position back away from recognizing this fact by granting the equal 'right' to hold and practice such views? These are questions worth pursuing, though Rawls does not take them up. Nor, I am afraid, can we.[15]

A second point worth noting about the liberty of conscience argument is that Rawls believes 'the reasoning in this case can be generalized to apply to other freedoms, although not always with the same force' (p. 206). His 'intuitive idea is to generalize the principle of religious toleration to a social form, thereby arriving at equal liberty in public institutions' (pp. 205–6, n. 6). But there is a serious question whether this generalization can be made. We have seen that the argument for liberty of conscience depends at steps (3) and (6)–(10) on appeals to the importance of obligations of conscience. If such obligations are granted a special importance, however, then there is serious question whether other types of wants, beliefs and preferences, which form the basis for desiring freedom of expression and certain personal freedoms, can claim a similar importance. But if there is no claim that a special importance accrues to demands of conscience, and all chosen features of life plans can give rise to interests which agents in the original position may feel they have to protect, then the First Principle rapidly mushrooms to include far more than what we might pick out as 'basic' liberties. Freedoms spring up whenever the seeds of desire are planted.

Another complaint about the generalizability of arguments for religious toleration might also be made, in this case challenging

[15] Cf. Gerald Dworkin's paper, 'Non-Neutral Principles', reprinted in this volume, which pursues this issue in some depth. See also Milton Fisk's and T. M. Scanlon's discussions of freedom of thought, also in this volume.

the attempt to treat religious and moral obligations on the same plane. One might say that the relativism implicit in religious toleration is acceptable because it is fairly harmless, at least when we are concerned with religious practices narrowly construed. But the acceptance of relativism with regard to other kinds of obligations, say moral obligations, may prove to be a more risky business since these obligations affect a wide variety of social interactions. This asymmetry, however, bodes ill for using religious toleration as the intuitive model for justifying other freedoms, even other freedoms of conscience.

Thirdly, it is also worth noting that, although Rawls views equality of liberty of conscience as a 'settled question', the structure of his argument can at best establish only a provisional agreement on equal distribution. If his argument succeeds, it shows only that equality is a preferred distribution principle when compared to majority will or to the utilitarian principle. Nothing in Rawls' argument precludes abandoning equality for another principle. Perhaps a principle weaker than equality but stronger than the utilitarian principle would be chosen if such a principle better meets all of Rawls' empirical constraints on choices in the original position. I shall briefly touch on one plausible alternative to equality shortly.

Before showing that choosing equal worth of liberty of conscience is rational if choosing equal liberty is, it is worth being clear just what worth of liberty is in this case. After all, examples of unequal worth of liberty for other basic liberties, like liberty of expression or political participation, are familiar and the subject of real concern. On the other hand, unequal worth of liberty of conscience seems less familiar and more abstract. Nevertheless, examples of unequal worth of liberty of conscience are readily found. For example, some religions view time-consuming, expensive acts, like pilgrimages, as obligations of the truly faithful. Inequalities in wealth clearly affect the ability to meet such demands and so would create inequalities in the worth of liberty of conscience. If this example is not compelling to us, it is probably because few of us feel compelled to make pilgrimages.

Other examples may be more relevant to our experience. Some religions consider it a matter of religious obligation to avoid killing or violence of any kind, including military service. It is a matter of historical experience in the United States, however, that conscientious objection to military service on religious grounds places a greater burden on lower- than upper-class objectors. The costs of legally establishing or defending objector status, and of facing up to hostile attitudes and discriminatory job practices toward pacifists are substantial and affect different classes differentially. This example is even clearer in the case of 'moral' or 'political' conscientious objectors to the Vietnam war, since these grounds for objector status were not legally recognized and often entailed defying induction, an outcome with far higher cost to working- or lower-class objectors than to upper middle-class objectors. Similarly, differences in wealth make it far easier for upper-class than lower-class opponents of South African apartheid to emigrate when refusal to live under such laws is felt to be a demand of conscience. No doubt, unequal worth of liberty of conscience may not seem as important to many as unequal worth of other liberties, but that may be because liberty of conscience is not as important as other liberties. In any case, we are forced to consider the worth of liberty of conscience because Rawls makes liberty of conscience the focus of his argument for other liberties. We come, then, to our first relative rationality argument.

Showing that equal worth of liberty of conscience would be chosen in the original position requires an argument analogous to (1)–(12), Rawls' argument for equal liberty of conscience. The crux of Rawls' argument is that equality is the only distribution principle which recognizes the importance individuals place on demands of conscience. Because these demands are so important, it is rational for persons in the original position to avoid risking any interference with meeting them. Thus, it is rational to reject majority will [steps (7)–(9)] and the principle of utility [step (10)] in favor of equality of liberty because these principles may yield obstacles, like adverse majority opinion or unfavorable utility calculations, which would interfere with meeting obligations. But whatever the principles of rationality guiding choice here, they do not distinguish among the source of obstacles; they simply require we avoid them. So, the desired analogue follows easily. If it is rational to reject principles producing obstacles like adverse opinion and unfavorable calculations because these obligations of conscience are so important, then it seems equally rational to reject inequalities in wealth and powers if they create similar obstacles.

By Rawls' definition, however, inequalities in the ability to meet demands of conscience, when caused by unequal wealth or powers, just *are* inequalities in worth of liberty of conscience. So the rational choice for those behind the veil of ignorance would be to rule out principles, like the Second Principle, which may reduce the worth of liberty for some. In short, if it is rational to choose equal liberty of conscience for the reasons Rawls gives, then it is equally rational to choose equal worth of liberty of conscience.

The priority of equal worth of liberty of conscience over other primary social goods can also be established. In step (3), Rawls argued that once a given index of the primary social goods was reached, even the worst-off members of society would prefer increments in their liberty to meet demands of conscience to further increments in the index. Thus rational agents would assign priority to liberty of conscience. But it seems equally rational to reject a higher index in favor of increments in the ability to meet demands of conscience. The alternative to this conclusion is the paradox of preferring liberty to other primary goods while at the same time preferring those goods to the ability to exercise the liberty, or at least to exercise it equally.[16]

Rawls' discussion of compensation for inequalities in the worth of liberty seems to verge on this same paradox since, on one interpretation, it fails to respect the *priority* of worth of liberty over other social goods. Inequalities are to be allowed only if they help the worst-off.

> The lesser worth of liberty is, however, compensated for since the capacity of the less fortunate members of society to achieve their aims would be even less were they not to accept the existing inequalities whenever the difference principle is satisfied [p. 204].

If we assume that the 'difference principle' referred to is the Second Principle, then there is an immediate problem. The Second Principle employs an index which includes all primary social goods *other than* basic liberties. It allows inequalities in individuals' indices only if the inequalities act to maximize the

[16] Rawls does assume that the level of the index at which preference for liberty emerges is a level at which the means for exercising liberty effectively exists. But the point of our discussion of equal worth of liberty in the cases of liberties of expression and political participation is that the 'effective' exercising of a liberty may require near equality in the ability to exercise the liberty. 'Effectiveness' may well be relative in the sense that the greater effectiveness of some persons' exercise of liberty renders ineffective lesser degrees of effective exercise of the liberty.

index of the worst-off. If worth of liberty is not included among the goods indexed, then Rawls appears to be authorizing a trade-off between it and the primary social goods which are indexed, since he claims that the lesser worth of liberty of the worst-off is compensated for by maximization of their index. But as we have just seen by the analogue to Rawls' step (3), once a certain index is achieved, increments in it are not as valuable to persons as are increments in the ability to meet demands of conscience. Therefore, such increments would not be accepted as compensation for lesser worth of liberty of conscience. Nor can this problem be avoided simply by counting worth of liberty itself as one of the primary goods included in the index. Since it would still be one among several goods indexed, worth of liberty will not necessarily be maximal whenever the index is maximal. Thus, maximization of even this expanded index would not be acceptable in the original position as compensation for lesser worth of liberty.

One way to try to compensate lesser worth of liberty while respecting its priority over other primary social goods would be to introduce a special Liberty-Restricted Difference Principle (LRDP), distinct from the Second Principle. This principle would permit inequalities of worth of liberty only if they act to maximize the worth of liberty of those with the least worth of liberty. As stated, the principle seems to capture Rawls' intention when he says the 'difference principle' will act to maximize 'the capacity of the less fortunate members of society to achieve their aims . . .' (p. 204). Its advantage over the Second Principle is that it prohibits direct trade-off of worth of liberty for a higher index of other primary goods. At the same time, it benefits from the general rationale Rawls offers for choosing the Second Principle in the original position.

Unfortunately, appeal to the Liberty-Restricted Difference Principle does not help to reconcile the First Principle demand for liberty with the Second Principle effect of generating unequal worth of liberty. To solve that problem by appeal to the LRDP requires an assurance that the inequalities in worth of liberty justified by the LRDP are exactly the same inequalities which are caused by the Second Principle. Unfortunately, there is no reason to believe they will be the same. For one thing, the worst-off groups may not be identical from the point of view of the two principles. The worst-off representatives, for the purposes of the Second Principle, are those with the lowest index of primary social goods. The worst-off representatives, for purposes of the

LRDP, are those with the least worth of liberty. Unless we assume that every variation in the index of primary social goods has a corollary variation in the worth of liberty, which there is no reason to assume, then the worst-off representatives will not be the same individuals (or classes).

But even if the worst-off individuals or classes happen to coincide, the inequalities justified by the two principles will most likely not coincide anyway. Second Principle inequalities of wealth and power which maximize the index of primary social goods may well not maximize worth of liberty for those with the lowest index. Indeed, what is most likely is that worth of liberty is especially sensitive to *relative* differences in the index of primary social goods and is not a simple monotonic function of it.

An example might help illustrate the point. Assume, for the moment, what we are often told, that granting corporation owners particularly high indices of primary social goods maximizes the indices of workers. The assumption is that profit incentive acts to increase investments and, thereby, allows for more jobs, higher wages, and greater tax monies for welfare. Also, grant what I have argued earlier, that significant inequalities of wealth and powers cause, in cases like this, significant inequalities in the worth of liberty. Then, the higher index of primary social goods enjoyed by the worker or welfare recipient does not necessarily produce more worth of liberty, as Rawls apparently assumes (cf. p. 204). Rather, the very inequality of wealth and powers which, we are assuming, acts to *increase* the index of the worst-off individuals can at the same time act to *decrease* his worth of liberty. The increased index of the corporation owner may give him substantial competitive advantage over those with lower indices. This effect is decisive where worth of liberty is affected by comparative access to those resources and institutions such as qualified legal counsel or the mass media, which are needed for the effective exercise of liberty. The result is that the worst-off, despite their increased indices, may be in a relatively worse position to effectively exercise their liberty. Their increased index is worth relatively less when it comes to exercising liberty because extra, even decisive, advantage has been ceded to the best-off.

The fact that the LRDP and Second Principle yield different sets of inequalities of worth of liberty is not the only problem with appeal to this new difference principle. First, let us leave aside the whole issue of conflict with the Second Principle. Next, let us assume that persons in the original position would choose to regulate worth of liberty in accordance with the LRDP. Their reasoning would be analogous to the reasoning for the Second Principle. That is, there is no reason to accept less worth of liberty for the purposes of maintaining equality if accepting certain inequalities in worth of liberty might lead to greater worth of liberty. It now becomes obvious that we can turn our relative rationality argument in the reverse direction. Just as it is rational to distribute worth of liberty according to the LRDP, why should it not be rational to distribute liberty itself according to a liberty-restricted difference principle instead of insisting on equal liberty.[17] In Rawls' original argument for equal liberty of conscience, equality of liberty was chosen because it was less risky than the principle of utility or than majority will. But the LRDP seems no riskier than equality. So if it is rational to choose LRDP for worth of liberty, then it seems equally rational to choose it for liberty itself, since it suffers none of the disadvantages that led to rejecting other distribution principles.

This reversal is not completely acceptable to Rawls. Equal liberty of conscience, we have seen, is viewed by Rawls as a 'settled question', a 'fixed point' in our moral judgments. Therefore Rawls is not likely to abandon the equal distribution in favor of a special difference principle. But if this is so, consistency demands, according to our relative rationality argument, that he also abandon the equal distribution of worth of liberty. Consequently, he must reject the LRDP here as well. In short, the LRDP provides no way around the relative rationality argument in the case of liberty of conscience.

Earlier I noted that Rawls' argument for equal liberty of conscience does not consider distribution principles weaker than equality but stronger than the utilitarian principle or majority will. The LRDP is just such a principle and because of that, creates the tension we have just seen in Rawls' position. On the one hand, Rawls wants equal distribution to be a fixed point. On the other, his argument can at best establish that equality of distribution is preferred to riskier principles. Against an equally non-risky principle, like the LRDP, Rawls has yet to show that equality is the rationally preferred choice.

Our first relative rationality argument establishes that choosing

[17] Thomas Nagel objects to this possibility. See his remark in 'Rawls on Justice' (p. 14), this volume.

equal worth of liberty of conscience is rational if choosing equal liberty of conscience is. This argument shows that considerations internal to Rawls' theory prevent him from reconciling possible conflicts between the First and Second Principles by appeal to his distinction between liberty and worth of liberty. A second relative rationality argument can be applied to the liberties of political participation, the only other liberties for which Rawls presents an extended argument for equality of distribution (although, in this case, Rawls uses his argument as much to show the general priority of liberty as to show why it should be equal). The argument for equal participation liberties or 'the participation principle', as Rawls calls it (p. 221), is based on the importance of self-respect, which is classified as a primary good. Although, as we have seen, Rawls intends the equal liberty of conscience to be a model for other liberties, there are problems with generalizing that argument. Because the central primary good of self-respect is at the heart of this second argument for equal liberty, it may provide more promise of being extended to cover other liberties.

Once again, since our purpose is not to analyze it, we shall content ourselves with a rough paraphrase of Rawls' argument.[18] Our paraphrase is composed of points found in Sections 36, 63, 67, and especially 82:

(1) Since without self-respect 'nothing may seem worth doing' (p. 440), self-respect is an important primary social good, basic to all life plans. (2) When the index of primary goods is at a certain level, most urgent needs of the worst-off will be met (p. 542) and 'the fundamental interest in determining our plan of life... assumes a prior place' (p. 543). (3) At this point, self-respect becomes crucially important and parties in the original position would want 'to avoid at almost any cost the social conditions that undermine self-respect (p. 440) or increase risks to self-respect. (4) Similarly, at this point, parties in the original position would reject further increases in the index in favor of increases in self-respect or at least in favor of eliminating conditions that undermine self-respect. (5) Self-respect (a) could be based on socio-economic status, or 'income share' (p. 544), as it is in current

[18] A detailed discussion of this argument can be found in Henry Shue's paper, 'Liberty and Self-Respect', *Ethics*, 85 (forthcoming), which came to my attention too late to be included in this volume. Shue's analysis of the self-respect argument is similar in outline to my own, though it differs in details. Shue is not concerned with the effects on liberty of unequal wealth.

societies, or (b) it could be based on 'the public recognition of just institutions' and 'the publically affirmed distribution of fundamental rights and liberties' (p. 544), especially the liberties of political participation. But, (6) basing self-respect on socio-economic status is risky; since the Second Principle allows inequalities in the index of primary goods, some persons would have less self-respect than others. What is worse, (7) those with less self-respect have no acceptable compensation. All they have in return for their lower index is the assurance that it is maximal. But the fact that their index is maximal does not mean that their self-respect is; self-respect is based on the *relative* level of the index, not on its absolute level. Moreover, the higher index itself is not acceptable compensation by step (4). So, (8) basing self-respect on the 'publically affirmed distribution of fundamental rights and liberties' would be less risky than basing it on income share, provided that the distribution were equal; unequal distribution would be subject to similar objections to those mentioned in step (7). (9) The liberties most relevant to enhancing self-respect, since they imply one's value to others, are those which recognize as equal the contribution each party can make to determining public policy and action. Therefore, (10) parties in the original position would choose to secure self-respect by the public affirmation of the status of equal citizenship for all.

Some points about this paraphrase of Rawls' argument are worth noting. First, the claim that self-respect is such an important primary goal, appealed to in step (1), depends on a general psychological theory which Rawls argues for elsewhere (cf. Sect. 63 and 67). Discussing this theory would take us too far afield, but discussion is certainly warranted. The general role of self-respect described in premise (1) is what may make this argument more generalizable than Rawls' argument for liberty of conscience. Second, the explicit claim that self-respect emerges as the central primary good only when the index has reached a certain level [steps (2) and (3)] is not made explicit in Rawls' own exposition. I believe, however, it is compatible with his intentions. If Rawls believed that self-respect was always to be viewed as the most important primary social good, he would have given it a more central role in discussions of how the index is constructed and how the Second Principle is to be applied. Third, Rawls may oversimplify the possible bases of self-respect when he suggests the contrast which I characterize in step (5). Though socio-economic status no doubt plays a significant role in determining self-respect, it hardly is the whole story, as Rawls would readily agree. But then the other important bases of self-respect would have to be discussed before we could simply reject disjunct (a) of step (5) and opt for disjunct (b). Finally, step (4) of my reconstruction is not explicit in Rawls' discussion. Yet it is extremely important. Without it, it becomes impossible to show that socio-economic status is not an acceptable basis for self-respect. That is, it is impossible unless we adopt an even stronger premise, one claiming that self-respect must be distributed equally. But Rawls does not make the stronger premise explicit either. Indeed, one wonders how one could guarantee such a distribution, so it is unlikely Rawls would feel committed to equality of self-respect. Therefore, I use the weaker premise (4) since it catches the intention behind steps (2) and (3).

We can now sketch our relative rationality argument for equal worth of citizenship liberties. To carry through the argument, we must assume what was claimed earlier, that Second Principle inequalities in wealth and powers may cause significant inequalities in the worth of these liberties. As we have seen, Rawls cannot rule out this possibility. Indeed, the distinction between liberty and worth of liberty was introduced to cope with it. Also, we must be clear what the core of Rawls' own argument is. Rawls' argument for equal citizenship liberties depends on three claims, that public affirmation of the equal liberties could act as a social basis for self-respect, that enhancement of self-respect would be equal because the liberties are equal, and that this arrangement, viewed from behind the veil of ignorance, minimizes the risk of having relatively low self-respect, making it rational to choose equal citizenship liberties. Are there analogous claims that could be made for worth of liberty? If we can find these analogues, the rest of the relative rationality argument is elementary and we need not take space to spell it out.

One similarity is obvious. Inequalities in worth of citizenship liberties, that is, inequalities in the ability of parties to influence and participate in the political process, would be no less 'publically known' than the equal liberties themselves. But if these inequalities in worth of citizenship liberties are publically known, do they have an effect on self-respect? Could public knowledge of inequalities in worth of citizenship liberties act to undermine the self-respect of those with less worth of liberty?

It seems plausible to say they would, that public recognition of unequal liberty to exercise the 'affirmed' basic liberty is just as likely to undermine self-respect as public recognition of unequal liberties themselves. For example, consider those who are worst-off, as determined by their index of primary goods. They know that those far better off than they not only enjoy a higher index, but also have greater worth of citizenship liberties. They know, for example, that those better-off are more able to have their views and interests put forward in the mass media, are better able to select candidates, and are more effective in influencing office holders. It is likely, then, that their self-respect would be diminished. The mechanism here seems identical to the one Rawls cites in arguing for equal basic liberties: 'This subordinate ranking in the public forum experienced in the attempt to take part in political and economic life and felt in dealing with those who have a greater liberty, would indeed be humiliating and destructive of self-esteem' (pp. 544–5).

Thus, it seems that public knowledge of worth of liberty can act as a basis for self-respect. Moreover, unequal worth of liberties can enhance or diminish self-respect depending on how much worth of liberty one has. What is worse, it is hard to see how the well-ordered society could succeed in guaranteeing that the affirmation of equal liberties would successfully serve as the basis of self-respect but prevent knowledge of unequal worth of liberty from playing any role. The problem is that many people keep their eyes on the doughnut and not on the hole. They would reject the idea that their self-respect would be enhanced and secured by the public affirmation of equal liberties which they know they cannot exercise equally with others. Parties in the original position presumably would also keep their eyes on the doughnut. From the original position, they would believe it as rational to guarantee equal worth of citizenship liberties as they would to guarantee equal basic liberties themselves. Thus all three of Rawls' core claims about the relation between self-respect and equal citizenship liberties have their analogues for the equal worth of those liberties.

It should also be clear, from step (4) in our reconstruction, why we cannot simply compensate those with less worth of citizenship liberties, and therefore, with possibly less self-respect, by reassuring them that their index of primary goods is maximal. Once a certain level of the index is reached, it is not rational to prefer further increments in it to increments in self-respect. Since unequal worth of liberty diminishes self-respect for some, from behind the veil of ignorance it is rational to secure maximal self-respect through maximally equal worth of liberty. Equal worth of citizenship liberties gains its priority over other goods through its causal relation to self-respect, and self-respect enjoys priority by step (4).

As in the previous relative rationality argument, the compensation problem cannot be averted by appeal to a special, Liberty-Restricted Difference Principle (LRDP) applied to worth of citizenship liberties. Such an LRDP would justify inequalities in worth of citizenship liberties only if they acted to maximize the worth of liberty of those with the least worth of liberty. In other words, it would be justified to grant some people greater worth of citizenship liberties, if their having made it possible for those with less worth of liberty to have more than they otherwise would have had. The problem here is that the inequalities justified by the LRDP are not likely to coincide with the inequalities in worth of liberty caused by Second Principle inequalities in wealth and powers. As in the first appeal to an LRDP, the worst-off classes are not likely to coincide unless we make the strong assumption, which is probably not true, that every variation in the index has a corollary variation in the worth of liberty. More important, for the same reasons as before, it is unlikely that the worth of citizenship liberties is a monotonic function of the index. It is more likely that it is affected more by relative differences in the index than by absolute levels of the index.

If, as in the first relative rationality argument, we set aside the problem of reconciling the effects of the LRDP and the Second Principle, and assume it is rational to adopt the LRDP for worth of citizenship liberties, then we run into the same problem we did earlier. The relative rationality argument can be run in reverse. Instead of arguing from the rationality of equal liberty to the rationality of equal worth of liberty, we could argue from the rationality of unequal worth of citizenship liberties to the rationality of unequal citizenship liberties themselves. Rawls does not say that equal citizenship liberties have the status of a 'fixed point', but he surely seems committed to them:[19] 'When it is the

[19] One qualification is in order here. Rawls does seem to suggest at points that certain inequalities in citizenship liberties might be justifiable.

> The priority of liberty does not exclude marginal exchanges within the system of freedom. Moreover, it allows although it does not require that some liberties, say those covered by the principle of participation, are less essential in that their main role is to protect the remaining freedom [p. 230].

This remark seems to ignore the special relation between self-respect and

position of equal citizenship that answers to the need for status, the precedence of the equal liberties becomes all the more necessary' (p. 545). So, in general, Rawls would want to resist applying an LRDP to basic citizenship liberties. To be consistent, he would be then forced to reject appeal to the LRDP for worth of citizenship liberties as well.

V

Our two relative rationality arguments, using considerations internal to Rawls' theory, show that choosing equal worth of liberty is just as rational in the original position as choosing equal basic liberty. They show this for liberty of conscience and liberties of participation (citizenship liberties), the only two cases in which Rawls argues explicitly that the liberties are to be distributed equally. But this result means that Rawls' distinction between liberty and worth of liberty cannot be used to reconcile the First and Second Principles, as might have been hoped.

Initially, the distinction between liberty and worth of liberty looked like it might work. It made it seem that the First Principle demand for equality would not be undermined by Second Principle inequalities of wealth and powers, since, by definition, these inequalities did not affect liberty but only worth of liberty. If we could justify these inequalities in worth of liberty by application of a difference principle, perhaps the Second Principle itself, then the conflict would disappear.

Unfortunately, as the relative rationality arguments show, there is no way to accept unequal worth of liberty in the original position. Unequal worth of liberty cannot be compensated for by increases in other primary goods, since the reasons for granting priority to equal basic liberties apply to equal worth of liberty

equal participation liberties appealed to in Sect. 82. So does the following passage:

> The passengers of a ship are willing to let the captain steer the course, since they believe he is more knowledgeable and wishes to arrive safely as much as they do . . . the ship of state is in some ways analogous to a ship at sea; and to the extent that this is so, the political liberties are indeed subordinate to the other freedoms that, so to say, define the intrinsic good of the passengers. Admitting these assumptions plural voting may be perfectly just [p. 233].

Cf. Brian Barry's remarks on this point in *The Liberal Theory of Justice*, p. 145.

with equivalent strength. No other difference principle, even one restricted to maximizing (the minimum worth) of liberty, will work either. The 'fair' inequalities that might result from a Liberty-Restricted Difference Principle, for example, will not in general coincide with the effects of the Second Principle. Since it is these Second Principle effects which need reconciling with the First Principle, the LRDP gives no relief from the force of the relative rationality argument.

The distinction between liberty and worth of liberty thus fails Rawls in two ways. First, it has no satisfactory rationale. The special exclusion of economic factor from constraints definitive of liberty seems arbitrary, as we showed in Section III. Second, it fails to accomplish the task that motivated its introduction. That is, it fails to reconcile the First and Second Principles. What is worse, as was shown in Section IV, it fails for reasons internal to and important to Rawls' theory. These internal reasons, however, also have an import not restricted to Rawls' theory. Showing that equal liberty and equal worth of liberty are equally rational choices in the original position goes part of the way toward showing why equality of basic liberty seems to be something merely formal, a hollow abstraction lacking real application, if it is not accompanied by equality in the ability to exercise liberty. Further, since equality in the ability to exercise liberty is directly affected by the distribution of wealth and powers, our discussion of relative rationality has another consequence not restricted to Rawls' theory. It shows that a strong egalitarian sentiment in the political sphere may not be so isolatable as Rawls and earlier theorists had hoped from strong egalitarian demands in the social and economic sphere.

Perhaps this last point will be better understood if we look a bit more carefully at where Rawls stands as a result of our argument. Rawls seems to have two main alternatives. One is to attempt reconciling the First and Second Principles by refusing to allow any Second Principle inequalities which undermine the First Principle by making worth of liberty unequal. This strategy could be justified by resting very heavily on the priority of the First Principle. Since liberty has priority over other social goods, no trade-off can be allowed between worth of liberty and the index of primary goods. Rawls can accept our contention that significant Second Principle inequalities in wealth and powers can cause inequalities in worth of liberty, yet respond by ruling out all such significant inequalities.

Throughout *Theory of Justice,* Rawls uses examples which make it seem that fairly significant inequalities are compatible with justice as fairness. Perhaps the most striking example is the attempt to leave it an open question whether or not inequalities resulting from private ownership of the means of production are compatible with the Second Principle. Operating on the supposition that they are, Rawls describes in some detail a constitutional democracy which has as its basis a private ownership economy. If Rawls follows this first alternative, however, many inequalities which might have been justified by the Second Principle taken in isolation will probably fail the test of compatibility with the First Principle.

In a sense, a more far-reaching egalitarianism may be forced on us as a result of the two principles of justice than we at first expected, and certainly one more far-reaching than Rawls' examples indicate.[20] Rawls, being primarily interested in the argument for the principles themselves, might be willing to roll with the punch. All this means is that his system is not compatible, as a matter of empirical fact, with as diverse a set of social systems as he might have hoped.

But even if Rawls is willing to accept this result, there remains something of a surprise in it. What I have shown is that it is the First Principle, rather than the Second, which carries the egalitarian punch. It is the First Principle, even more than the Second, which is likely to force strong egalitarianism with regard to primary social goods other than liberty. As we have seen, however, Rawls' conjunction of the First and Second Principles is only a contemporary version of earlier attempts to conjoin equality in the political sphere with various social and economic inequalities. The thrust of our argument is that this historical attempt has also consistently underestimated the egalitarian force of the demand for equality in the political sphere.

There is, of course, another alternative. Rawls could reject the claim that significant economic and social inequalities cause inequalities in liberty or worth of liberty. But the attempt to reject this claim would involve Rawls in the 'policy sociology' he had clearly hoped to ignore while developing his 'ideal'

[20] The result is reminiscent of Engels' remark that one form of the proletarian demand for equality arises 'as the reaction against the bourgeois demand for equality, drawing more or less correct and more far-reaching demands from this bourgeois demand . . . in this case, it stands and falls with bourgeois equality itself.' Cf. *Anti-Duhring* (New York: International Publishers, 1966), pp. 117–18.

theory (cf. pp. 226–7). Nevertheless, the alternative does remain as a challenge. The serious point of social theory which Rawls, as well as the earlier liberal theories, would have to answer can be put succinctly: can a maximally extensive and equal system of liberties be successfully achieved without ruling out all significant inequalities of wealth and power?[21] I believe not.[22]

[21] The Marxist would rephrase the question as follows: does the demand for equal liberties make sense except when couched as the demand for the abolition of classes? As such, the question becomes a central focus of debate between liberal and marxist political theory.

[22] I would like to express my thanks to Hugo Adam Bedau and John Rawls for many helpful criticisms they have made of an earlier draft of this paper. A version of this paper was read at the American Philosophical Association Western Division Meetings, Chicago, April 1975.

VIEWS FROM THE SOCIAL SCIENCES

12 Rawls versus Bentham: An Axiomatic Examination of the Pure Distribution Problem[1]

A. K. SEN*

1. *Introduction*

In this paper I would like to compare and contrast the decision rules yielded respectively by the Rawlsian 'maximin' conception of justice[2] and by classical utilitarianism. Much of the discussion will take place in the context of a pure distribution problem, typified by the exercise of justly dividing a cake among n persons, which brings out some of the differences sharply.

In Section 2 a set of axioms are presented which the various choice rules may be expected to follow, and it is examined which of these axioms are satisfied respectively by the Utilitarian, the Rawlsian and other choice rules. The presentation in Section 2

[1] I have benefited from the comments of Partha Dasgupta and Norman Daniels. This is a slightly revised version of a paper published in *Theory and Decision*, Vol. 4 (1974).

* London School of Economics.

[2] In this paper I shall not be concerned with the contractual conception of fairness developed by Rawls [John Rawls, 'Justice as Fairness', *Philosophical Review* 67 (1958); *ibid., A Theory of Justice* (Harvard University Press, 1971)], and the justification of the maximin rule in terms of choices in the 'original position'. I have tried to argue elsewhere [A. K. Sen, *Collective Choice and Social Welfare* (Holden-Day and Oliver & Boyd, 1970), chap. 9] that the contractual conception may be more readily acceptable than the maximin rule as such. See also J. Harsanyi, 'Cardinal Welfare, Individualistic Ethics and Interpersonal Comparisons of Utility', *Journal of Political Economy* 63 (1955) and P. K. Pattansik, *Voting and Collective Choice* (Cambridge University Press, 1971).

is informal, but the axioms are more formally stated in Section 3 in which the results presented in Section 2 are fitted with proofs. If the reader is bored by formalities, he can easily move from Section 2 directly to Section 4, where the results are discussed again in completely informal terms.

The main conclusion is that Bentham and Rawls capture two different aspects of interpersonal welfare comparisons—both necessary and neither sufficient as a basis of ethical judgment. The utilitarian procedure is based on comparing *gains and losses* of different persons (e.g. 'person 1 gains more from this change than person 2 loses'), and is completely insensitive to comparisons of *levels* of welfare (e.g., 'person 1 is better off than person 2'). The Rawlsian procedure does exactly the opposite and is based on comparisons of *levels* only without making essential use of comparisons of *gains and losses.* In so far as both types of information can influence our ethical judgments, each approach must be recognized to be essentially incomplete. In my judgment Rawls' contributions are best appreciated as a welcome corrective to Benthamite blindness to comparisons of levels, and not as a complete theory in itself.

2. *Choice Rules and Axioms of Ranking*

There are n people rather austerely christened l, . . ., n. There is a fixed homogeneous income (cake) to be distributed among them. Each likes more and more of income but the gain from an additional unit goes down as he gets richer and richer. We take his welfare to be a function of his own income only and it increases at a diminishing rate as he gets more and more. The problem is to rank all possible distributions of the cake according to some rule of choice.

The Utilitarian rule (henceforth, UR) is to maximize the sum of individual welfares. The simplest version of the Rawlsian maximin rule (henceforth, MR) is to maximize the welfare level of the worst off person. The lexicographic version of the Rawlsian maximin rule (henceforth, LMR) is to follow MR, but if the worst off persons in two distributions are equally well off, then to maximize the welfare of the second worst off person. If the worst off persons are equally well off and so are the second worst off persons in two distributions, then maximize the welfare of the third worst off. And so on, under LMR.

Three axioms on rules of choice are now introduced.

THE SYMMETRY PREFERENCE AXIOM (SPA)

If everyone has the same welfare function, then any transfer from a richer man to a poorer person, which does not reverse the inequality, is always preferable.

THE WEAK EQUITY AXIOM (WEA)

If person i is worse off than person j whenever i and j have the same income level, then no less income should be given to i than to j in the optimal solution of the pure distribution problem.[3]

THE JOINT TRANSFER AXIOM (JTA)

It is possible to specify a situation in which j is [slightly] better off than k (the worst off person), and [strongly] worse off than i, such that some transfer from i to j [sufficiently large], even though combined with a simultaneous transfer [sufficiently small] from k to j, leads to a more preferred state than in the absence of the two transfers.

The Symmetry Preference Axiom simply stands in favour of a reduction of inequality if the persons have identical 'needs'. The rationale of this can take various forms, e.g., avoidance of arbitrary discrimination, or preference for equality of welfare levels of different persons, or for equating the welfare value of an additional dollar for everyone, or simply a preference for symmetry.[4] The Weak Equity Axiom demands that a person who is more deprived in non-income respects should not be made to receive less income as well. The Joint Transfer Axiom suggests that an inequality increasing transfer (from k to j) can be outweighed by a sufficiently large inequality decreasing transfer (from i to j). That is, some trade-offs are permitted. The words in square brackets are not needed in the statement of JTA and have been included only to motivate the axiom. While SPA is concerned

[3] The Weak Equity Axiom was defined in A. K. Sen, *On Economic Inequality* (Clarendon Press and Norton, 1973) in a somewhat more demanding form, requiring that person i should receive *more* (and not merely no less) income than j under the circumstances specified.

[4] On the last, see particularly S. Ch. Kolm, 'The Optimal Production of Social Justice' in J. Margolis and H. Guitton, eds., *Public Economics* (Macmillan and St. Martin's Press, 1969).

with single transfers, JTA is concerned with pairs of transfers.

The following results are true and are proved in Section 3 below.

(T.1) The Utilitarian Rule violates the Weak Equity Axiom for some set of permissible individual welfare functions.

(T.2) The Maximin Rule violates the Symmetry Preference Axiom and the Joint Transfer Axiom for some set of permissible individual welfare functions.

(T.3) The Lexicographic Maximin Rule can violate the Joint Transfer Axiom for some set of permissible individual welfare functions.

(T.4) There exist choice rules that can satisfy all three axioms (SPA, WEA and JTA) for all permissible individual welfare functions.

Readers uninterested in formal presentation and proofs can go directly to Section 4 from here, omiting Section 3.

3. *Formal Presentation*

The share of income of person i is y_i, for $i = 1, \ldots, n$. The problem is to rank all vectors y, i.e., $(y_1, \ldots, y_n)$, subject to:

$$\sum_{i=1}^{n} y_i = Y > O \tag{1}$$

$$\forall i: y_i \geqslant O. \tag{2}$$

Person i's welfare W_i is a monotonically increasing and twice differentiable function of his income y_i and is strictly concave.

$$W_i = W_i(y_i), \text{ with } W_i' > O \text{ and } W_i'' < O \tag{3}$$

The W_i functions can vary from person to person but are interpersonally fully comparable (see Sen, *Collective Choice*, Chapter 7). That is, different persons' welfare levels can be treated as being exactly as comparable as the welfare levels of the same person.

The following notation will be used in addition to standard symbols of algebra: → for 'if-then'; ↔ for 'if and only if'; & for 'and' (conjunction); v for 'or' (alternation); ∼ for 'not' (negation); ∀ for 'for all' (the universal quantifier); and ∃ for 'for some' (the existential quantifier). Further, R is the binary relation of 'at least as good as', P that of 'better than', and I that of 'indifferent to'.

$$xPy \leftrightarrow [xRy \,\&\, \neg yRx]. \tag{4}$$

$$xIy \leftrightarrow [xRy \,\&\, yRx]. \tag{5}$$

UR states that:

$$xRy \leftrightarrow \sum_i W_i\,(x_i) \geqslant \sum_i W_i\,(y_i). \tag{6}$$

MR states that:

$$xRy \leftrightarrow \operatorname*{Min}_i W_i\,(x_i) \geqslant \operatorname*{Min}_i W_i\,(y_i). \tag{7}$$

In distribution x, call the worst off person xl, and generally the i-th worst off person xi. (In case of ties in the poverty ranking, take the tied persons in either order.) Similarly, yi is the i-th worst off person in distribution *y*.

LMR states that:

$$\begin{aligned} xRy \leftrightarrow [W_{x1} > {}_{y1}\,\}\mathrm{v}\{\, W_{x1} = W_{y1} \,\&\, W_{x2} > W_{y2}\}\ \mathrm{v} \ldots \\ \mathrm{v}\ \{\forall i\text{: } i \leqslant n-2\text{: } W_{xi} = W_{yi} \,\&\, W_{x(n-1)} > W_{y(n-1)}\} \\ \mathrm{v}\ \{\forall i\text{: } i \leqslant n-1\text{: } W_{xi} = W_{yi} \,\&\, W_{xn} \geqslant W_{yn}\}] \end{aligned} \tag{8}$$

The axioms are now formally defined.

Symmetry Preference Axiom (SPA):

$$\begin{aligned} &[\{\forall i,j,y\text{: } W_i(y) = W_j(y)\} \\ &\&\ (y_i < x_i \leqslant x_j < y_j\} \,\&\, \{x_i - y_i = y_j - x_j\} \,\&\, \\ &\{\forall k \neq i, j\text{: } x_k = y_k\}] \rightarrow xPy \end{aligned} \tag{9}$$

Weak Equity Axiom (WEA):

$$\begin{aligned} &[\{\forall y\text{: } W_i(y) < W_j(y)\} \,\&\, \\ &\{\forall y\text{: } xRy\}] \quad \rightarrow \quad x_i \geqslant x_j. \end{aligned} \tag{10}$$

Joint Transfer Axiom (JTA):

$$\begin{aligned} &\exists\ x,y,\delta_1,\delta_2\text{: } [(y_i > y_j > y_k) \,\&\, (x_i \geqslant x_j \geqslant x_k) \\ &\&\ (\forall y_r\text{: } y_r \geqslant y_k) \,\&\, (\delta_1,\ \delta_2 > O) \,\&\, (\forall r \neq i,j,k\text{: } \\ &x_r = y_r) \,\&\, (x_i = y_i - \delta_1) \,\&\, (x_k = y_k - \delta_2) \,\&\, \\ &(x_j = y_j + \delta_1 + \delta_2) \,\&\, xPy]. \end{aligned} \tag{11}$$

Now the proofs of (T.1) – (T.4):

Proof of (T.1):
Consider $W_i(.) = mW_j(.)$ with $O < m < 1$ and $W_i(.)$ always positive. If $\forall y$: $x R y$, then under UR:

$$\sum_i W_i(x) = \underset{y}{\text{Max}} \sum_i W_i(y_i) \tag{12}$$

In view of the strict concavity and twice differentiability of each W_i, this implies that:

$$W_j'(x_j) = W_i'(x_i) = mW_j'(x_i). \tag{13}$$

Since $m < 1$, and $W_j'' < O$, it must be the case that $x_j > x_i$. Thus UR violates WEA.

Proof of (T.2):
Consider the antecedent in the statement of SPA, and take a case in which $\exists y_k$: $y_k < y_i$. Since $\forall k \neq i, j$: $x_k = y_k$, and $y_i > \leqslant x_i < y_j$, evidently

$$\text{Min } W_i(x_i) = \text{Min } W_i(y_i).$$

Hence xIy. Thus MR violates SPA.

Now consider JTA, if all conditions within the square brackets are satisfied except xPy, then clearly

$$\underset{i}{\text{Min}}\, W_i(x_i) \leqslant \underset{i}{\text{Min}}\, W_i(y_i).$$

Hence yRx according to MR. But this rules out JTA.

Proof of (T.3):
The same reasoning holds for LMR as in the case of MR in the latter part of the proof of (T.2).

Proof of (T.4):
An example will suffice and many examples do exist. We pick the choice rule of minimizing that widely used measure of inequality, the Gini coefficient, which we know can be written as:[5]

$$G = 1 + (1/n) - (2/nY)\,[ny_{y1} + (n-1)y_{y2} + \ldots + y_{yn}], \tag{14}$$

[5] See P. Dasgupta, A. K. Sen, and D. Starrett, 'Notes on the Measurement of Inequality', *Journal of Economic Theory* 6 (1973), p. 186. Note in that paper $Y=1$, and G is taken to be the negative of the Gini coefficient, i.e.,— G here. For a detailed discussion of the Gini coefficient, see Sen, *On Economic Equality*, pp. 29–34.

in which y_{yi}, as defined in the context of LMR, is the income of the i-th worst off person in the distribution y.

This amounts to the choice rule of maximizing W given by:

$$W = \sum_{i} (n + i - i)\, y_{yi} \qquad (15)$$

The unique optimum is given by $y_i = y_j$ for all i, j. Evidently WEA is satisfied.

It is clear that SPA is satisfied, since:

$$W(x) - W(y) = (x_i - y_i)\omega, \qquad (16)$$

where ω is the difference between the worst off ranks of person j and person i, which is positive, and so is $(x_i - y_i)$. Finally, for JTA, consider the effect on W of a joint transfer which does not alter the ranking of poverty of the persons. Let r(i), r(j) and r(k) be the worst off rank positions of i, j and k respectively. Obviously, $r(i) > r(j) > r(k)$.

$$W(x) - W(y) = \delta_1 [r(i) - r(j)] - \delta_2 [r(j) - r(k)] \qquad (17)$$

We can easily get $W(x) > W(y)$, by choosing:

$$\delta_1 > \delta_2 [r(j) - r(k)] \;/\; [r(i) - r(j)]$$

This completes the proof of (T.4).

But, of course, minimization of the Gini coefficient is an unappetizing rule from many points of view. Indeed, it passes WEA and SPA trivially, since it is defined independently of the individual welfare functions. It is an operation on individual incomes without referring to welfare levels. While the statements of the conditionals required under WEA and SPA involve individual welfare functions, these references occur only in the respective antecedents. The rule of minimizing the Gini coefficient satisfies WEA and SPA only because it renders the consequents true *irrespective* of the truth value of the antecedents concerned with individual welfare functions such that an automatic fulfilment of the consequences by the choice rules yielded by the Gini inequality measure will be ruled out.[6]

A more robust example can be constructed by devising a rule which sums some strictly concave transform of individual welfare

[6] In fact the Gini coefficient does not satisfy the stricter form of WEA proposed in Sen. *On Economic Inequality*.

functions—the concavity being sufficient to guarantee WEA in addition to SPA and JTA.[7]

4. *Discussion*

While neither classical Utilitarianism (UR) nor the Rawlsian maximin rules (MR and LMR) can satisfy all three of the axioms SPA, WEA and JTA, choice rules do exist that satisfy all three. In fact, as was shown, even minimizing a standard measure of inequality, *viz.*, the Gini coefficient, is such a rule.

The uses made of interpersonal comparisons of welfare in the different approaches are worth contrasting. For this I shall use an analytical framework that I have presented elsewhere.[8]

Non-comparability:
Welfare numbers of different persons cannot be compared *in any way.*

Level comparability:
Welfare levels of different persons can be compared but not differences between levels of different persons. E.g., it makes sense to say that A is better off than B, but none to say that A's welfare gain in moving from x to y is greater than B's gain (or loss) from the movement.

Unit Comparability:
Welfare level differences of different persons can be compared but not the levels themselves.[9]

Full Comparability:
Both levels and differences can be compared.

[7] *Ibid.*, pp. 20–2.

[8] Sen, *Collective Choice and Social Welfare*, chap. 7 and 7*.

[9] Under level comparability $W_i(x) > W_j(y)$ makes sense, but not $[W_i(x) - W_i(y)] > [W_j(y) - W_j(x)]$. Under unit comparability the latter makes sense but not the former. Note that level comparability holds if individual welfare functions are all 'ordinal' (order homomorphic to the real numbers) without being 'cardinal' (group homomorphic), but are entirely comparable. Unit comparability holds if individual welfare functions are all 'cardinal' and any change of unit of the numerical representation of the welfare function of one person must be combined with a change of unit in the same ratio for all persons, but an arbitrary constant can be added to any individual's welfare function without a similar constant being added to the welfare functions of others. On all this, see Sen, *Collective Choice and Social Welfare*, chap. 7*.

What are the requirements of comparability of the three approaches discussed in the earlier sections?

(1) Utilitarianism (UR): Unit comparability, or full comparability.
(2) Rawlsian maximin rules (MR or LMR): Level comparability, or full comparability.[10]

A distinctive feature of the Rawlsian maximin rules is their concentration on welfare levels, whereas Utilitarianism concentrates on welfare differences only. If unit comparability holds, MR or LMR cannot even be formulated, but UR can be used, since all it requires is to sum the welfare differences of all the persons in moving from x to y. And x is preferred, or y, or both equally, according as the sum is positive, negative, or zero, respectively. On the other hand, it is easily seen that if level comparability holds UR cannot be formulated whereas MR or LMR can flourish. The real conflict arises only if full comparability is assumed, when MR, LMR and UR can all be used. This was the framework used in the earlier sections of this paper.

5. *Concluding Remarks*

Critical comments on Rawls' maximin rules have mainly been concerned with their extreme nature concentrating only on the worst off individual ignoring the rest (or the k-th worst off, if the more worse off persons tie in the poverty scale, under LMR). This certainly is a significant aspect of Rawls' conception of justice, and this does differ sharply from other approaches such as utilitarianism. On the other hand, there is another, possibly more serious, aspect of the contrast between the two approaches, and this concerns the concentration on *levels* of welfare under the Rawlsian approach in contrast with the concentration on welfare *differences* in the utilitarian scheme of things. The contrast can be easily seen if it is asked: Under what circumstances should a transfer of income from person i to person j be recommended under the two approaches? Under UR, such a transfer should take place if and only if the welfare *gain* of j is greater than the welfare *loss* of i from the transfer. Under MR, it should take place if and only if i has a higher *level* of welfare than j who is the worst off person (and i stands to gain *something* from the transfer,

[10] Some uses of UR, MR or LMR are possible even when the required type of comparability is only partial, but not under all circumstances; see Sen, *Collective Choice and Social Welfare*, chap. 7 and 7*.

it does *not* matter how much). The extreme nature of the MR (or LMR) criterion in concentrating on the welfare level of the *worst off* person (or the k-th worst off, in the case of ties, under LMR) can be removed in a more general approach which could still retain the concentration on welfare *levels* as opposed to the exclusive concern with marginal gains or losses in the utilitarian approach. The Weak Equity Axiom is an example of a partial rule that is not extremist in the sense in which MR and LMR are, but which uses exactly the same type of information as the Rawlsian criteria.

Finally, it is reasonable to argue that in making ethical judgments on distributional issues (and in other types of social choices as well), one is typically concerned *both* with comparisons of levels of welfare as well as with comparisons of welfare gains and losses. It is not surprising that the utilitarian approach and the maximin approach both run into some fairly straightforward difficulties since each leaves out completely one of the two parts of the total picture. Given the powerful hold that utilitarianism has had on thinking on public policy for centuries, it is understandable, and in many ways entirely welcome, that Rawls has concentrated totally on the other half of the information set. But a more complete theory is yet to emerge.

13 Justifying Justice: Problems of Psychology, Politics and Measurement in Rawls[1]

BENJAMIN R. BARBER*

John Rawls' *A Theory of Justice*, as befits a work of such magisterial grandeur, has attracted a great deal of essentialist criticism: attacks calling into question its fundamental ideological and philosophical premises. Its rigid egalitarianism, its bourgeois predilections, its spirited defense of liberalism, and its partiality

[1] A grant from the Rutgers University Research Council for a more general project helped to make this critique possible. I am also grateful to Brian Barry, Quentin Skinner and Gordon Schochet for their comments on an early draft of the paper.

* Professor of Political Science, Rutgers University.

to fully developed, capitalist societies have all come under assault, and properly so. But my intention here is not to rehearse or elaborate these radical charges; rather, I want to raise certain questions about *A Theory of Justice* in Rawls' own terms—accepting his premises but examining his reasoning by his own stated criteria. It is my view that even in this limited perspective the Rawlsian theory of justice is wanting.

Rawls would like to persuade us that two intuitively attractive fundamental rules of justice, the equal liberty rule and the difference principle (with its fair equality for opportunity corollary), can be both philosophically justified by abstract rational argument, and concretely corroborated by appeal to their congruence with intuitive notions of man's sociability and the good.

I believe that my arguments here will show that the abstract justificatory appeal to the original position is unsatisfactory in certain ways, and that it raises problems of comparison and measurement not adequately disposed of by Rawls. I want also to show that the appeal to congruence is founded on an inadequate political and historical sociology which in turn creates further problems for the argument from the original position. In sum, I hope to show that while Rawls has lit his candle at both ends, he has gotten neither end to burn.

I

Rawls denies that he is making a Cartesian appeal to the original position as a source of necessary first principles from which the balance of his argument can be regarded as a mere deduction (pp. 577–8). Nevertheless, the original position occupies a critical role in his theory of justice. Technically, it functions as a hypothetical point of mutual disinterest that satisfies the requirements of an Ideal Observer in adducing the notion of justice as fairness. Because men in the original position are not yet particular men with particular notions of the good, Rawls is able to develop a proceduralist definition of justice uncontaminated by substantive first principles (the bane of institutionism). Yet because men in the original position are *potential* particular men with *potential* particular fates they will not be satisfied with non-particular or non-individuating notions of aggregate utility (the bane of mean utilitarianism). In brief, because men in the original position cannot determine who they will actually be they can be counted on to make disinterested and thus fair rules; but because they also anticipate living as actual particular men they will reject rules which sacrifice the welfare of particular men to the general good.

The original position also serves, rather like the notion of the state of nature in the earlier contractarians, as a hypothetical context for the definition of essential man stripped of all contingent particularity. Justice as fairness is not 'at the mercy, so to speak, of existing wants and interests. It sets up an Archimedean point for assessing the social system without invoking *a priori* considerations' (p. 262). Man's nature in the original position consists then in rationality and a generalized interest, not in particular desires, aims and aspirations.[2]

I want to suggest that Rawls' attempt to de-particularize the original position is not in fact very successful, and that in consequence its pre-moral (pre-substantive) character cannot be upheld. Men in the original position are defined by rough equality and freedom, by a general knowledge of the laws of nature and society, and by rationality—the capacity to anticipate consequences (Hobbes' ratiocination). They do not have particular interests but they do have a generalized interest in whatever particular interests they may acquire. Rawls argues that these limited conditions account for the emergence of the rules of justice as fairness. I believe that they do not, that additional assumptions about men that contaminate the original position need to be made if the rules of justice as fairness are to be regarded as the inevitable choice of rational men in the original position.

The first point that needs to be made concerns the meaning of interest. Rawls knows that while he can strip men in the original position of particular interests and particular desires he cannot leave them bereft of interest and desire altogether or they will cease to be men at all. 'Human actions' do, after all, 'spring from existing desires' (p. 568). Indeed, the judgments men make in the original position about alternative rules of justice are made 'solely on the basis of what seems best calculated to further their interests' (p. 584). Apparently men in the original position have interests, but not particular interests; they comprehend and

[2] 'It is not our aims that primarily reveal our nature', Rawls writes, 'but rather the principles that we would acknowledge to govern the background conditions under which these aims are to be formed and the manner in which they are pursued' (p. 560).

presumably feel the power of desire but are ignorant of which desires they will actually have.

Now there is a considerable question in my mind about whether it is possible to conceive of men as having a hypothetical knowledge of what it means to have interests and desires without having particular interests and particular desires. Mutually disinterested men might turn out to be uninterested men, men incapable of comprehending the meaning of interest. Rawls suggests as much when he concedes that 'some may object that the exclusion of nearly all particular information makes it difficult to grasp what is meant by the original position' (p. 138). At the level of psychology it seems possible that particularity is built into the notion of interest and that it cannot be cut away without rendering interest unintelligible.

Rawls seems to regard the idea of 'primary goods' as a reponse to this difficulty. Although men in the original position are not permitted to have specified, substantive ends, they are allowed through ratiocination to share a common interest in a set of common means. These common means are the primary social goods that can be thought of as instrumental to the pursuit of any and all particular aims, interests and ends. Although men in the original position remain mutually disinterested with respect to interests as particular ends, they understand that 'in general they must try to protect their liberties, widen their opportunities, and enlarge their means for promoting their aims whatever these are' (p. 143). Thus, they naturally (i.e., rationally) attempt to 'win for themselves the highest index of primary social goods, since this enables them to promote their conception of the good most effectively whatever it turns out to be' (p. 144). It is presumably in this sense only that men in the original position make calculations to 'further their interests'.

Yet this does not really answer the question of whether interest is intelligible at all in the absence of particularity; for the interest men take in primary goods is presumably only explicable in terms of the potential interest they have in particular ends. Moreover, Rawls draws the category Primary Good in terms so generous that its instrumental status seems critically compromised. Primary goods turn out to encompass not only the obvious instrumentalities like opportunities, powers, income and wealth, but also rights and liberties, and self-respect. The latter is a good so self-evidently contrary to the instrumental spirit that it is difficult not to conclude that it is a substantive first principle, an end-in-itself, smuggled into the original position under cover of the supposedly prudential primary goods.[3]

But let me for the moment accept that the device of primary good does meet the difficulties of rendering the notion of interest in the original position intelligible. Is Rawls then justified in claiming that the choice men in that position will supposedly make in favor of justice as fairness is a choice uncontaminated by any substantive, *a priori* idea of the good? or by the particular psychologies that attend particular men's experience in particular social systems? In sum, can it be safely assumed that the parties in the original position are not 'influenced by different attitudes towards risk and uncertainty, or by various tendencies to dominate or to submit, and the like?' (p. 530). I think not. In the relevant section (Sect. 26), Rawls introduces an 'analogy' that appears to go well beyond the minimal conditions portrayed earlier as definitive of the original position, namely the maximin rule for choice under uncertainty (p. 152).[4] The maximin rule 'tells us to rank alternatives by their worst possible outcomes' (pp. 152–3), tells us how to act as if our particular place in society were to be assigned by our enemies. Under these assumptions, Rawls believes it is rational to 'adopt the conservative attitude' expressed by maximin (p. 153). Rawls is at pains to persuade us that while the rule is neither self-evident, nor usual, nor generally applicable, it is uniquely suited to the peculiar conditions of the original position. Indeed, quite propitiously, 'the original position has been defined so that it is a situation in which the maximin rule applies' (p. 155), and it is to the maximin rule that the logic of the rules of justice as fairness apparently conforms. In guaranteeing themselves as much liberty as is compatible with an equal liberty for other men (the equal liberty rule), and in guaranteeing that whatever inequalities exist will be to the advantage of the least advantaged member of society (the difference principle), the parties in the original position are doing

[3] Rawls appreciates the contrast sufficiently to defer, for the sake of 'simplicity', his discussion of self-respect to the section on the Aristotelian principle and the full theory of the good (Part III). In the sections where primary goods are treated as facilitators of interest in the original position, self-respect is prudently and completely ignored (see p. 92).

[4] For other kinds of critical discussion of the maximin rule see Kenneth J. Arrow, 'Some Ordinalist-Utilitarian Notes on Rawls' Theory of Justice', *The Journal of Philosophy*, 10 May (1973); and David Lyons, 'Rawls versus Utilitarianism', *The Journal of Philosophy*, 5 October (1972). Rawls appears to back away slightly from his views on risk aversion in his 'Reply to Lyons and Teitelman', *The Journal of Philosophy*, 5 October (1972), pp. 556–7.

no more than following a strategy of minimal risk—are establishing rules of justice designed to protect them given the worst possible outcome for themselves in actual societies. This is, of course, the essence of maximin, a strategy which in Rawls becomes the vital bridge linking the rules of justice with the conditions described by the original position.

It is my view that there is nothing in the original position that suggests maximin as the only rational or most rational solution to the problem of choice under uncertainty; that, moreover, the question of which strategy would be most rational cannot be settled without further knowledge about attitudes towards risk and uncertainty, towards freedom and security, not given by the formalistic conditions of the original position; and that, finally, to treat adequately these attitudes Rawls must import into his original position covert special psychologies of the kind it was explicitly designed to exclude. Rawls in fact leaps from the original position, where men are prevented by the veil of ignorance from knowing what their particular statutes will be, to the unwarranted conclusion that this uncertainty will produce in them a rational preference for minimizing risks. Yet, as Rawls acknowledges, this assumes not merely that particular prospects are uncertain but that they are unpromising; not simply that particular statuses will be assigned by lot but that they will be assigned by enemies; not only that shares in the cake will be chosen in a random order but that they will be chosen last. It might equally well be assumed that friends will assign statuses, that men (a particular man) will get to choose first. It is no less rational, although suggestive of a different and less conservative temperament to be sure, for men to pursue, say, a moderate risk strategy whose aim would be to create the possibility of somewhat greater gains than afforded by maximin even at the risk of somewhat greater possible losses. Indeed, the scarcity built into all contractarian views of society (Rawls is no exception on this point) enhances the attractiveness of gambling strategies that, should an individual win, permit him far greater benefits than allowed by an austere egalitarianism—particularly if he regards his losses as comparatively insignificant as measured by the alternative (an unattractively austere minimum below which maximin guarantees he will not fall). Lotteries function precisely on this basis. Given still more radical assumptions about attitudes towards risk, it can be contended that some men may choose rationally to risk starvation and even death for the chance—even against the odds—to be very rich or very powerful. War is an extreme but hardly irrational example of this win-all/lose-all strategy. The development of capitalism is scarcely thinkable in the absence of high-risk attitudes in the face of uncertainty. A consideration of actual historical developments and concrete institutions as they manifest special psychologies may in fact suggest that the no-risk predilection for security may be atypical of human choice in the face of uncertainty.

There seem to be a number of psychological reasons for this. For one thing, the 'satisfactory minimum' afforded by the maximin rule may not be 'satisfying' at all by the criteria of maximin satisfaction. Avoiding pain or penury or powerlessness may not be measurable on the same scale as achieving (and enjoying) pleasure, wealth or omnipotence. This possible asymmetry may in turn reflect fundamental psychological disparities between the need for security expressed in the fear of pain, in anxiety, in the longing for serenity, and perhaps even in the death-drive, and the need for self-expression manifested in the quest for freedom, for spontaneity, for domination and for self-fulfillment. Rawls' focus on primary goods as instrumentalities for both avoidance and achievement ends (security and self-expression) blinds him to these kinds of possible disparities.[5] Consequently, he can opt for security (via maximin) in the original position without realizing that in doing so he is implicating a substantive special psychology. It is very odd in a philosophy of justice that champions the priority of liberty over all other goods and orders the rules of justice in accordance with this priority, to discover that a conservative special psychology predisposed towards security has been installed where rationality is supposed to be.

There is another, related difficulty in trying to guarantee a threshold of minimum satisfaction. Rawls argues that the interpersonal index defined by 'expectations of primary social goods' (p. 92) provides an acceptable standard for the kinds of interpersonal comparison required of a theory of justice. Yet primary goods as distinctive as freedom, power, wealth and self-respect can hardly be regarded as satisfying in some unitary way, except from the perspective of a rudimentary hedonism to which Rawls does not appear to subscribe. Freedom may seem more satisfying

[5] Rawls agrees with Santayana that 'we must settle the relative worth of pleasure and pain' (p. 557), but does not raise the issue in the context of primary goods and leaves its resolution to the discretion of 'subjective individuals'.

than survival—thus, the gospel song: 'Before I'd be a slave, I'd be buried in my grave/Go home to my Lord and be free . . .' Or a poverty-based self-respect may seem more satisfying than minimal economic welfare bought at the cost of an ignominious obeisance to bureaucracy. Certain Christian ascetics may even complain that the Rawlsian standard deprives them of the austerity and struggle for survival they regard as necessary to their otherworldly beliefs (see below).

Rawls may want to reply that because primary goods are means rather than ends such objections are not pertinent. But this response only raises again the question of whether the primary social goods can be thought of exclusively as means when they function as an interpersonal index of comparative expectations of particular men.

If these kinds of criticism have any truth, then those commentators who have tried to read Rawls as typically bourgeois in his outlook on human nature, who have suggested that the original position is informed by capitalist market biases, are mistaken.[6] On the contrary, Rawls' inclination towards a risk-free maximin strategy in the face of uncertainty suggests biases that are profoundly conservative—anti-capitalist in their thrust and to some degree anti-liberal in their spirit. They reveal a primary concern with secur¨y and the achievement of minimal conditions for individual welfare. The egalitarianism in which they issue is purely prudential, a device to ensure that the self-interested man will not be worse off than anyone else. Despite the considerable preoccupation with good life plans and the Aristotelian standard of excellence towards the end of the book, self-expression and self-fulfillment are not the major aims of the two rules of justice. Although freedom as a means is the first (and lexically prior) principle of procedural justice, freedom as an end receives little attention.

Rawlsian man in the original position is finally a strikingly lugubrious creature: unwilling to enter a situation that promises success because it also promises failure, unwilling to risk winning because he feels doomed to losing, ready for the worst because he cannot imagine the best, content with security and the knowledge he will be no worse off than anyone else because he dares not risk freedom and the possibility that he will be better off—all under the guise of 'rationality'. Recall that Rawlsian

[6] See, for example, Steven Lukes, 'An Archimedean Point', *Observer*, 4 June (1972).

men choose for minimal equality for the least advantaged not out of altruism or benevolence or social responsibility, but solely in order to protect themselves in the pursuit of their interests, whatever those interests turn out to be.[7] Surely more spirited, aggressive, optimistic men—freed of the constraints of morality and altruism as they supposedly are in Rawls' original position —might choose to pursue their interests more vigorously, less cautiously. Nor would they be any less rational for doing so, assuming that they had weighed and accepted the risks involved.

Rawls is faced with a dilemma: if he wishes to preserve egalitarianism he must contaminate the original position, for the rules of justice that precipitate egalitarianism are generated not by pure rationality but by the special psychology of no-risk planning under conditions of uncertainty. If on the other hand he wishes to preserve the pristine formalism of the original position, the two rules of justice as fairness and the egalitarianism they produce cease to be the inevitable choice of rational men in the original position, and justice as fairness becomes only one of many rational options. Yet the entire argument of *A Theory of Justice* precludes the surrender of either formal rationality or egalitarianism. Hence the dilemma.

Rawls has made clear, it is true, that the 'original position is not intended to explain human conduct except insofar as it tries to account for our moral judgments' (p. 120), and may on this score deem these psychological remonstrations beside the point. However, it is Rawls, I have argued, who has introduced special psychology into the pristine pre-particularity of the original position. To dismiss my arguments in support of the special psychology of moderate or extreme-risk strategies is to dismiss his own preferences for a low-risk strategy: all such strategies are cut from the same cloth and are part of a fabric that has nothing to do with rationality. In the absence of these kinds of preferences the two rules of justice simply are not defeasible as the inevitable choice of rational men in the original position. Admitting them means lifting the veil of ignorance partially and thus robbing the original position of its defining character. The argument from the original position seems crippled by the very crutch that makes it ambulatory.

[7] 'The theory of justice assumes a definite limit on the strength of social and altruistic motivation. It supposes that individuals and groups put forward competing claims, and while they are willing to act justly, they are not prepared to abandon their interests' (p. 281). Also see Part IV, below.

There is a more decisive reply to this line of criticism implicit in Rawls' suggestion that the original position is defined in such a way that it can be regarded as the original position (as rational) only insofar as it issues, by way of maximin, in the two rules of justice as fairness. Rationality, in this perspective, does not issue in but is defined by maximin. Although there is considerable ambiguity in Rawls on the point, a number of passages appear to support the implication: 'we want to define the original position so that we get the desired solution', he writes at one point (p. 141). 'The original position', he later notes, in what is probably the strongest statement of the matter in the book, 'has been defined so that it is a situation in which the maximin rule applies' (p. 155). But there is little cause to dwell on this line of defense. It is not a position to which Rawls can really afford to commit himself, and his ambivalence on the point elsewhere suggests he does not mean to do so. If the original position is defined simply as the position that issues in the two rules, it becomes analytic with respect to those rules and Rawls' entire justificatory enterprise is rendered truistic. If the two rules are necessarily entailed by the original position, the elaborate examination of alternative strategies pursued in Chapter 3 becomes a deception, and the extended debate with utilitarianism is made superfluous. Rawls is quite clear, however, that ideally men are 'to choose among all possible conceptions of justice' in the original position' (p. 122). Moreover, since he regards determining the rational preference between justice as fairness and the principle of average utility as 'perhaps the central problem in developing the conception of justice as fairness as a viable alternative to the utilitarian tradition' (p. 150), I think it fair to assume he does not really mean to argue that the original position is rigorously analytic with respect to the conditions under which rational preferences are developed. This appears to leave my original criticism unanswered.

II

As the difficulties of establishing comparable indicators for the satisfaction potential of different primary social goods suggest, there are serious problems of interpersonal comparison and measurement in Rawls' theory of justice. Rawls comments critically on the problem as it affects advocates of average utility theory (pp. 92–3) in order to argue that the contract doctrine makes it possible to 'abandon entirely' the thorny question of 'measuring and summing well-being' (p. 324). Although Rawls' approach does clearly avoid certain summing difficulties since it does not require determining collective well-being or average utility at all, it does not elude the difficulties of establishing ordinal scales of interpersonal comparison for critical terms like satisfaction, primary good and least-advantaged/most-advantaged.

Rawls leans very heavily on the notion 'least-advantaged'. The crucial difference principle thus reads: 'social and economic inequalities are to be arranged so that they are . . . to the greatest benefit of the least advantaged' (p. 83). According to Rawls, this formulation requires no 'accurate interpersonal comparison of benefits . . . it suffices that the least favored person can be identified and his rational preference determined' (p. 77); which in turn requires only that we compare the 'expectations' of individuals as defined by the index of primary social good 'which a representative individual can look forward to' (p. 92). No cardinal judgments need be made; only simple ordinal rankings are necessary.

The matter is not really quite so straightforward. As Rawls acknowledges, the construction of the index does generate problems, some of which are lodged in the category 'primary good'. Rawls asks, 'how are the different primary social goods to be weighed?' (p. 93). His strategy seems to be to narrow the category primary good by stipulation to an operationalizable core. Self-respect is deferred to the final section of the book where measurement is not an issue. Liberty, Rawls intimates, is not a problem because of its lexical priority over the other primary goods: since 'The fundamental liberties are always equal, and there is a fair equality of opportunity, one does not need to balance these liberties and rights against other values' (p. 93). This reduces the problem to identifying those with 'the least authority and the lowest income' (p. 94), presumably a manageable enterprise.

Unfortunately, manageability seems to have to be purchased at the price of meaning: quite aside from Rawls' own doubts about the lexical priority of liberty,[8] it can be doubted that crude indi-

[8] Rawls concedes that the precedence of liberty comes into play only after 'a certain level of wealth has been attained' (p. 542), and that below this threshold, liberty may not only have to be weighed against but perhaps subordinated to the other primary goods in whose absence freedom has no

cators like income are sufficient to measure so complex a notion as justice, particularly in modern industrial democracies. In the United States recently, blacks, white middle-class students, women, the rural poor, blue-collar workers and even the 'long-suffering' middle class have vied with one another for the title 'least-advantaged'. Depending on whether wealth, dignity, life purpose, political power, self-importance, employability or some other indicators are used, each of these groups can make its case. Income suggests only one dimension, and not necessarily the most salient dimension, of the issue. Whom then is to be regarded as least-advantaged: the prosperous black or the poor white? The unemployable, self-deprecating wealthy suburban housewife or the self-respecting, overburdened welfare mother? The over-taxed, undervalued assembly line worker or the alienated, anomic college drop-out? Rawls provides no criteria by which such social judgments can be made; yet surely in relatively affluent societies these questions are at least as relevant to the problems of justice as gross indicators like income.

Nevertheless, Rawls ultimately rests his entire technical case for the interpersonal viability of his ordinal measures on wealth and income alone. Not even authority and power, noted in passing, are given serious attention.[9] Comparability has been won only by gutting the category primary good and leaving a shell called income behind.

Even if we put these objections aside and accept that income may provide an approximate measure of the least-advantaged, there remains a serious deficiency in the Rawlsian argument. In establishing the superiority of the difference principle over the principle of average utility (Sect. 13), Rawls utilizes a graph plotting the relative income of a representative least-advantaged and representative most-advantaged man. The argument expressed in this graph, however effectively it may challenge the mean utilitarians,[10] is remarkable for its apoliticity and ahistoricity.

meaning. Depending on where the threshold is established, even Marx might be comfortable with such a viewpoint!

[9] By passing over power, Rawls evades the critical problems of definition and measurement that have attended its conceptualization among sociologists and political scientists. See, for example, Robert Dahl, *Who Governs* (New Haven, 1961), P. Bachrach and M. Baratz, 'The Two Faces of Power', *The American Political Science Review*, LVI (1962), pp. 947–52, and J. R. Champlin (ed.), *Power* (New York, 1971).

[10] There seems to be some doubt about this as well: see Scott Gordon, 'John Rawls' Difference Principle, Utilitarianism, and the Optimum Degree of Inequality', *The Journal of Philosophy*, 10 May (1973), and C. B. Mac-

Both Rawls and his critics seem to take for granted the social setting within which incomes covary in order to air their differences concerning the hypothetical point of maximum equality on a fixed, time-blind curve. Thus, on Rawls' graph (Graph A) where X_1 depicts the income of a representative most-advantaged man and X_2 the income of a representative least-advantaged man, 'a' is the point of maximum equality (pp. 76–9) on a contribution curve OP showing the relative position of X_1 and X_2.

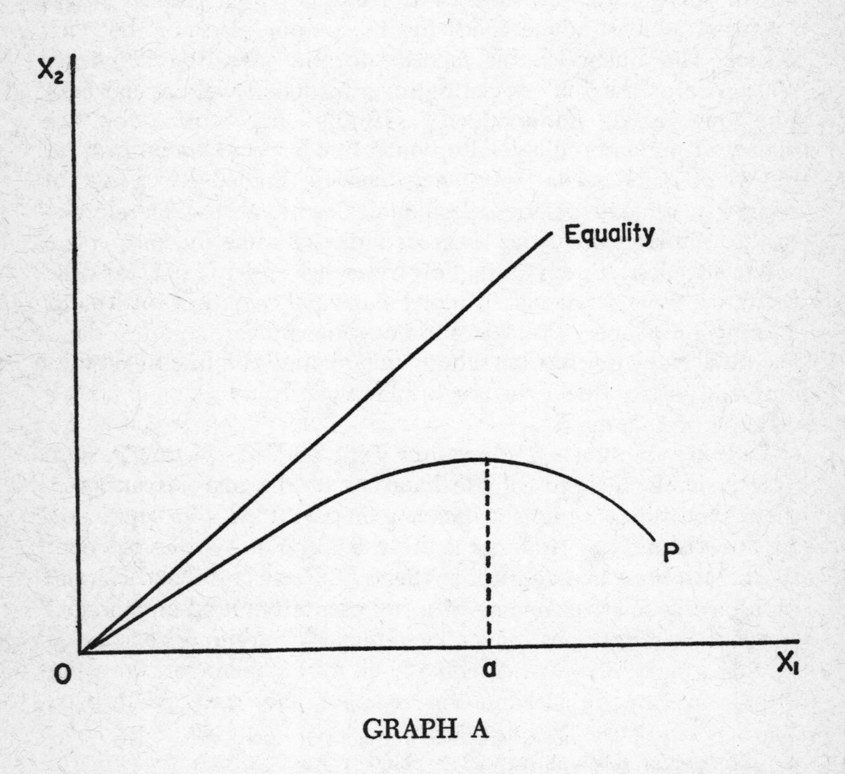

GRAPH A

I do not wish to quarrel with Rawls' technical discussion debating the location of 'a' on the static curve OP but to query the shape of OP itself, not only—as Scott Gordon has it—as 'an empirical matter in the realm of positive economics',[11] but as an

Pherson, 'Revisionist Liberalism', in his *Democratic Theory: Essays in Retrieval* (Oxford, 1973), pp. 87–94.

[11] Gordon, *ibid.*, p. 279.

empirical matter in the realm of historical development and social theory. Rawls' contribution curve OP assumes that 'the social cooperation defined by the basic structure is mutually advantageous' (p. 77). However, if OP is plotted in the context of some general theory of historical development or some particular economic theory, its conveniently symmetrical shape may be radically modified and the rendering of a point of maximum equality made correspondingly difficult. To make this clear let me portray two possible historical situations in which Rawls' abstract and static curve OP turns into a writhing snake hostile to Rawls' handling techniques.

Let us assume that X_1 represents (through representative individuals) the capital-owning class and X_2 the working class. Envision, then, two crucial threshold periods in recent industrial history: a period of rapid, disruptive unionization (exemplifying problems of incremental change), and a period of socialist revolution on the Marxist-Leninist model (exemplifying problems of radical, structural change). Under conditions of rapid unionization (Graph B), contribution curve OP would be radically deflected from any normal shape, leaving the point of maximum equality uncertain—ambiguous over time. If 's' is understood as the point in historical time where unionization is initiated through (say) a General Strike, a dramatic decline in the expectations of the working class (X_2) can be anticipated, its degree and duration depending on the reaction of the capital-owning class (X_1). At some point 'u', however, presupposing that the strike succeeds, the union shop is recognized by the bosses and a period of rapidly rising expectations for the working class ensues, to some extent at the expense of the owners. Following this phase, some point 'b' of maximum equality is reached that far exceeds the original 'maximum' at point 'a'.

What is telling about this portrait is that it confronts the advocate of justice as fairness with several alternative points of 'maximum' equality, between which it may not be possible to choose without reference to considerations of time, place and putative laws of development not given by the two rules he is essaying to honor. Viewing OP in its early phase, he might conclude that 'a' was indeed the point of maximum equality, particularly if he examined the projections for the segment between 'a' and 'u' that augured only ill for the expectations of X_2. He might, on the other hand, count on the eventual success of unionization, *if* his theory of development anticipated the phase from 'u' to 'b', and

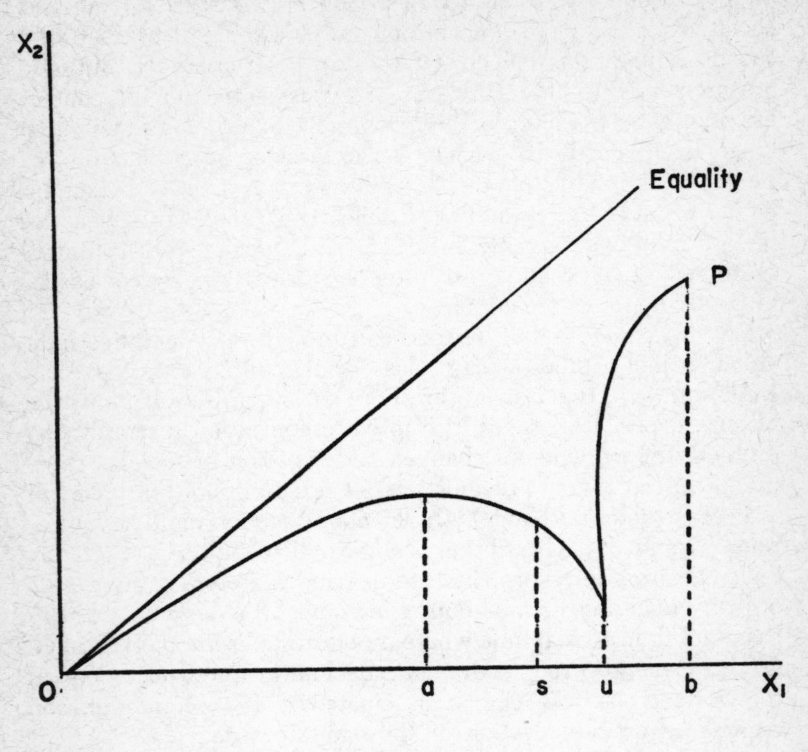

GRAPH B

sacrifice the apparent equality of 'a' for the greater equality of the slightly less certain but much more equitable 'b'. To do so he would of course be acting under conditions of uncertainty where risks might be taken in vain or success achieved at too great a cost, say following a five-year General Strike that destroys the economy in order to achieve a degree of equality only slightly greater than 'a'.

The point is the judgment concerning the point of maximum equality cannot be made in a timeless void, as Rawls seems to think it can. It must be reached in the context of a particular time and particular place where the relations between X_1 and X_2 conform to some developmental laws—laws which may or may not be generally known and understood. Rawls' strategy here becomes not merely undialectical and ahistorical but, in conjunction with the predeliction for maximin, anti-historical—predisposed to inertia. Under the conditions portrayed in Graph B, maximin would dictate forgoing a possible, but historically uncertain 'b' equality in return for a guaranteed 'a' equality. 'Just' men might thus, against their instincts to be sure, find themselves obliged by justice as fairness to act as strike-busters and scabs—a most melancholy and ironic consideration.

The situation reflects even greater uncertainty under conditions corresponding to the Marxist-Leninist theory of socialist revolution. Once again (Graph C) there is an early period of apparent maximum equality 'a' followed by a period of declining expectations during which the disappointments and grievances of the working class (X_2) feed revolutionary sentiment. In time,

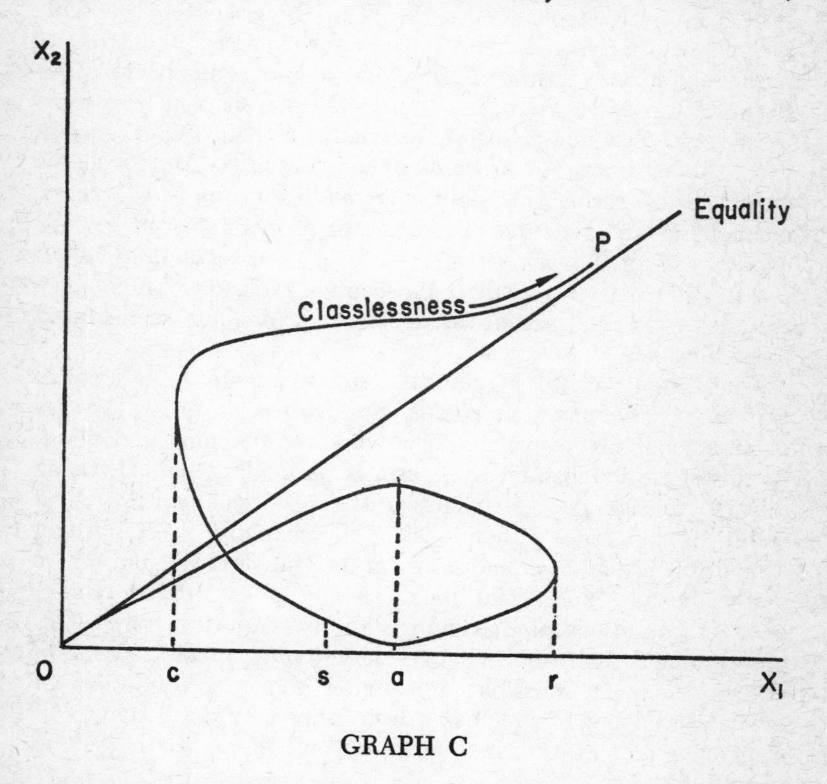

GRAPH C

the working class under the leadership of a revolutionary party overthrows the capital-owning class (X_1) and seizes its representative institutions (at point 'r'). During the initial period of disruption and civil war, this leads to the diminishing of both X_1s' and X_2s' expectations. But following the successful establishment of a revolutionary regime (at point 's') X_2s' expectations once again begin to grow, while X_1s' continue to decline as the expropriation of the expropriators becomes a reality. During the ensuing period of the dictatorship of the proletariat (the segment from 's' to 'c') the capital-owning class virtually disappears, and OP veers towards total equality; the class system itself is liquidated and X_1 and X_2 become indistinguishable, Everymen chosen at random in a classless society.

Now this projection creates further difficulties for Rawls:[12] by adumbrating fundamental structural changes in the relations between the two classes and, eventually, liquidation of the class system entirely, it challenges the static Rawlsian picture with a dynamic that seems quite beyond the compass of justice as fairness. For the ahistorical Rawlsian would again be tempted to prefer the certainty of 'a'—that is, the present reality of capitalist society (or feudal society or whatever the 'present' society was!)—to the risks and costs of some eventual P. Moreover, his anti-historical formalism might even, as history pressed forward, compel him to favor the restoration of property to the capital-owners during the period from 's' to 'c'; for, at least after OP crossed the axis of equality, the capitalist class would qualify technically as the 'least-advantaged' class. Maximin seems fated to such ironies.

I am not trying to suggest that Rawls intends to rationalize capitalism. The maximum rule is conservative *vis-à-vis* all modes of change: it favors propinquity, whatever the time and place. In this sense its abstract formalism is particularly salient, being adverse to risk *per se*, whatever the historical context. As a result, the difference principle is simply incapable of establishing a point of maximum equality except in a timeless vacuum where neither history nor social theory need be confronted. Under abstract conditions of maximum stability, enduring tranquillity and minimal structural change, identifying a static point of equality may be possible; but under the real conditions of evolving societies forever being transformed by the dynamics of economics and history, it is quite beyond the static projections of maximin.

Rawls may possess the conceptual apparatus to deal with some

[12] As would any projection based on a theory of structural change or revolution; the credibility of the Marxist-Leninist model is obviously not at issue.

of these objections. He recognizes that defining expectations 'solely by reference to such things as liberty and wealth' can only be 'provisional', and that it is eventually necessary 'to include other kinds of primary goods' that raise 'deeper questions' (pp. 396–7). A careful examination of the place of self-respect in identifying the least-advantaged, set within the more general discussion of the good and the Aristotelian principle, would presumably precipitate a much richer understanding of injustice than is afforded by a raw, ahistorical comparison of incomes over some limited period.[18] But self-respect can neither be operationalized in the fashion of income, nor rendered philosophically determinate in the manner of the two rules. It thus stays entirely in the background during the operational analysis of the difference principle, acting indirectly to leaven the otherwise flat ingredients of the primary good loaf. Consequently, inequality is construed in narrow pecuniary terms that neglect the psychological and political roots of exploitation in relationships conditioned by domination and submission, depersonalization and alienation, or socialization and public opinion. Rawls' formalism simply immunizes him to sociological modes of understanding. Even were he, say, to extend his contribution curves over a historical period encompassing unionization, it seems unlikely that his focus on equalizing income would nourish in him a sensitivity to the social and psychological consequences of unionization—for example, the standardization and routinization of work, the legitimation of wage-capital relations through the amelioration of conditions, the subordination of the creative ownership of labor power to the just distribution of its anonymous fruits, and so forth.

The evidence that Rawls' analysis is insufficiently political as well as egregiously ahistorical is in fact overwhelming. There is no need to mediate it through the technical arguments in favor of the difference principle. Hence, while my intention in this section was to indicate inadequacies in Rawls' criteria of interpersonal comparison, I want in the next to confront directly his apoliticity.

[18] Because the Aristotelian principle suggests a qualitative ranking of our activities and life plans corresponding to their complexity and the degree to which they fulfill our capacities, it gives to self-respect a moral depth it lacks as an instrumental primary good (see Sect. 65).

III

Rawls did well to avoid Godwin's title *On Political Justice* in labeling his enterprise. His formalism, his preoccupation with economic models and metaphors, and his predilection for abstract reasoning all combine to deter him from setting his emergent theory of justice in the context of historical or political reality. His tendency is, when theory meets practice, to eschew practice—dismissing questions of 'political sociology' and disclaiming any intentions of developing a 'theory of the political system' (pp. 226–7). His examples are often trivial, usually apolitical. Choosing between a trip to Paris and a trip to Rome is permitted to stand as an example of fundamental life choices (p. 412). Terms suggestive of modern man's political dilemmas—racism, alienation, nationalism, socialization, indoctrination—are hardly to be found; nor does the material face of politics characterized by power, command, authority and sanction (as against the ideal face described by legitimacy, obligation and justice) show itself. When political terms do occasionally appear, they appear in startlingly naive and abstract ways, as if Rawls not only believed that a theory of justice must condition political reality, but that political reality could be regarded as little more than a precipitate of the theory of justice. This raises a serious philosophical question about the relationship between the original position and what we can call the historical position—between the normative theory of justice and historical reality. I will examine this dilemma in the final section of the essay. Here I am anxious to illustrate the charges of apoliticity I have leveled at Rawls. Take the following propositions:

1. It is a 'mistaken view that the intensity of desire is a relevant consideration in enacting [just] legislation' (p. 230; also see p. 361).
2. 'The liberties of the intolerant [i.e. tolerating the intolerant] may persuade them to a belief in freedom' (p. 219).
3. 'There ought to be a cooperative, political alliance of minorities to regulate the overall level of dissent' (p. 374).
4. 'Throughout the choice between a private-property economy and socialism is left open; from the standpoint of the theory of justice alone, various basic structures would appear to satisfy its principles' (p. 258).

5. 'Each person must insist upon an equal right to decide what his religious obligations are' (p. 217).

I would like to suggest that a theory of justice that permits assertions of this sort, notwithstanding the caveats and contexts that attend them, is *prima facie* a theory of limited utility in addressing the injustices of actual political systems.[14] Let me make the point by commenting briefly on the cited passages.

1. If Rawls argued that intensity of desire was not a *sufficient* consideration in enacting just legislation, or not constitutive of the definition of justice, there could be little objection. But in pursuit of philosophical immunity for his notion of justice to particular wants and desires, he is constrained to claim intensity is irrelevant. This seems a foolhardy contention at best. Although intensity of desire hardly defines the just cause, history teaches that it can often reveal it. Liberal democratic systems are thus designed to permit an intense minority to resist (the filibuster) or even to overcome (the electoral primary) a lethargic majority. Moreover, the intensity question is absolutely crucial to the stability of democratic systems, and Rawls evinces a deep concern for stability: the 'well-ordered' society, he asserts significantly, is a stable society (p. 398). Now almost all social scientists agree that neglecting intensity can destabilize democracy;[15] how then can Rawls be committed to stability as an integral feature of his theory of justice, yet at the same time be indifferent to intensity? Presumably only by refusing to countenance political sociology.

2. Rawls is able to cite neither psychological nor historical evidence suggesting that tolerating the intolerant renders them more tolerant and liberty-conscious. The sorts of data deployed by those who are interested in evidence indicate that the intolerant are largely beyond rationality and thus quite incapable of grasping what is at stake in the very idea of toleration.[16] The

[14] Other perhaps than as a practical 'rule of thumb' in making clear-cut ethical decisions in completely unambiguous circumstances; see Peter Caws, 'Changing Our Habits', *The New Republic*, 13 May (1972), p. 24.

[15] Robert Dahl has argued, for example, that asymmetrical patterns among minorities and majorities where the latter employ the majoritarian principle to lazily and indifferently obstruct the will of the former can lead to the radical destabilization of the precarious democratic balance; see *A Preface to Democratic Theory* (Chicago, 1956), pp. 90–123. Also see W. Kendall and G. W. Carey, 'The "Intensity" Problem and Democratic Theory', *The American Political Science Review*, LXII, 1, March (1968).

[16] See Theodore Adorno et al., *The Authoritarian Personality*, 2 Vols. (New York, 1950); S. M. Lipset, *Political Man* (New York, 1960), especially

intolerant defer to imagined superiors with the same irrationality that they bully imagined inferiors. Fanatical true believers often seem to regard toleration by others as a compliment to their own rightness rather than as a reprimand to their intolerance. We need not, however, falsify Rawls' contention in order to make clear that, in the absence of political sociology, it remains not only unproven, but unexamined.

3. This may be an attractive idea, but I cannot think of a single historical case that confirms its viability under actual conditions of dissent in democracies. In the instances that come to mind—Czarist Russia in its 'parliamentary' phase prior to the revolution, Weimar Germany, America in the sixties, for example—internecine conflict among dissenting minorities has been an inherent feature of political dissent. Indeed, competition among dissenters has often been more violent and protracted than between the dissenters and the system they purport to oppose in common, and for understandable reasons. The scarcity of access to the system (via media impact, for example) compels claimants on its attention to out-shout, out-radical and otherwise out-do one another. Their belief in contradictory solutions to common problems makes them natural competitors in vying for revolutionary legitimacy. It is little wonder then that in recent years in America blacks, women, students and others have each claimed to be the 'most oppressed' group in American society, have each tried to characterize themselves as the true 'niggers of the world'. Telling such competing minorities, as Rawls does, that while their complaints may be legitimate *an sich*, taken in conjunction with the grievances and demands of other dissenting groups they simply overburden the capacity of the system for tolerance and change, seems a good deal less than just; particularly in a theory of justice that claims immunity from contingency.

4. This assertion is so striking that citing it almost obviates the need to comment on it. Albeit he acknowledges that 'a doctrine of political economy must include an interpretation of the public good which is based on a conception of justice' (p. 259), Rawls nonetheless wishes to leave the choice between capitalism and socialism 'open'. This, to me, is like developing a geometry in which the question of whether parallel lines meet is left open, or generating an aesthetic that refuses to take sides on questions of taste. Given the intimate interdependence of

Chapter 4; and S. A. Stouffer, *Communism, Conformity and Civil Liberties* (New York, 1955).

political and economic institutions in the West, and given the undeniable culpability of capitalism in the history of Western injustice, a theory of justice that sees nothing to choose between capitalism and socialism is either extravagantly formalistic to the point of utter irrelevance, or is a badly disguised rationalization for one particular socio-economic system, namely 'property-owning democracy'. Either way, from the standpoint of politics and sociology, it is deeply inadequate.[17]

5. Rawls' discussion of religion seems quite nearly as naive and incomplete as his treatment of economic systems. The passage cited contains a clue to his misperceptions in its strange supposition that men 'decide' on their religious obligations—presumably much the way they decide on whether to travel to Paris or Rome. This point of view follows from Rawls' positing of the priority of the just to the good: for the just is then also necessarily prior to the religious. That is to say, religion is admitted into Rawlsian psychology only after men, stripped of all religious particularity, have chosen the rules that are to condition whatever religious beliefs they end up holding. Religion is thrust into the realm of contingency and particularity, lexically subordinate to the settling of fundamental procedural questions. Religious ontology becomes secondary to rationalist epistemology.

Unfortunately, the religious view of the world begins with precisely the opposite assumptions. It presupposes the priority of the good to the just, the subordination of the epistemological to the ontological ('believe that ye may know'), the insignificance of contingency in the face of faith. The religious believer simply would not comprehend, let alone accept, the secular-skeptical premises of the original position. If he refused to acknowledge the 'liberty' of a non-believer, he would be acting not out of disrespect but out of the conviction that no man can possibly be free who does not believe. A Christian believer might not regard the perils of being in a persecuted minority during 'this life' as a sufficient warrant to compromise any aspect of his otherworldly faith. Indeed, he might even reject the rules of justice as a deprivation that would compel him to live without the austerity and suffering to which God has intentionally condemned men.

[17] Rawls remarks in passing that 'some socialists have objected to all market institutions as inherently degrading' (p. 280), but apparently dismisses socialism as too dependent 'on the strength of social and altruistic motivation' (p. 281). Altruism is an issue that comes back to haunt him, however, as becomes evident in the discussion below (Part IV).

If Rawls were to say to such a man, 'but think, my friend, what if you turn out to be a Confucian? or an atheist?' he could only reply 'all the worse for my immortal soul!' Put bluntly, there is simply no room in the afterlife for the original position: the only thing original in the Christian way of thinking is sin.

Not that there is any room in the original position for the afterlife—or for sin. Rawls' treatment of religion simply assumes that religious beliefs have no ontological or epistemological bases in truth whatsoever. The Impartial Observers' perspective is that of the agnostic, the skeptic; religion is thus reduced to dimensions where it can be encompassed and assimilated by justice as fairness. That is to say, it is rendered both trivial and contingent.

The problems raised by these inadequacies in Rawls' political sociology go well beyond political naiveté and inadvertent ahistoricity. To some extent his apolitical abstractness is intentional. Some of his sharpest critics have defended him on this point. Stuart Hampshire has labeled as 'unfair' (although accurate) the charges that Rawls 'pays no attention to the dominant powers of corporations, unions, the military machine, the secret police, the mass media . . .' since he intends only to provide 'a theoretical reconstruction of the notion of justice, and not of all the virtues of a good society and a good life; and he is not concerned with the practicalities of political science or with the theory of democracy'.[18] Such remonstrations might appear to blunt the line of criticism developed in this section. However, I believe I can show that Hampshire underestimates the deep ambiguity that surrounds Rawls' attitude towards, and account of, political reality. Rawls is never able quite to decide what sort of reality the rules of justice are intended for. On the one hand, it is a reality that manifests instinctively virtuous inclinations in congruence with and thus corroborative of the procedural principles of justice established by the original position. On the other hand, it is a reality very much like our own where the sublime verities of the original position are forever confronted by competition, conflict, inequality, war and even slavery. This disturbing ambivalence sets Rawls off in several contrary directions, plunging him into dilemmas from which he never altogether succeeds in extricating himself.

[18] Stuart Hampshire, 'A New Philosophy of the Just Society', *The New York Review of Books*, 24 February (1972), p. 39.

IV

Rather like Hobbes' model of the state of nature, the original position contains two moments, reason and interest, whose thrust tends in contrary directions. It is characterized, on the one hand, by rationality and collaboration, by the stipulated tabling of passions like rancor and envy, by a point of view that suggests a beneficent even mutualist interpretation of human nature. On the other hand, it is also characterized by interest psychology, by the drive for power and liberty as instrumentalities of personal ambition, and by an explicit individualism that treats all social collaboration as a means to individual ends. The motivational input in the original position is apparently individualistic, the social output (the rules of justice) a good deal more collaborative. Reason itself totters in its meaning between a consequentialist Hobbesian ratiocination and a deontological Kantian 'practical reason'.

This same ambiguity creeps into Rawls' portrait of the context into which the rules developed in the antiseptic original position are introduced, the context I have called the historical position. When Rawls suggests that in the original position men 'decide solely on the basis of what seems best calculated to further their interests' (p. 584), he seems to construe the rules of justice as typically liberal control mechanisms for the accommodation of conflicting interests in a real historical position that is essentially competitive. At one point justice itself is defined as 'the virtue of practices where there are competing interests and where persons feel entitled to press their rights on each other' (p. 129). If this is all to be taken seriously, the rules of justice must be understood as proximate criteria of minimal justice for conventional interest-oriented, power-responsive, conflict-ridden real societies rather than as absolute criteria of pristine conduct in ideal 'well-ordered' societies. As such they can presumably be deployed in the *ad hoc* fashion recommended by Peter Caws and others, justice being little more than a matter of choosing 'between several unjust, second best, arrangements' (p. 279). Although Rawls differs with the utilitarians on the question of aggregating utility, he shares with them from this perspective a common ambience in his hedonistic focus on interest, his instrumentalist view of reason, his manipulative approach to political institutions, and his pluralist preoccupation with conflict resolution.

Kantian though he may deem himself, and his protestations that his theory is 'ideal-regarding' rather than 'want-regarding' notwithstanding,[19] Rawls here seems to equate the historical position with what he calls 'private society'—an association 'not held together by a public conviction that its basic arrangements are just and good in themselves, but by the calculations of everyone . . . that any practicable changes would reduce the stock of means whereby they pursue their personal ends' (p. 522). It is precisely the point of the basic arrangements reached in the original position that they are not 'just and good in themselves' but decided 'solely on the basis of what seems best calculated to further [men's] interests' (p. 584).

Yet these are conclusions Rawls cannot accept and anyone who has read *A Theory of Justice* will recognize the above to be a one-sided picture. However Hobbesian his consequentialist reasoning may seem, his instincts remain Kantian; Rawls the Kantian understands that the argument from the original position must be corroborated, in keeping with the soft logic of congruence, by the evidence of man's instincts and intuitions in natural societies. It is perfectly obvious that 'private society' is not the model of natural society with which Rawls aspires to make the rules of justice congruent. He is at pains to deny that the 'contractarian view [is] individualistic'.[20] Although he anticipates that critics will contend, as I have above, that 'the contract doctrine entails that private society is the ideal', he is insistent that 'this is not so, as the notion of the well-ordered society shows' (p. 522). The well-ordered society, far from being a 'private society', is 'a form of social union' (p. 527); and social union is the form of association that corresponds with man's 'social nature' and embodies his 'shared final ends', the form of association where men naturally 'value common institutions and activities as goods in themselves' (p. 522). Now this construction, with its mutualist tone and implications of natural sociability, clearly contradicts the Hobbesian understanding of Rawls' state as an artifice of individual interests. It is more deontological than teleological, more mutualist than individualistic, more ideal-regarding

[19] Brian Barry argues that Rawls' theory is not convincingly 'ideal-regarding' at all, in his 'Liberalism and Want Satisfaction', *Political Theory*, I, 2, May (1973). Barry's general position, the most sweeping critique of Rawls to date, can be found in his *The Liberal Theory of Justice: A Critical Examination of 'A Theory of Justice' by John Rawls* (Oxford, 1973).

[20] In his 'Reply to Teitelman', *Journal of Philosophy*, 5 October (1972), p. 557. Also see *A Theory of Justice*, p. 584.

than want-regarding, more harmonistic than competitive. With it, Rawls provides an intuitive foundation in natural instinct for the procedural rules adduced without the mediation of instinct in the original position. Although 'the notion of respect or of inherent worth of persons is not a suitable basis for arriving at [the rules of justice]', these rules will 'be effective only if men have a sense of justice and do therefore respect one another' (p. 586). The rules of justice, it would seem, operate effectively only in an already 'nearly just society' where there already exists 'a constitutional regime and a publically recognized conception of justice' (pp. 385–6).

What this argument amounts to is an appeal by Rawls for secondary support to the kinds of intuitive, *a prioristic* notions necessarily excluded from the primary theory. The argument from the original position with its rather uncertain foundations in Hobbesian ratiocination is thus shored up by Kantian braces drawn from *a priori matériel* not admissible in the original position. It is in this sense that Rawls tries to burn his candle at both ends . . . the more light the better. Justification, we will recall, 'is a matter of the mutual support of many considerations, of everything fitting together into one coherent view' (p. 579).

The trouble is the two parts of the argument do not fit—the candle will not burn at either end. When Rawls sanitizes the original position he is careful to 'assume a definite limit on the strength of social and altruistic motivation' (p. 281). There is an important 'difference between the sense of justice and the love of mankind', the latter being 'superogatory, going beyond the moral requirements and not invoking the exemptions [in the name of private interest] which the principles of duty and obligation allow' (p. 476). In other words, too much altruism would obviate the need for a rationalist theory of justice altogether: 'A society in which all can achieve their complete good,' Rawls warns, 'in which there are no conflicting demands and the wants of all fit together without coercion, is a society in a certain sense beyond justice' (p. 281); for in 'an association of saints agreeing on a common idea, if such a community could exist, disputes about justice would not occur' (p. 129). But surely the notion of 'social union' or of the 'well-ordered' 'nearly just' society where men share 'final ends' and 'value common institutions and activities as goods in themselves' (see above) leans precisely in the direction of an 'association of saints agreeing on a common idea'. The intuitive foundations appealed to by Rawls to shore up his prudential theory ultimately render that theory superfluous; the theory of justice is convincing only in a context that is itself beyond justice!

Rawls does try to avoid this startling irony by construing social union as a form of association more mutualist than private society yet less utopian than an association of saints. The footing on this middle ground is not, however, very secure. As if on the top of a ridge separating the valley of private society and perfect union, Rawls seems continually in danger of plunging down one side or the other, able to traverse the terrain comfortably only by descending to the valley of private society and interest theory (the Hobbesian path), or sliding down the other side into the valley of perfect union and superogatory mutualism (the trans-Kantian trail, as it were). No doubt he would like to remain on the ridge where his perspective can encompass the Hobbesian and Kantian valleys at a glance. The point is he cannot, and the Hobbesian and Kantian in him are never really made compatible. The argument from congruence, far from reconciling these contradictions, only brings them into relief.

The conclusion that seems to me unavoidable is that the original position, even if the objections raised in Section I can be put aside, is incomplete, perhaps even untenable in the absence of intuitive arguments positing man's natural sociability and instinctive sense of justice; yet the rules of justice that issue from the original position are rendered superfluous in the presence of such arguments. If, on the other hand, the historical position is made equivalent to private society, then Rawls must be classified as a radical individualist in a manner he explicitly rejects, and the argument from congruence loses its force.

Finally, it seems that what Rawls has taken to be the smooth channel of congruence running between the perilous Scylla of utopianism and the stark Charybdis of realism is an impassable rapids in whose currents the theory of justice founders. The theory is a magnificently constructed vessel, however; and it will undoubtedly take more than some white water to sink it.

14 Constitutional Welfare Rights and *A Theory of Justice*

FRANK I. MICHELMAN*

How does *A Theory of Justice* bear upon the work of legal investigators concerned about recognition, through legal processes, of claims to be furnished with basic requisites of human fulfillment in society—subsistence, health care, education and the like—or with the money these things cost?

It is natural to ask this question. An apparent main purpose of Rawls' book is to advance and clarify discourse about claims respecting distribution of social goods including material social goods—income, wealth, and what you can buy with them. Rawls' theorizing seems aimed at bringing out and systematizing moral notions supposed to be already implanted in the bulk of his readership. His suggested principles of just distributions, while relying in part on a notion of 'pure procedural justice' (as embodied in a free market economy), demand in addition the constraint of outcomes to assure provision to all of a 'reasonable social minimum' and 'fair equality of opportunity'. Moreover, the principles and the arguments supporting them proceed visibly from the broad traditions of western individualistic democratic liberalism within which our characteristic notions of legal order and doctrine (including the notion of a constitutional legal order) have arisen.

One can thus well imagine that constitutional lawyers and scholars, seeking or weighing legal definition, recognition, and enforcement of specific welfare rights, would eagerly take to Rawls in search of a principled account of such rights—one which could be used to support or explain such legal events (actual or desired) as inclusions of specific welfare guaranties in a constitution or (more spectacularly) determinations by the judiciary that such guaranties are already present in the spacious locutions of, say, section one of the fourteenth amendment.[1]

* Professor of Law, Harvard University.

[1] 'No state ... shall ... deprive any person of life, liberty, or property,

By a specific welfare guaranty (or, as I shall sometimes say, an 'insurance right') I mean a right to provision for a certain need—on the order of shelter, education, medical care—as and when it accrues. I put the matter this way even though, as will appear, it is easier to find in Rawls support for the creation of more broadly formulated rights to guaranteed money income at some fixed or calculable rate (e.g., not less than half the median income), or—more broadly still—of rights against excessive or unnecessary inequality in wealth of income. But insofar as one is asking what can be found in Rawls to help promote or explain judicial discovery of welfare rights in the existing Constitution, the special focus on insurance rights seem at least provisionally in order. However strange may seem the nomination of welfare insurance rights as early candidates for direct judicial recognition as components of due process of equal protection,[2] most lawyers will instinctively feel that constitutional minimum-income rights are less likely still.

A number of questions, arising in various perspectives and at various levels, would have to be answered in reaching a final appraisal of what Rawls' theory may have to contribute to a conception of constitutional welfare rights. Although this essay deals with only one of these questions, it will help to chart them all in their broadest outlines.

To begin with, we have to note that Rawls is mainly concerned with what he calls 'ideal theory'—the selection, defense, and working out of principles of justice for and in the supposed context of a society which is 'well-ordered' or in a state of 'near justice'. The crucial assumption of ideal theory is that the principles of justice (whatever ones are under examination) are generally and explicitly acknowledged, accepted, and on the whole applied in the society. It is on that assumption that the

without due process of law; nor deny to any person . . . the equal protection of the laws.' See, e.g. Michelman, 'Foreword: On Protecting the Poor Through the Fourteenth Amendment', *Harvard Law Review* 83 (1969), pp. 14–16. The cited article should be compared with the vigorous criticism of it found in Winter, 'Poverty, Economic Equality, and The Equal Protection Clause', *Supreme Court Review* 41 (1972).

[2] Such recognition might take other and subtler forms than broad judicial orders to legislative and administrative authorities to establish and operate social-welfare programs. *See* Michelman, 'In Pursuit of Constitutional Welfare Rights: One View of Rawls' Theory of Justice, *University of Pennsylvania Law Review* 121 (1973), pp. 1013–15 [hereinafter cited as 'Welfare Rights']; Michelman, 'Formal and Fraternal Aims to Procedural Due Process' in *Nomos XI: Due Process*, J. Pennock, ed. (1974).

implications and consequences of the principles are worked out and appraised.

This feature of Rawls' work immediately suggests that one might first try to discover whether and to what extent the ideal theory itself (that is, the theory as worked out for a society supposed to be well-ordered by the very criteria it implies) suggests welfare rights that are judicially enforceable. We would then have to consider whether our society is well-ordered (or potentially well-ordered) in the indicated sense, and if not, what then follows. For a theory which suggests justiciable welfare rights for a society which is well-ordered might lead to different conclusions for one which is not. Conversely, and perhaps more important, it may turn out that even if the theory does not indicate justiciable welfare rights under ideal conditions, it suggests them as a way of coping with non-ideal conditions.

Focusing first on ideal theory, one would encounter another major division of inquiry. We want to know whether the principles of justice have a substantive content pointing to welfare rights and, if so, what specific shape these rights might take; but we need to ask also whether the theory contemplates that any such rights would be a part of the written constitution and (a different though related question) whether they would be enforceable in the face of legislative quiescence by judicial action. A like duality of 'substance' and 'procedure' would help organize pursuit of the theory into its non-ideal applications.

Lurking behind all this is still another major enquiry. Even if one decided that Rawls' theory suggests, for our society (well-ordered or not as the case may be), the constitutionalization and judicial enforcement of welfare rights, one would have to consider what special claims this particular theory might have to govern or influence the decisions of our judges.

Of all these questions, the only one I deal with here is that of the substantive welfare-rights implications of the ideal theory of justice as fairness.[8] In approaching this question, it is necessary first to consider the level of abstraction at which the theory is formulated. Rawls proposes to address only

> the basic structure of society, or more exactly, the way in which the major social institutions distribute fundamental rights and duties and determine the division of advantages

[8] I have tried to deal with the other questions in 'Welfare Rights'—from which the present essay is adapted (with some new material added).

> from social cooperation . . . The intuitive notion here is that this structure contains various social positions and that men born into different positions have different expectations of life determined, in part, by the political system as well as by economic and social circumstances . . . It is these inequalities . . . to which the principles of justice must in the first instance apply. These principles, then, regulate the choice of a political constitution and the main elements of the economic and social system [p. 7].

The notion of the basic structure may be crucial to the welfare-rights search, for it seems to suggest a response to one important line of objection to welfare insurance rights—that is, that such rights signify redistribution from the prudent and industrious to those who have culpably failed to grasp opportunities to provide for their own security. The notion of the basic structure suggests the possibility that society now contains a correctible shortage of economically secure positions (or an excess of insecure ones); and welfare insurance rights can then be appraised as a possible corrective device.

At the same time, the focus of justice as fairness on the basic structure may pose difficulties for the welfare-rights search. This focus means that the theory's principles of justice are selected for their capacity to serve as rather abstract, broad-gauged constraints against which to test more specific and circumstantially contingent proposals at the constitutional and legislative levels. And insofar as these constraints would allow variation in the specific content of welfare rights, according to societal circumstances and how other available choices (regarding the basic structure) are made, a specific catalogue of welfare rights cannot directly appear.

We turn now to our specific search for the welfare-rights implications of the principles of justice. It will be convenient to consider separately the difference principle, the opportunity principle,[4] and the liberty principle in that order. Eventually we shall have to consider the bearing of the theory as a whole.

The Difference Principle

Within the general framework of a free-market system, the

[4] That is, part (b) of the second principle of justice (see p. 302).

difference principle is said to imply an obligation to furnish each person with a 'social minimum', in order that the residual market determination of income and wealth may be considered just. Rawls apparently contemplates that this claim can be fully satisfied by general monetary transfer schemes such as negative income taxes (see p. 275), which take no specific account of particular needs to be covered. If this is correct, the difference principle's welfare-rights implications will be exhausted by a minimum-income claim which, I have suggested, will furnish no judicial purchase on the detection of welfare rights in the existing Constitution.

Our study of the difference principle's insurance-rights implications will proceed in the following way. We shall first pursue the matter on the assumption that the prospects of the least advantaged ('the bottom'), which the difference principle requires to be maximized, are to be defined in terms only of the primary social goods of income and wealth (rights and liberties evidently being taken care of by the liberty principle). On this assumption we shall compare the difference principle with that of average utility, which Rawls plainly regards as the difference principle's most plausible competitor (see pp. 161–6), to see which is more receptive to a notion of welfare insurance rights. We shall then ask whether, under the difference principle, the bottom's expectations are to be maximized also with regard to the primary social good of self-respect or 'a sense of one's own worth' (the most important primary good of all, as Rawls ultimately says); and if so, what difference this makes.

WELFARE RIGHTS IMPLICATIONS OF AVERAGE UTILITY

The difference principle asks whether the tax-transfer structure, with its associated incentive structure, can be altered so as to improve the situation of the bottom.[5] The average-utility principle asks whether any possible alteration can increase total

[5] Treating tax-transfer arrangements as the only component to be manipulated may be an oversimplification, since even if these arrangements are optimal for the bottom given the rest of the basic structure as it stands, some possible alteration of other sectors of the basic structure, or some general revision of the structure itself might improve the bottom's situation. I am simply assuming, subject to further investigation, that if even the oversimplified inquiry made here fails to yield justiciable welfare rights, they would not be derivable by a more complex inquiry, entailing still further debatable choices.

consumer satisfaction (assuming no change in population).[6] For average utility, at least, a notion of 'basic needs' might be relevant insofar as we make certain assumptions about the relationships between the intensity of a person's desire for additional quanta of certain goods and the 'amounts' of those goods he already has. These assumptions are but variations and refinements of the assumption of 'declining marginal utility of income' which utilitarians have traditionally made in arguing for some degree of income-equalizing tax-transfer activity.[7] They posit that persons now enjoying little or none of certain goods would be willing to pay, for certain increments to their existing low levels of enjoyment, amounts exceeding production costs by more than the normal market rate of profit.[8] Borrowing from economic jargon, we might call this excess a 'consumer's surplus' arising under nondiscriminatory pricing when dollars are spent to satisfy urgent needs arising out of deprivations, absolute or relative, not widely experienced in the economy.[9]

Utilitarians may well believe that income transfers from the top to the bottom which enabled the bottom to cover the normal cost of such high-surplus increments would, even allowing for disincentive effects which reduce total material output, seem likely to raise the total of consumer satisfaction; and if so, such

[6] Rawls' definition of average utility is more refined but leads to the same thing: 'To apply this conception to the basic structure, institutions are set up so as to maximize the percentage weighted sum of the expectations of representative individuals. To compute this sum we multiply expectations by the fraction of society at the corresponding position' (p. 162).

[7] For an elaboration and criticism of the notion of declining marginal utility of income, see W. Blum and H. Kalven, *The Uneasy Case for Progressive Taxation* (1953), pp. 56–63.

[8] Let a, b, c and d stand for an individual's progressively higher levels of enjoyment of some good. If a is a very low level, b is barely enough to be of any use, and c rather high, the value of b minus a may greatly exceed that of d minus c, even though there are plausible scales (e.g., production costs or physical quantities) under which the two differences are the same. Indeed, there may be no increment to a which has the same value as any increment to c. Increments to a which are smaller than b minus a may all be virtually worthless, and increments to a at least equal to b minus a, of great value; whereas increments to c seem always to have some marginal value but never as much as b minus a. Consider all this in the case of food (allowing increments to reflect improvements not only of quantity but of mix, variety, quality, refinement of preparation, rarity, delicacy, etc.).

[9] See A. Marshall, *Principles of Economics* 124 (1890); Hicks, 'The Rehabilitation of Consumers' Surplus', *Review of Economic Studies* 9 (1941), reprinted in K. Arrow and T. Scitovsky, eds., *Readings in Welfare Economics* (1969), p. 325.

transfers would be required by average utility.[10] The argument depends on a willingness to speculate about interpersonal utility comparisons; but of course the basic utilitarian ethic can hardly rule out action based on such speculations, insofar as they are believed true. One can thus imagine utilitarians trying to specify for any given economy a critical list of minimum service levels regarding certain needs, such that the anti-productive effects of intensified tax-transfer activity required to enable those at the bottom to cover the normal costs of this basic set would be just offset by the welfare gained through shifting resources to the satisfaction of these hitherto unmet basic needs. This would be the point where total consumer satisfaction is at its maximum and average utility is constant. Through this line of thought it is at least conceivable that under average-utility reasoning a social minimum could be exhaustively defined or calculated by reference to the normal costs of satisfying a basic set of minimum levels of service to particular needs.[11]

Of course this will seem an outrageously impractical recipe for calculating a complete social minimum, especially considering that the contents of the basic set of needs must undergo continuing redefinition to keep abreast of changes in total output, distribution, social practice, and 'tastes'. And even disregarding its impracticality, what has been described is only a method for *calculating* a minimum, not a method for *delivering* it. While the calculation is hinged to specific needs, delivery might well be in unrestricted cash. The calculation proceeds from general assumptions about most people's perceptions of their basic economic priorities; but even if one thought that such general

[10] This ignores the claims of future generations and the associated notion of a 'just savings principle', discussed at pp. 284–93. Like the oversimplification discussed in note 5 *supra*, this oversimplification seems allowable in view of our inquiry's limited purpose.

[11] To be precise we ought to speak of basic packages of interrelated needs and service levels, rather than collections of mutually independent basic needs and levels. The statement that I would give everything I had in return for treatment of my tuberculosis breaks down if all I have is just enough to eat. A more precise statement might be that I would give all I had to be both cured and fed. Again, how much shelter (in the form of artificial heat) I deem essential may depend on what diet and clothing are available to me; and how much education I deem essential might depend on what level of subsistence is assured me in any event. Thus there seem to be various sets of value-interchangeable basic packages. See Tribe, 'Policy Science: Analysis or Ideology?, *Philosophy and Public Affairs* 2 (1972), pp. 91–2. It would seem that in looking for that package whose provision will satisfy average utility we should focus on that package in each such set which is cheapest to produce.

assumptions might be accurate enough to provide the best available approach to calculating the universal social minimum guaranty that would maximize total consumer satisfaction, one would probably still choose to make some or all distributions in cash so as to avoid unnecessary sacrifices of welfare in the cases of recipients whose actual priorities vary from those of the general model.

Given the difficulties of calculation, the elusive and presumptuous nature of the necessary suppositions, and the possible wastefulness of in-kind distribution, it seems unlikely that utilitarians would try to compose their *entire* social minimum by packaging and pricing various constitutional insurance rights. Yet some select members of the basic set of needs might seem so obviously universal and durable as to warrant recognition as insurance rights, were it not for a fundamental problem of principle: under average-utility notions, transfers (no matter how calculated) are undertaken for the sake of a maximizing interest ascribed to the populace as a whole, and not for the sake of any acknowledged claim of justice or right on the part of the disadvantaged claimant as an individual. This being so, it is quite unclear why any policy favoring such transfers should be advanced in the guise of 'rights'—meaning demands that are to be met despite a currently opposing legislative will—especially when it is supposed, as an ideal theory, that the legislature is the authentic voice of a morally enlightened population.

THE DIFFERENCE PRINCIPLE COMPARED WITH AVERAGE UTILITY

More than anything else, what appears to distinguish the theory of justice as fairness from that of average utility is that the former *does* establish distributive claims for the disadvantaged to press on their own behalf, and not simply as happenstance advocates for the general public interest (cf. pp. 27–8, 31, 160; see also p. 182). And so one turns to the possibility of extending the 'basic set' method of defining and calculating a social minimum, or part of it, to the minimum implied by the difference principle. But this, it turns out, cannot be done. One *can* say that the difference-principle social minimum must be at least as large as the average-utility social minimum, because all changes designed to achieve the latter minimum must increase the bottom's income and all such changes are required by the difference principle.[12] But this is not at all the same thing as saying that the amount of the difference principle minimum would ever be calculated, even in part, by asking what particular needs the bottom was able to satisfy at various income levels. The special relevance of basic needs for average utility is that the abnormally high consumer's surplus they imply can overbalance the antiproductive effects of tax-induced disincentives so as to yield a net increase in total consumer satisfaction. But increases in total consumer satisfaction hold, as such, no interest for the difference principle. That principle simply pursues any increase in the bottom's income, whether or not large enough to yield a net rise in consumer satisfaction in the face of tax increases and associated incentive and production losses. Income-transfer activity is simply to be intensified just up to the point where any further intensification lowers total output so much that the bottom's absolute purchasing power begins to fall even as its relative share of total consumer satisfaction continues to rise. Under the difference principle, that is all there is to it. There can be no implicit insurance-rights package because there is no concern for what the bottom spends (or is able to spend) its income on. Income is a form of general power to achieve one's aims whatever they are—a primary, an elemental, social good, of which the bottom simply wants and is entitled to as much as can possibly be furnished it.

So we have arrived at this point: Under average utility there is a difficult argument for relating a social minimum to basic needs (or articulating a minimum in terms of social insurance for basic needs), but no argument for erecting the minimum so conceived into an individual right; whereas under the difference principle an individual right seems to be implied, but it is a judicially ungraspable income right rather than a set of insurance rights.

DOES A MINIMUM INCOME GUARANTY ADEQUATELY CARRY OUT THE CENTRAL CONCEPTION OF JUSTICE AS FAIRNESS?

Quite apart from any possible interest in judicial enforcement, this income-focused version of the difference principle seems troublesome. A precept for the distribution of material social goods which ignores claims regarding basic needs as such, and

[12] This reflects the conservative attitude (or maximin strategy) which, Rawls argues, is appropriate for parties in the original position and should lead them to prefer the difference principle over average utility (pp. 152–6).

is sensitive only to claims regarding the 'primary social good' of money income, will for many of us seem incomplete and thus not fully in harmony with our 'considered judgments'.

It may be asked how anyone's considered judgments can be offended by a principle that demands the maximum possible income for the bottom, merely because that principle exhibits no concern for satisfying any particular needs the bottom may be supposed to have. If certain needs remain unsatisfied even after the difference principle is satisfied, nothing more can be done to meet them. How, then, can considered judgments be offended?

One answer is that a considered judgment is composed not only of a conclusion regarding the rightness of a given act or outcome, but also of a reason or reasons, however indistinct or inarticulate, for that conclusion. I further suggest that the reasons may have practical importance not only in indicating what is to be done in the particular case (if the conclusion is that correction is required), but also in conditioning our more general response to situations of that type.

Suppose a number of persons are seen to have incomes so low that they cannot obtain the basic health care required by a person with no special medical problems. Two observers might agree that the situation demands correction, but disagree about the reasons. The first may say that if the lowest incomes are so low that their recipients cannot obtain basic health care out of private income, that is strong evidence that a higher minimum income could be guaranteed. It is hard to know what remedial action this observer could demand except that the legislature, with such economic-policy advice as it can garner, immediately consider whether the bottom's income might not be raised by some means consistent with the liberty and opportunity principles. The necessary analysis is likely to prove time-consuming and complex. A decision to stand pat, on the ground that no increase in the bottom's income is in fact possible, will almost certainly not be demonstrably wrong. The first observer could strongly believe that it is wrong, without questioning the legislature's good faith. At this point, the first observer could make no further demands.

The second observer says that the existing situation is troubling not only because it suggests that the bottom's income is not as high as it could be, but also because persons have a right not to be barred from basic health care by monetary shortage—recognizing as a natural limitation that the right is exhausted once it can be shown that the bottom's income is as high as it can possibly be, or that there is no way to free the health-care interest from this risk without displacing the risk onto some other interest deemed at least equally important. The second observer, then, can demand that the legislature enact a health-insurance program, and the burden of persuasion will then be on those who disagree to show that the natural limitation supports their position.

Thus the difference in reasons given for the same conclusion may have an important effect on attitudes with which the problem of correction is approached, and therefore on the legislative outcome; and this possibility exists although we assume, consistent with the suppositions of ideal theory, that the legislature always acts in good faith.

One reason, then, for feeling troubled about the insensitivity to needs inherent in Rawls' notion of primary social goods—a feature of the theory which apparently plays an important role in masking insurance rights out of a difference-principle social minimum—is the possible effect on attitudes toward the plight of representative persons whose incomes are so low that they experience specifically detectable, severe deprivations. But the sense of trouble may also have quite different sources: it may be a reaction against Rawls' insistence that the proper object of solicitude in a scheme of distributive justice is a worst-off *representative* person—that is, a member of worst-off *class* who supposedly has needs and other attributes typical for that class, rather than a motley crew of badly off individuals each with his own distinctive needs. For if the classes which representative persons represent are conceived, as Rawls indicates, as socio-economic or victim classes (see pp. 98, 99), then the difference principles allows the possibility of guaranteeing to each member of the worst-off class an income—the highest which can be arranged for all members of the class—which enable typical members who are, say, diabetics unable to afford both insulin and a basically healthy, daily diet.

'PRIMARY SOCIAL GOODS' AND AGGREGATION OF 'THE DISADVANTAGED'

The resort to an aggregative conception of the 'worst-off' whose expectations are to be maximized, and the measurement of those expectations in terms of universally valued primary goods (such as income and wealth), seem to be two aspects of a single moral notion. If we really thought that the only need people had was the undifferentiated need for more income, then under Rawls' 'general conception' of justice as fairness[13] the 'worst-off' would simply be the class of those whose incomes were below the highest guaranteed minimum income of which the economy was capable. Conversely, if individualized needs on the order of basic health are considered discrete components of expectations, it is hard to see how any simple notion of worst-off representative persons can satisfy the precept of maximin which characterizes the general conception. Represented classes would have to be defined by categories of need pertaining to the several components of expectations (basic goods), measuring need by existing shortfall from some standard minimum service level (health might be such a good and diabetics a class); weights would have to be assigned to the need categories; 'worse-offness' would consist of the total weights borne by an individual over a period;[14] and transfers for that period would then have to be made in accordance with the 'lexical difference principle' (see pp. 82–3). If one could speak at all of a standardized 'social minimum' guaranty, its form would have to be that of a package of output-measured[15] minimum levels of service to various basic needs—the broadest list of needs and/or highest levels of service that the economy is able to guarantee to all, taking into account tax-generated disincentives. The guaranteed minimum—or any part of it—could be provided either through social services freely available to all, or by cash distributions designed to assure each person of ability to bring himself up to all the minimum levels through open-market purchases.[16] This sort of social minimum

[13] 'All social primary goods—liberty and opportunity, income and wealth, and the bases of self-respect—are to be distributed equally unless an unequal distribution of any or all of these goods is to the advantage of the least favored' (p. 303).

[14] To speak of 'worse-offness' as the total of the weights is only a very crude first approximation. The combinatorial relationships of the various needs are bound to be much more complex than the continuously additive relationship suggested by a simple summing of the weights—involving gaps, groupings, lexicalities, etc. Compare note 11, *supra*.

[15] By output measurement I mean a performance standard—how well can the patient hear? how well can the student calculate?—as distinguished from a measurement of the resources devoted to a given case.

[16] Presumably we would use the latter alternative when we were prepared also to use a 'means test' designed to require a person to spend private income on achievement of the minimum levels, insofar as he was able. Some of the factors bearing on this question are discussed in Michelman, 'Welfare Rights', at 1011 and nn. 141, 142. See also B. Barry, *The Liberal Theory of Justice* (1973), pp. 114–15.

would somewhat resemble the needs-focused average utility social minimum earlier described. There is, however, a crucial difference which perfectly typifies the essential distinction between average utility's concern with aggregate satisfactions and the concern of justice as fairness with individual rights: for average-utility purposes, it may be quite appropriate to guarantee each person a standard amount of cash—the amount which would enable a typical person to get himself up to all the basic levels;[17] whereas the maximinimum I am now describing would attempt to assure each person of enough cash and services to get *that person* up to the basic levels.

Rawls contends that persons in the original position, having to choose the principles by which they will judge features of their society's basic structure, should rationally exhibit a conservatism so pronounced as to lead them to a maximin strategy in choosing these principles. If so, why would they not choose principles which generate a social-minimum guaranty taking cognizance of individual basic needs, rather than a guaranty which knows only the 'primary social good' of income?[18] It is true that a basic needs-focused minimum would be quite impractical unless great latitude for approximation and oversimplification were allowed in its construction. It is also true that a needs-focused approach would require an objective or collectivized method of identifying and defining need, so that the resulting standard minimum package would never be just the package which each person (or any person) would have preferred for himself. But Rawls' own theory of the three stages of increasingly

[17] The questions for the average utilitarian would be: (i) whether a more individualized investigation of ability to satisfy high-surplus needs would garner more in primary resource-allocation benefits than it would cost in administration expense, and (ii) how much 'extra' can be spent on bringing persons with abnormally costly problems up to the standard levels, consistently with the precept of maximizing *average* welfare.

[18] B. Barry, *supra* note 16, at pp. 56–8. Rawls cites three features of the original position which he believes make a maximin strategy rational for the parties. The first and third, but not so clearly the second, suggest that the parties would want a minimum guaranty geared to the costs required by individuals to satisfy their basic needs. The three features are: *First*, 'there must be some reason [namely, total and insuperable ignorance] for sharply discounting estimates of [likelihoods of . . . possible circumstances]'; *second*, 'the person choosing has a conception of his good such that he cares very little . . . for what he might gain above the minimum . . . that he can . . . be sure of by following the maximin rule'; *third*, 'the rejected alternatives [but not the preferred one] have outcomes that one can hardly accept'. The problematical bearing of the second feature will play a part in my discussion which follows.

specific social choice—the original position, the constitutional convention, and the legislative stage—shows that these are not fatal objections. Obviously the specific contents of a needs-focused minimum package cannot themselves be a 'principle of justice' to be chosen in the original position; but perhaps the parties could choose a principle obligating the legislature periodically to undertake the design and implementation of a basic needs-focused minimum guaranty having the general shape and structure I have indicated.[19] Whether they would prefer such a principle to the aggregative difference principle which generates a legislative obligation to arrange for the highest possible minimum income seems to depend on how they would estimate the probabilities of (a) substantial convergence at the legislative stage of individual perceptions that certain needs are basic; and (b) the legislature's ability accurately to detect the convergent perceptions and feasibly to make provision for them. Offhand it seems that these probabilities would be rated good enough for highly risk-averse parties to have preferred a more needs-sensitive approach to the social minimum than is dictated by the notion of income as a primary social good.[20]

There remains, to be sure, a persuasive objection to a social-minimum guaranty made strictly responsive to needs. I think it is an objection which goes to the very heart of Rawls' moral intuition. Exploring this objection may suggest that justice as fairness is not so neatly and thoroughly at odds with average utilitarianism as Rawls would sometimes have us believe[21] (e.g., pp. 3, 14, 15, 22, 30) and ascribing the objection to parties in the original position will require a sharp qualification on the hypothesis of their rational aversion to risky choices in this

[19] There would, then, be nothing in the argument directly specifying any judicially cognizable welfare insurance rights. Yet the conception of a moral duty incumbent on the legislature to come forward with a package of such rights may have important implications for judicial action. See note 2, *supra*.

[20] These probabilities might be thought to be affected by the availability of a persuasive, philosophical account of the notion of basic needs. For one such account see C. Fried, *Difficulties in the Economic Analysis of Rights, with Application to the Case of Health* (to be published). See also C. Fried, *Medical Experimentation—Personal Integrity and Social Policy* (1974), ch. 4. Agreement on a principle committing the legislature to assure provision for each person's basic needs would not require any unattainable 'agreement on how to estimate happiness as defined, say, by men's success in executing their rational plans, much less on the intrinsic value of these plans' (p. 95).

[21] But this is not at all to suggest that justice as fairness is a utilitarian theory, or that it is not clearly distinguishable for utilitarianism (cf. p. 316).

situation. This objection is simply that satisfying a highly disadvantaged person's basic needs will sometimes be possible only at exceedingly large cost; and a commitment to pay such costs whenever required will apparently force an unacceptable lowering of the minimum assurances which can be extended to the disadvantaged generally. The moral intuition at work here evokes a hybrid of maximin and average utility: it calls for something like the highest attainable level of average provision for the group of the disadvantaged.[22]

But such a precept no more points to a strictly income-focused social minimum than to a strictly needs-focused minimum. It points to some sort of hybrid or compromise, whose possible contours seem susceptible of loose description: the hybrid guaranty might consist of (i) an array of standard minimum service levels respecting various basic needs;[23] (ii) an array of assured fallback treatments (sometimes consisting of no treatment) to be provided in cases of shortfall from the minimum standard levels which are unacceptably expensive to 'cure';[24] (iii) an array of compensatory assurances of super-standard treatment in other needs categories, to be provided in certain cases where fallback treatment comes into play;[25] and (iv) a residual minimum income guaranty. The specific content of these various, interconnected assurances would be chosen so as to produce as 'high' a guaranteed minimum—assuring as full and complete

[22] Cf. p. 98: 'The expectation of the lowest representative man is defined as the average taken over this whole [lowest socio-economic] class.' See also p. 101 (principle of redress to be balanced against other principles such as improving average standard of life).

[23] For example, a standard of hearing acuity.

[24] Suppose a case of hearing impairment which is medically treatable at enormous expense, and which can also be fully compensated for (so far as auditory sensibility is concerned) by an inexpensive mechanical hearing aid. Granting that anyone, given the choice, would prefer unimpaired natural hearing to mechanically assisted hearing (and assuming that some standard of good hearing is a component of the minimum standard for health care), we would still define our guaranty so as not to differentiate between natural and mechanically assisted hearing.

[25] Suppose that after provision of any guaranteed 'fallback' treatment a person remains functionally deaf. Although we have now done 'everything we can' in the way of responding to that person's basic needs *through medical care*, we may not yet have done everything we can—everything the highest possible social minimum would do—to respond to those needs. For it might well be possible to guarantee for such cases a degree or kind of special education or training, over and above the minimum guaranteed for normal persons, which would significantly compensate for the hearing disability.

satisfaction of the basic needs of everyone (though not of each one), and beyond that as large a minimum share of residual goods for all—as circumstances would permit.

Now it must be recognized that Rawls nowhere insists that a guaranty of this form would be an inappropriate way for the legislature to carry out its obligation under the difference principle. Conceivably Rawls would be troubled about legislatures presuming to determine for persons what their needs are (cf. p. 250); but he indicates nothing stronger than a policy preference for a guaranteed-income form of social minimum, based on an apparent assumption that in our society such an income guaranty would cover all basic needs—save, perhaps, in those cases where satisfying them would be extremely costly—assuming that the level of the guaranty were properly selected.[26] The question, then, is not why Rawls rejects the hybrid form of social minimum —for he does not—but rather why he has not favored a principle of justice (a version of the difference principle) which would *dictate* the hybrid form and so perhaps allow us to assert rights to have specific needs provided for. The answer may lie in the extreme difficulty—it may be the impossibility—of articulating such a principle in a form suitably abstract, and at the same time sufficiently structured, for adoption in the original position. In order to explain, I want to go back over some of the territory we have covered in somewhat more systematic fashion.

Let us say that goods are either basic or residual. Basic goods are those providing a foundation upon which a person builds his pursuit and enjoyment of other (residual) goods, or those that partially constitute the self which does the pursuing and enjoying of residual goods. A person's access to basic goods thus radically controls the worth to that person of residual goods.[27] Now to speak of a good as foundational or constitutive in this way means that a person's demand respecting that good is a demand for the good itself, not a demand for resources convertible into

[26] Rawls' references to the social minimum seem to equivocate between concern with needs and concern with holding down inequality of income. Compare pp. 276–7 with pp. 285–6. One comes away with the impression that the minimum is to be set with a view both to assuring fulfillment of certain basic needs and to accomplishing a desired, but highly contingent and ineffable, degree of income equalization. Yet it also seems that the former goal is not a direct one, but rather left to be satisfied as a presumed consequence of pursuing the latter goal.

[27] The notions and terminology are adapted from C. Fried, *Economic Analysis of Rights, supra* note 20. Rawls treats self-respect, or 'the bases of self-respect', as basic in this sense (see p. 397).

the good.[28] To see whether a person has the good we look to performance levels, not resource inputs. And conversely, it seems that to speak of goods as residual is to say that with respect to them what a person wants is precisely the resources—income, wealth, powers, opportunities—which an adequately constituted and grounded self can use to realize its own plans and inclinations.

We can imagine a race of beings whose foundational and constitutive requirements, or basic needs, are automatically provided for by the natural order and for whom, accordingly, all distributive goods are residual. For them a clear and simple mandate issues from the general conception of justice as fairness: provide the highest possible guaranteed minimum income. Now although this mandate matches Rawls' own social-policy derivation from the difference principle, it is clear that he does not mean to assume away humanity's subjection to basic needs. For if all distributable goods were residual, there would be no need to worry, as Rawls does, about choosing between an individualized 'lexical difference principle' and some aggregative version, or about how to define the worst-off group when an aggregative version is to be used. These questions become problematic only when individuals are conceived to differ in their basic needs.

To see how these questions arise, we can imagine a race for whom all goods are basic and none residual. Perhaps for members of this race well-being consists solely of fitness achieved in the biological function of one's body, so that the only consumption goods are food and medical care. Individuals work (some as physicians) to produce resources and incomes, but with no other object than to obtain for themselves higher and higher levels of fitness in bodily function. Since the costs of maintenance at any given level of fitness are different for different individuals, the aim of the general conception cannot in this society be met by a simple precept of maximizing the guaranteed minimum money income. It looks as though we shall have to repair to the lexical version of the difference principle: Place all members in a rank order from most fit to least fit; give treatment to the least fit (using resources transferred from any or all of the others) until his fitness level matches that of the member next above him; now treat these two until their level has been

[28] See Fried, *Economic Analysis of Rights, supra;* Tribe, 'Policy Science: Analysis or Ideology?' *Philosophy and Public Affairs* 2 (1972).

raised to that of the member next above them; and proceed in this manner until the attempt to redistribute more fitness away from those at levels above the floor thus far achieved would so severely impair incentives and production as to be unavailing.

The intuitive moral objection to this method can best be captured by example. Suppose the society has twenty-one members named, A, B . . . U, whose fitness levels before redistribution range along a scale from plus-10 to minus-10. A part of the schedule of resource costs required to bring about various improvements looks like this:

Member to Be Raised	*From Level*	*To Level*	*Cost*
U	minus-10	minus-9	x
U	minus-9	minus-8	$2x$
T	minus-9	minus-8	$7000x$
U	minus-8	minus-7	$2x$
S	minus-8	minus-7	$3x$
U	minus-7	minus-6	x
S	minus-7	minus-6	$\cdot 5x$
R	minus-7	minus-6	$3x$

Assume there is no doubt that the economy can sustain the redistribution needed to bring U, S, and R up to minus-6—and, indeed, to raise them and Q, P, and O even higher. Suppose, though, that the cost of raising T to minus-8 would exhaust the economy's redistributive capacity. Insistence on use of the individualized lexical difference principle then would mean denial to U, S, and R—who are very badly off indeed in comparison with most members of society—of substantial improvements which society is fully capable of providing, for the sake of raising T from minus-9 to minus-8. It seems wrong to permit T's misfortune to have such devastating effect on the prospects of others who are very badly off. Our intuition (I speak for those who share it with me)[29] looks in the direction of an aggregative version of the general conception which would maximize the prospects of some 'least advantaged' *group*, but distribute within that group according to vague and indeterminate precepts—like 'the greatest needs satisfaction for the greatest number'—hard to distinguish from the familiar litany of average utility.

Left open, and calling for responses which it seems will have to be intuitive and arbitrary, are interrelated questions about

[29] As for whether Rawls is among them, see note 22, *supra;* cf. p. 157.

where to set the standard level of fitness defining the 'disadvantaged' to whom transfers are to be made, and how to determine when treatment of an 'ailment' suffered by a member of the disadvantaged group is too costly to 'waste' transfer proceeds on. But we should notice one important special case in which these questions will have at least partially determinate answers. Suppose there is some level L along the fitness scale, such that once a person falls below L it matters not how far below L he falls. A person falling further below L is conscious of a worsening condition, of increasing misery, but he does not care about it. Where such is the case, it seems clear that at least all those at fitness levels below L should be regarded as disadvantaged, and that the transfer proceeds should be applied so as to minimize the number who remain below L—which means spending them first on those who can be raised to L most cheaply.[30]

Now we have to deal with human beings, for whom there are both basic and residual goods to be distributed. To keep things from becoming unduly complicated, we can suppose there is only one basic good. We want to fashion a social minimum guaranty—a regularized program for periodic transfers—responsive to the fact that there are these two types of good, and responsive also to a general conception of justice which commits us to doing the best we can for the disadvantaged, without sacrificing the prospects of some of the disadvantaged to the inordinately expensive needs even of those who are extremely disadvantaged. It would be nice if there were some way of lending a trenchant structure to the legislative choices entailed in fashioning such a social minimum program—especially nice if the necessary choices could be reduced to the filling in of values for a few parameters, where each parameter stood for some morally cognizable dimension of the problem. Only if something like that were so would the way be open to argue for a right to have basic needs specifically provided for, so as to overcome Rawls' apparently considered view that a minimum-income guaranty provides an appropriate way of (implicitly) defining a disadvantaged group, and distributing basic as well as residual goods within that group, in a manner consistent with the general conception of justice. If the legislative choices involved in fashioning the complex social minimum could be structured into something like the assignment of values to a set of parameters, then one could argue that this structure for legislative choice—

[30] See B. Barry, *supra* note 16, at p. 98.

together, perhaps, with some procedure for setting upper and lower bounds on the parameter values—is the controlling principle that would be adopted in the original position or constitutional convention.

But it is not easy, to say the least, to dream up the structure or parameters we want. I shall discuss here only the apparently easiest case, in which it might seem that the problem could be solved rather handily—that is, the sort of case mentioned earlier in which there is, on the basic-good side, a level of service L below which life, or residual goods, become virtually worthless. Alas, the very fact that there *are* residual goods deprives us of our determinate solution. It is true that for anyone below L on the basic-good side residual goods have no value—or, as we can say, an unfulfilled basic need exists. But it seems equally true that for anyone lacking residual goods satisfaction of basic needs has no point—yielding only a foundation with no possibility of a superstructure, or a self with no possibility of realizing its aims. Extreme shortage of residual goods thus seems to signify disadvantage no less severe than does existence of basic need. There is no clear gain in satisfying the basic needs of some when the cost is eliminating the residual goods of others. Thus the determinate solution of redistributing so as to minimize the number of unsatisfied basic needs will not work. Even in this seemingly most manageable of cases, there is no readily apparent, structured procedure which can guide a decision as to how many should be how severely deprived of residual goods for the sake of reducing to what level the number of cases of unsatisfied basic needs. Only in what may be the trivially special case in which the economy can fully meet each person's basic needs and also provide a significant volume of residual goods for all does a determinate solution emerge—that of providing for each person's basic needs, and on top of that arranging for the highest possible minimum (residual) income. (For situations in which this extraordinarily benign condition does not hold, one might want to say that at least fulfillment of basic needs should be assured in all cases for which the costs are not abnormally high, and on top of that the highest possible minimum provided. But then one might as well proceed directly to the minimum-income guaranty, and forget about basic needs whose satisfaction is assumed to be equally costly for all.)

Thus Rawls' failure to arrive at a constitutional mandate to provide for basic needs seems to reflect a utilitarianesque indeterminacy intrinsic to his general conception of justice. We should note, though, that this is not the only way of interpreting the aggregative, income-focused approach to the question of maximizing the prospects of the least advantaged. This approach seems translatable into a postulate about compensation which may hold some appeal: In effect, the social minimum is treated as compensation to those who must accept the least advantageous positions defined by the basic structure, compensation which it might be felt is owing just on account of society's insistence on having a basic structure which defines such disadvantageous positions. But it is only disadvantage crystallized by society's choice of a basic structure which is thus compensated for. Rawls says that his principles 'attempt to mitigate the arbitrariness of natural contingency' (p. 96), but this is true for his aggregative, income-focused version of the difference principle only insofar as the disadvantages imposed by natural contingency are reflected in curtailed earning power or assignment to inferior positions in the basic structure. Yet the problem faced by a diabetic is not that his earning capacity is impaired; his problem is that it costs him more than it costs others to satisfy his basic needs.

The postulate of compensation (through the social minimum) only for disadvantage arising from the basic structure's being what it is seems harmonious with Rawls' thesis that the difference principle represents a fair basis on which to ask for the cooperation of the least advantaged in the social scheme of cooperation (p. 103).[31] The postulate may have other philosophical merit. It may even be one which we can imagine parties in the original position accepting, though not without revisiting their supposed rational adoption of a maximin strategy. Their acceptance of the postulate seems to mean that they are extremely conservative in dealing with the risk that the socially generated basic structure will define some very low socio-economic positions, but not especially conservative about the risk that naturally generated impairments will be very expensive to overcome. I do not stop to inquire what intuitively appealing moral notions might correspond with this ascription of sensitivities to the contracting parties.

[31] For criticism of this thesis, see R. Nozick, 'Distributive Justice', *Philosophy and Public Affairs* 3 (1973), pp. 85–94.

SELF-RESPECT

Perhaps we can get where we—and, it sometimes seems, Rawls as well—would like to go including the primary good of self-respect among those as to which the bottom's prospects are to be maximized.

Rawls' definition of self-respect makes clear that it is what I have been calling a basic good:

> We may define self-respect . . . as having two aspects. First of all, . . . it includes a person's sense of his own value, his secure conviction that his good, his plan of life, is worth carrying out. And second, self-respect implies a confidence in one's ability, so far as it is within one's power, to fulfill one's intentions. When we feel that our plans are of little value, we cannot pursue them with pleasure or take delight in their execution. Nor plagued by failure and self-doubt can we continue in our endeavors. It is clear then why self-respect is a primary good. Without it nothing may seem worth doing, or if some things have value for us, we lack the will to strive for them. All desire and activity becomes empty and vain, and we sink into apathy and cynicism. Therefore the parties in the original position would wish to avoid at almost any cost the social conditions that undermine self-respect [p. 440].

I shall assume without detailed discussion that the notion of self-respect as a moral entitlement is perfectly capable of implying some conception of a minimum insurance-rights package. Rawls so implies when he refers to a 'psychological condition' in which 'persons lack a sure confidence in their own value and in their ability to do anything worthwhile', and notes that

> many occasions arise when this psychological condition is experienced as painful and humiliating. The discrepancy between oneself and others is made visible by the social structure and style of life of one's society. The less fortunate are therefore often forcibly reminded of their situation, sometimes leading them to an even lower estimate of themselves and their mode of living [p. 535].

But there is initial difficulty in feeding this view into applications of the difference principle, because by itself it does not seem to fit the difference principle's 'more is better' attitude. Self-respect is a basic good, and with respect to it one would want to satisfy a minimum, not to get more and more.[32] Rawls states that 'in applying the difference principle we wish to include in the prospects of the least advantaged the primary good of self-respect; and there are a variety of ways of taking account of this value consistent with the difference principle' (p. 362). But in fact he intimates only one way in which self-respect might be counted *in applications of* the difference principle (though, as we shall see, there are clearly many ways to take account of self-respect *without contradicting* the difference principle). Rawls says:

> To some extent men's sense of their own worth may hinge upon their institutional position and their income share... [But] with the appropriate background arrangements, [the] inclinations [to social envy and jealousy] should not be excessive, at least not when the priority of liberty is effectively upheld. But theoretically we can if necessary include self-respect in the primary goods, the index of which defines expectations. Then in applications of the difference principle, this index can allow for the effects of excusable envy [i.e., injuries to self-respect]; the expectations of the less advantaged are lower the more severe these effects. Whether some adjustment for self-respect has to be made is best decided from the standpoint of the legislative stage where the parties have more information about social circumstances and the principle of political determination applies [p. 546].

Rawls is, however, far from persuaded that any adjustment in the difference principle's application will ever be required in the interest of self-respect.[33] And he does not give us any direct

[32] The possibility of a basic needs-related self-respect continuum seems to be recognized at p. 546. Elsewhere, however, Rawls seems to differentiate self-respect, viewed as a basic kind of good, from income and wealth, viewed as residual kinds of good (pp. 396–7, 440, 543–4).

[33] A need for such adjustments also to some extent mars the structural nicety of justice as fairness. Says Rawls:

> [A]n equal division of all primary goods is irrational in view of the possibility of bettering everyone's circumstances by accepting certain inequalities. Thus the best solution is to support the primary good of self-respect as far as possible by the assignment of the basic liberties that can indeed be made equal, defining the same status for all. At the same time, distributive justice as frequently understood, justice in the relative

statement of what form such an adjustment could take. Self-respect aside, the difference principle has already, presumably, led to the highest possible guaranteed income floor. There is no way the bottom's purchasing power can be increased. The only available adjustment, and the one apparently implied by the quoted passage, is to intensify tax-transfer activity beyond the 'optimal' point which maximizes the bottom's income, so that the top's standard of living will be reduced to a level such that the bottom's standard, although lower than that available at the optimal point, is relatively high enough to confirm its self-respect. The supposition must be that insofar as self-respect may be undermined by low standards of living, this is primarily a matter of relative deprivation.

So in suggesting that adjustments for self-respect may be required in the course of applying the difference principle, Rawls must be supposing the possibility of cases in which more intense tax-transfer activity is required to exhaust the possibilities of confirming the bottom's self-respect then would maximize its income. Since in the Rawlsian vision, income is virtually worthless without self-respect,[34] the question of adjusting the difference principle's application to take account of self-respect boils down to a conceptually simple idea: If it is necessary, in order to confirm the bottom's transfer activity self-respect, to increase tax-transfer activity beyond the point where the bottom's income is maximized, we are to do so.[35]

We have thus brought the basic good of self-respect within the maximizing structure of the difference principle, by showing that on its account maximization of the bottom's prospects may require tax-transfer activity which is suboptimal with regard to the bottom's income. In so doing, we have come as far as we

shares of material means, is relegated to a subordinate place. Thus we arrive at another reason for factoring the social order into two parts as indicated by the principles of justice. While these principles permit inequalities in return for contributions that are for the benefit of all, the precedence of liberty entails equality in the social bases of esteem.

[34] See pp. 440, 535. It is true that Rawls speaks in passing (p. 362) of assigning relative weights to self-respect and income in arriving at an index of the bottom's prospects. But this view seems so sharply at odds with the subsequent, more comprehensive elucidation of self-respect that it should perhaps be disregarded.

[35] I make no attempt to square the central idea here, that relative deprivation may undermine self-respect, with Rawls' general disclaimers that envy plays or should play any part in his arguments for maximizing the bottom's prospects. It all depends on the notion, which to me remains difficult, that there are exceptional cases of 'excusable envy' (see p. 534).

can in pursuit of some clear implication in the difference principle of an insurance-rights package. Yet the quest has failed. Through a somewhat elusive notion of relative deprivation, we can see the possibility that an existing state of distribution might be perceived as unjust because of some people's inability to satisfy certain needs, considered in the comparative light of what others are able to enjoy. But as long as we are required to think of this 'bottom' as a socio-economic class, there are only two possible remedies for such a situation. The preferred remedy, expected to suffice in most cases, is to raise the bottom's real income to the maximum achievable level. As we have already seen, there are no insurance-rights implications in that approach. An additional possible remedy—the one brought to light by the introduction of self-respect as a primary social good—is taxation and transfer calculated to reduce both the top's and the bottom's real incomes, but to reduce the top's at a faster rate so as to narrow the relative-deprivation gap. But this is hardly a remedy that one would associate with the idea of insurance rights.

Though we have finished our exploration of the difference principle, we have not yet finished with the primary good of self-respect. Before continuing with our examination of it, it will be helpful to see what the opportunity and liberty principles might have to contribute to a notion of constitutional welfare rights.

The Opportunity Principle

This principle requires a degree of compensatory service to what we could call the 'priming' needs of each individual, as by provision of basic education. Rawls does not say that any education must be publicly provided in kind, or that any compensatory right regarding education cannot be satisfied by general monetary transfers. But by attaching this right to the lexically preferred opportunity principle, Rawls must mean that it has to be satisfied before the difference principle can be allowed to operate—that compensatory satisfaction of the opportunity interest is a prerequisite to allowing any income inequalities to arise in the marketplace. It seems to follow that no one may be precluded from the requisite education by income shortage; and this would be, then, an insurance right. Even so, it might not generate a judicially enforceable claim unless its extent could be more precisely determined.

Now although the expression 'fair equality of opportunity' seems to suggest an intuitionistic criterion of how far to cut into dispositive liberty[36] on behalf of opportunity, the theory as a whole may give stronger guidance. Rawls believes his contractarian construct should generate assurances of education to compensate for environmental accidents of background and upbringing, but not for genetic accidents of inheritance.[37] This distinction may not defy translation into social policy.[38] If in the present state of educational technology it does not yield claims so determinate as to be judicially definable, that it may someday yield them is perhaps not unimaginable.

The scope of welfare rights implied by the opportunity principle may not be limited to goods which are strictly 'educational'. Unaccompanied by subsistence or health or freedom from extreme environmental deprivation, how could educational offering effectuate fair equality of opportunity? The priority of the opportunity principle over the difference principle and that of dispositive liberty must mean that education's effective biological entailments, whatever they are, must be satisfied as a prior condition to reliance on the pure procedural justice of the market—even supposing that such reliance would tend to maximize the bottom's real-income expectations, thus satisfying the difference principle. If so, the catalogue of insurance rights would reach beyond educational goods and (at least with regard to persons of educable age) into welfare domains which we tend to associate with the difference principle and the social minimum. And these additional rights seem more amenable to judicial definition than the core education claim itself, and more so under the opportunity principle than under the difference principle. It seems likely that one could arrive at somewhat objective and nonrelativistic descriptions of the levels of subsistence, health, and environmental amenity necessary to make persons receptive to educational offerings.

[36] In 'Welfare Rights' I argue that Rawls' theory affirmatively recognizes and values a person's liberty to dispose over the proceeds of his transactions in the marketplace ('dispositive liberty'), though it ranks the principle of respecting such liberty as lexically inferior to the Difference Principle. 'Welfare Rights' at pp. 969–76.

[37] Any further compensation for disadvantage in the natural-talents lottery is to be of the less-than-fully-offsetting type implied by the difference principle (see pp. 73–5, 511).

[38] See *California General Assembly, Senate Select Committee on School District Finance* 1, Final Report (1962), pp. 39–43.

The Liberty Principle

Enjoyment of basic liberties, like enjoyment of educational opportunity, has fairly straightforward and objective biological entailments. Thus the right to provision of these may rank with liberty among the social priorities established by the theory of justice as fairness. Rawls so indicates when he says that the priority of liberty over satisfaction of material wants is not unqualified, that in some circumstances the lexically articulated conception of justice may have to give way to a 'general conception' which permits trade-offs among all social values—liberty along with material goods—as long as those trade-offs satisfy the difference principle (see pp. 62, 303). This line of thought leads to a division of material wants into those which are and are not basic:

> To be sure, it is not the case that when the priority of liberty holds, all material wants are satisfied. Rather these desires are not so compelling as to make it rational for the persons in the original position to agree to satisfy them by accepting a less than equal freedom . . . Until the basic wants of individuals can be fulfilled, the relative urgency of their interest in liberty cannot be firmly decided in advance. It will depend on the claims of the least favored as seen from the constitutional and legislative stages. But under favorable circumstances the fundamental interest in determining our plan of life eventually assumes a prior place [p. 543].[39]

Justice as Fairness as a Whole

The good of self-respect seems to play a dual role in the theory of justice as fairness. It is, we have seen, one of the primary goods, along with income and wealth, with respect to which the difference principle demands that the bottom's prospects be maximized. In this role self-respect may be said to be coordinate with income and wealth[40] and subordinate to rights and liberties, in the lexicon of primary social goods. But there is another perspective

[39] This passage should be compared with my discussion above.

[40] But more precisely it seems to be a condition of the enjoyment of income and wealth.

in which confirmation and nurture of self-respect are the end and objective of all the principles of justice taken together. In this perspective, self-respect is the preeminent social good, subordinate even to rights and liberties. The liberty principle, the opportunity principle, and the difference principle—each separately and all in their convergent impact—are elaborated and justified in terms of their tendency to instill and safeguard self-respect; and insofar as it is feared that one or another of these principles in isolation may be insufficiently solicitous of self-respect, it is shown how one or another of the other principles may counteract this concern (see pp. 534–48). Self-respect thus becomes a central element in the justification of the whole theory.

It follows that if welfare rights can be shown essential to self-respect, in any sense additional to those in which the difference, opportunity, and liberty principles severally imply such rights, the theory of justice as fairness implies them in this additional sense as well. Thus the difference principle implies welfare rights in the elusive form of whatever is necessary to prevent the undermining of self-respect by relative deprivation. The opportunity and liberty principles imply welfare rights as more objective, less relativistic biological entailments of opportunity and liberty. In addition, the central and preeminent good of self-respect may imply welfare rights reaching beyond those biological entailments, and not depending on notions of relative deprivation for their justification. Perhaps, for example, self-respect requires the opportunity to be creative in some medium which gives pleasure to others—artistic performance, crafts, sport, or whatever (see pp. 440–2)—though perhaps no one would say that he was degraded by not having had such opportunities, or that his educational offering or civil and political liberties were rendered worthless without it.

Professor Fried has suggested that it should be possible to '[elaborate] a comprehensive theory of rights within the terms of Rawls' general scheme'.[41] With one qualification, the foregoing discussion confirms that this should be possible for welfare rights, although it seems to require some verbal modification of the 'general schame' of lexically articulated principles of justice. Somewhere near the top of the priority list there may belong a 'welfare rights' principle, signifying that the questions of maximizing the bottom's position with regard to money income and

[41] C. Fried, Book Review, *Harvard Law Review*, 85 (1972), pp. 1691, 1697.

wealth, and even of protecting the bottom against the degrading effects of relative deprivation, are simply not to be addressed until provision for some articulated package of basic welfare needs has been secured. The qualification pertains to those cases in which satisfying basic needs would be ruinously expensive.[42] It seems certain that some such cases exist. Whether their existence means that there can be no specific welfare rights at all is a question I have not been able to answer.

To conclude: It is a mistake to think of the social minimum as an institutional feature linked specifically and peculiarly to the difference principle (whereas democracy or a bill of rights, say, would be associated with the liberty principle). The social minimum is an implication of justice as fairness taken as a whole theory. While the difference principle taken in isolation seems to have a simple, maximizing thrust, that is not true of the whole theory. The theory as a whole reflects a degree of risk aversion, imputing to representative persons a structured set of priorities under which the question of generally amplifying one's income simply is not reached until adequate assurance has been made for what one specifically needs in order that his basic rights, liberties, and opportunities may be effectively enjoyed, and his self-respect maintained.

[42] See pp. 335 ff, *supra.*

Selected Bibliography on John Rawls: Post 1971

Acton, H. B., 'Distributive Justice, the Invisible Hand, and the Cunning of Reason', *Political Studies*, 20 (Dec. 1972), pp. 421–31.

Altham, J. E. J., 'Rawls' Difference Principle', *Philosophy*, 48 (Jan. 1973), pp. 75–8.

Arrow, Kenneth J., 'Some Ordinalist-Utilitarian Notes on Rawls' Theory of Justice', *Journal of Philosophy*, 70 (10 May 1973), pp. 245–63.

Barry, Brian, *The Liberal Theory of Justice: A Critical Examination of the Principal Doctrines in 'A Theory of Justice' by John Rawls*. Oxford: Clarendon Press, 1973.

—, 'John Rawls and the Priority of Liberty', *Philosophy and Public Affairs*, 2 (Spring 1973), pp. 274–90.

Bates, Stanley, 'The Motivation to Be Just', *Ethics*, 85 (Oct. 1974), pp. 1–17.

Bentley, D. J., 'John Rawls: A Theory of Justice', *University of Pennsylvania Law Review*, 121 (1973), pp. 1070–8.

Bowie, Norman B., 'Some Comments on Rawls' Theory of Justice', *Social Theory and Practice* 3 (1974), pp. 65–74.

Braybrooke, David, 'Utilitarianism with a Difference: Rawls' Position in Ethics', *Canadian Journal of Philosophy*, 3 (Dec. 1973), pp. 303–31.

Brock, Dan W., 'Contractualism, Utilitarianism and Social Inequalities', *Social Theory and Practice*, 1 (Spring 1971), pp. 33–44.

—, 'Recent Work in Utilitarianism', *American Philosophical Quarterly*, 10 (Oct. 1973), pp. 241–76.

— 'The Theory of Justice', *University of Chicago Law Review*, 40 (1973), pp. 486–99.

Brown, D. G., 'John Rawls: John Mill', *Dialogue* (Canada), 12 (S 1973), pp. 477–9.

Choptiany, Leonard, 'A Critique of John Rawls' Principles of Justice', *Ethics*, 83 (Jan. 1972), pp. 146–50.

Cunningham, Robert L., 'Justice: Efficiency or Fairness?', *Personalist*, 52 (Spring 1971), pp. 253–81.

Dasgupta, Partha, 'On Some Problems Arising from Professor Rawls' Conception of Distributive Justice', *Theory and Decision*, 4 (Feb.–Apr. 1974), pp. 325–44.

Demarco, Joseph P., 'Some Problems in Rawls' Theory of Justice', *Philosophical Context*, 2 (1973), pp. 41–8.

Eshete, Andreas, 'Contractarianism and the Scope of Justice', *Ethics*, 85 (Oct. 1974), pp. 38–49.

Feinberg, Joel, 'Duty and Obligation in the Non Ideal World', *Journal of Philosophy*, 70 (10 May 1973), pp. 263–75.

Frankel, Charles, 'Justice, Utilitarianism and Rights', *Social Theory and Practice*, 3 (1974), pp. 27–46.

Friedmann, W. G., '*A Theory of Justice*: a lawyer's critique', *Columbia Journal of Transnational Law*, 11 (1972), pp. 369–79.

Gauthier, D., 'Justice and Natural Endowment: Toward a Critique of Rawls' Ideological Framework', *Social Theory and Practice*, 3 (1974), pp. 3–26.

Hare, R. M., 'Rawls' Theory of Justice, Part I, II', *Philosophical Quarterly*, 28 (1973), pp. 144–55; 241–52.

Harris, Charles E., 'Rawls on Justification', *Southwestern Journal of Philosophy*, 5 (Spring 1974), pp. 135–43.

Hart, H. L. A., 'Rawls on Liberty and its Priority', *University of Chicago Law Review*, 40 (1973), pp. 534–55.

Haskar, Vinit, 'Rawls' Theory of Justice', *Analysis*, 32 (May 1972), pp. 149–153.

Hicks, Joe H., 'Philosophers' Contracts and the Law', *Ethics*, 85 (Oct. 1974), pp. 18–37.

Kohlberg, Lawrence, 'The Claim to Adequacy of Highest Stage of Moral Development', *Journal of Philosophy*, 70 (Oct. 1973), pp. 630–46.

Lessnoff, M., 'John Rawls' Theory of Justice', *Political Studies*, 19 (March 1971), pp. 63–80.

— 'Barry on Rawls' Priority of Liberty', 4 (1974), pp. 100–14.

Levine, Andrew, 'Rawls' Kantianism', *Social Theory and Practice*, 3 (1974), pp. 47–63.

Lyons, D., 'On Formal Justice', *Cornell Law Review*, 58 (1973), pp. 833–61.

—, 'Rawls versus Utilitarianism', *Journal of Philosophy*, 69 (5 Oct. 1972), pp. 535–45.

MacCormick, N., 'Justice According to Rawls', *Law Quarterly Review*, 89 (1973), pp. 393–417.

Mandelbaum, Maurice, 'Review of *A Theory of Justice*', *History and Theory*, 12 (1973), pp. 240–50.

Merritt, G., 'Justice as Fairness: a commentary on Rawls' new theory of justice', *Vanderbilt Law Review*, 26 (1973), pp. 665–86.

Michelman, F. I., 'In Pursuit of Constitutional Welfare Rights: One View of Rawls' *Theory of Justice*', *University of Pennsylvania Law Review*, 121 (1973), pp. 962–1019.

Miller, Richard, 'Rawls and Marxism', *Philosophy and Public Affairs*, 3 (1974), pp. 167–92.

Mueller, Dennis C., Tollison, Robert D., Willet, Thomas D., 'The Utilitarian Contract: A Generalization of Rawls' Theory of Justice', *Theory and Decision*, 4 (Feb.–Apr. 1974), pp. 345–68.

Murphy, C. F., 'Distributive justice, its modern significance', *American Journal of Jurisprudence*, 17 (1972), pp. 153–78.

Nagel, Thomas, 'Rawls on Justice', *Philosophical Review*, 82 (April 1973), pp. 220–34.

Neilsen, Kai, 'A Note on Rationality', *Journal of Critical Analysis*, 4 (April 1972), pp. 16–19.

Norton, David L., 'Rawls's Theory of Justice: A "Perfectionist" Rejoinder', *Ethics*, 85 (Oct. 1974), pp. 50–7.

Nozick, R., 'Distributive Justice', *Philosophy and Public Affairs*, 3 (1973), pp. 45–126.

Parekh, B., 'Reflections on Rawls' Theory of Justice', *Political Studies*, 20 (Dec. 1972), pp. 479–83.

Pettit, Philip, 'A Theory of Justice?', *Theory and Decision*, 4 (Feb.–Apr. 1974), pp. 311–24.

Pollock, Lansing, 'A Dilemma for Rawls', *Philosophical Studies*, 22 (April 1971), pp. 37–43.

Reiman, Jeffrey H., 'A Reply to Choptiany on Rawls on Justice', *Ethics*, 84 (April 1974), pp. 262–5.

Ross, Geoffrey, 'Utilities for Distributive Justice', *Theory and Decision*, 4 (Feb.–Apr. 1974), pp. 239–58.

Scharr, John, 'Reflections on Rawls' Theory of Justice', *Social Theory and Practice*, 3 (1974), pp. 75–100.

Schwartz, Adina, 'Moral Neutrality and Primary Goods', *Ethics*, 83 (July 1973), pp. 294–307.

Scott, Gordon, 'John Rawls' Difference Principle, Utilitarianism, and the Optimum Degree of Inequality', *Journal of Philosophy*, 70 (10 May 1973), pp. 263–75.

Sen, Amartya, 'Rawls versus Bentham: An Axiomatic Examination of the Pure Distribution Problem', *Theory and Decision*, 4 (Feb.–Apr. 1974), pp. 301–10.

Shue, Henry, 'Liberty and Self-respect', *Ethics*, 85 (1974–5).

——, 'Justice, Rationality and Desire', *Southern Journal of Philosophy*, 13 (1975–6).

——, 'The Current Fashions: Trickle-downs by Arrow and Close-Knits by Rawls', *Journal of Philosophy*, 71 (13 June 1974), pp. 319–26.

Slote, Michael A., 'Desert, Consent and Justice', *Philosophy and Public Affairs*, 2 (Summer 1973), pp. 323–47.

Smith, James Ward, 'Justice and Democracy', *Monist*, 55 (Jan. 1971), pp. 121–33.

Sterba, J. P., 'Justice as Desert', *Social Theory and Practice*, 3 (1974), pp. 101–16.

Teitelman, Michael, 'The Limits of Individualism', *Journal of Philosophy*, 69 (5 Oct. 1972), pp. 545–56.

Wittman, Donald, 'Punishment as Retribution', *Theory and Decision*, 4 (Feb.–Apr. 1974). pp. 209–38.

Index